ADVANCED MANUFACTURING – AN ICT AND SYSTEMS PERSPECTIVE

BALKEMA – Proceedings and Monographs
in Engineering, Water and Earth Sciences

# Advanced Manufacturing –
# An ICT and Systems Perspective

*Editors*

Marco Taisch
*Department of Management, Economics and Industrial Engineering,
Politecnico di Milano, Milano, Italy*

Klaus-Dieter Thoben
*BIBA, University of Bremen, Germany*

Marco Montorio
*Department of Management, Economics and Industrial Engineering,
Politecnico di Milano, Milano, Italy*

Taylor & Francis
Taylor & Francis Group

LONDON / LEIDEN / NEW YORK / PHILADELPHIA / SINGAPORE

*Taylor & Francis is an imprint of the Taylor & Francis Group, an informa business*

© 2007 Taylor & Francis Group, London, UK

Typeset by Charon Tec Ltd (A Macmillan Company), Chennai, India
Printed and bound in Great Britain by Antony Rowe Ltd (CPI-Group), Chippenham, Wiltshire

Published by:  Taylor & Francis/Balkema
       P.O. Box 447, 2300 AK Leiden, The Netherlands
       e-mail: Pub.NL@tandf.co.uk
       www.taylorandfrancis.co.uk/engineering, www.crcpress.com

ISBN 13: 978-0-415-42912-2

Advanced Manufacturing – An ICT and Systems Perspective – Taisch,
Thoben & Montorio (eds)
© 2007 Taylor & Francis Group, London, ISBN 978-0-415-42912-2

# Table of contents

*Advanced Manufacturing – An ICT and Systems Perspective – Taisch,*
*Thoben & Montorio (eds)*
*© 2007 Taylor & Francis Group, London, ISBN 978-0-415-42912-2*

# Preface

Manufacturing has played, and continues to play, a vital role in the European economy and society, and it will remain a significant generator of wealth in the future. A strong manufacturing industry will indeed continue to be fundamental to creating stable employment inside the European economy.

For these reasons the ability to maintain and develop the competitiveness of the manufacturing industry is essential for Europe's prosperity. This is especially the case at the present time when the risk of de-industrialisation is becoming increasingly serious owing to the growing intensity of competition emerging from low-wage countries, outsourcing and off-shoring trends, and the brain-drain phenomenon.

In such circumstances it is important to share a common understanding and create awareness of the new challenges and opportunities for the next generation of manufacturing, and from this, to develop a strategic research agenda and stimulate new research initiatives.

This is the rationale behind the preparation and publication of this book on Advanced Manufacturing, which summarises the results of three years of work within the IMS-NoE project (Network of Excellence on Intelligent Manufacturing Systems).

The IMS-NoE, funded by the European Commission, has provided a sound support infrastructure and a stimulating forum for discussion on the future of manufacturing and promoting excellence in manufacturing-related research.

The Network consists of over 300 experts in manufacturing, coming not only from the European Union, where it originated, but also from other IMS Regions (Australia, Canada, Japan, Korea, Switzerland and the USA). The IMS-NoE owes its success to its multi-regional and multidisciplinary nature.

The IMS-NoE brought together the expertise and experience of hundreds of researchers, industrial managers and policymakers worldwide and has collected their ideas, forming them into the *IMS-NoE Vision on the Future of Manufacturing*. This was developed through the organisation of initiatives such as: IMS-NoE Special Interest Group brainstorming meetings (www.ims-noe.org/SIG.asp), the International IMS Forum 2004, a *survey on technology foresight projects*, a *Delphi study on ICT in manufacturing*, and a series of *workshops on ICT in manufacturing* (www.ims-noe.org/FP7.asp), which set the cornerstone for the ICT in manufacturing initiative in the fields of:

- ICT to support management of IPR in view of industrial outsourcing/insourcing strategies;
- New intelligent and networked products;
- The agile wireless manufacturing plant;
- New manufacturing technologies for miniaturised ICT;
- Strategies for the design and manufacturing of new products.

The IMS-NoE activities must be analysed in the global framework of the IMS (Intelligent Manufacturing Systems) program (www.ims.org). The IMS Phase I came to an end in 2005 and Phase II started in 2006.

This book presents the achievements of the IMS-NoE, and in accordance to the open and collaborative nature of the Network, integrates these with visionary contributions coming from other initiatives.

The result is a broad, but necessarily non-exhaustive, vision on the future of manufacturing, which is here analysed from a system management perspective and with a special focus on

ICT-related matters. Each contribution intends to present such a complex and multidisciplinary research domain from a specific perspective, while focusing on a particular research domain.

Therefore while *Part I: Advanced Manufacturing: Foresight and Roadmapping Initiatives* provides an overview of past and continuing technology foresight exercises in manufacturing, the remaining parts will focus on:

- Part II: Product Lifecycle Management
- Part III: Sustainable Products and Processes
- Part IV: Production Scheduling and Control
- Part V: Benchmarking and Performance Measures
- Part VI: Industrial Services
- Part VII: Human Factors and Education in Manufacturing
- Part VIII: Collaborative Engineering
- Part IX: Supply Chain Integration

The ideas presented are nevertheless not intended to be an end in themselves. The book, provocative in its nature, aims at building consensus and also to stimulate fresh discussions, which may lead to novel research initiatives in the future.

The editors wish to thank the contributors as well as the European Commission for their support.

Marco Taisch
*IMS NoE Project Co-ordinator*
Klaus-Dieter Thoben
*IMS NoE Partner*
Marco Montorio
*IMS NoE Project Manager*

*Part I*
*Advanced manufacturing: Foresight and*
*roadmapping initiatives*

The shift from product to service dominance in value chains does not imply that products in the future will not play a critical role in the value creation process. Many human needs are and will always be satisfied in a physical, rather than in a virtual way. What is radically and unavoidably changing is not the role of manufacturing, but the way artefacts are manufactured. The ability of the European production sector to remain competitive in the decades to come depends on its ability to adapt and face future manufacturing challenges.

To remain competitive in this rapidly changing context it is vital to understand where to go and how to change. In this respect several technology foresight exercises have been undertaken in Europe, Japan, USA, Korea, Australia, Canada, as well as in other countries, with the intent to understand how to support the manufacturing sector in the transformation from a physical resource-based to a knowledge-based paradigm.

Discrepancies among the different studies are symptomatic of the difficulty to delineate a clear vision on the future of manufacturing, especially given the quick and ever increasing pace of technology shifts. Nevertheless, many similarities can be identified and, building on them, it is necessary to delineate a clear strategic research agenda to support manufacturing.

This section presents technology foresight, visionary and roadmapping activities in manufacturing, which have been undertaken within the IMS-NoE project as well as in the framework of other initiatives, with the intent of identifying similarities and of reaching a common understanding of them.

The first two contributions present, respectively, a survey analysis and a Delphi study that were carried on within the IMS-NoE project.

The survey investigates past and continuing technology foresight initiatives in manufacturing, taking into consideration projects from several IMS regions (Australia, Canada, European Union, Korea, Japan and USA), and presenting a critical analysis of the developed visions.

The Delphi study presents a vision on the future of manufacturing, with a special focus on the impact that ICT will have on next generation production systems.

The third contribution deals with the results of the project *Manufacturing Visions – Integrating Diverse Perspectives into Pan-European Foresight (ManVis)*, which started in 2004. Based on several initiatives such as a Delphi-survey, workshops and interviews with experts, *ManVis* intends to support the continuing policy process of enhancing the competitiveness of European manufacturing industries.

*Advanced Manufacturing – An ICT and Systems Perspective – Taisch,*
*Thoben & Montorio (eds)*
*© 2007 Taylor & Francis Group, London, ISBN 978-0-415-42912-2*

# The future of manufacturing: Survey of international technology foresight initiatives

Marco Montorio & Marco Taisch

*Politecnico di Milano, Department of Management, Economics and Industrial Engineering, Milano*

ABSTRACT: This paper surveys several technology foresight initiatives in manufacturing. Research projects from Australia, Canada, European Union, Korea, Japan and USA have been investigated and a critical analysis of the developed visions has been undertaken to identify specific features and common elements.

*Keywords*: Technology foresight, Next-Generation Manufacturing, roadmap, vision.

## 1 INTRODUCTION

Manufacturing remains a key generator of wealth and is still at the heart of the economic growth in industrialised economies. But in recent years manufacturing in developed countries has undergone profound changes that are bringing it from a resource-based and centralised paradigm to a knowledge-intensive, innovation-based, adaptive, digital and networked one.

Given the prominence of manufacturing in developed economies, and being aware of the profound shift that it is now facing, the ability to maintain and develop the competitiveness of manufacturing through relevant R&D investments will be essential for the prosperity of industrialised countries. This is especially true at the present time when the risk of de-industrialisation is becoming increasingly serious, owing to outsourcing and off-shoring trends and to increased competition coming from low-wage countries, such as China and India.

In this complex and rapidly changing environment planning a suitable R&D policy for manufacturing will be crucial for industrialised countries to face the continuing transition while maintaining their competitive position.

## 2 MANAGING THE TRANSITION: THE ROLE OF TECHNOLOGY FORESIGHT

Recognising the importance of investing in R&D for manufacturing is the first step, but understanding which areas to invest in and which technologies to research is the heart of the problem. To plan R&D investments it is necessary to envision next generation manufacturing and to understand how enterprises need to change to face future market challenges while remaining competitive. In this sense, technology foresight represents the basis for decision making in R&D strategy.

According to Ben Martin (SPRU – Science and Technology Policy Research Unit, at the University of Sussex – 1995) technology foresight can be defined as an activity which: "involves systematic attempts to look into the longer term future of science, technology, the economy, the environment and society with a view to identifying emerging generic technologies and the underpinning areas of strategic research likely to yield the greatest economic, social and environmental benefits."

In the last decade great attention has been given to technology foresight in manufacturing and several initiatives have been launched in industrialised countries such as Australia, Canada, Europe, Japan, Korea and USA, with the intent to develop a vision on the next generation manufacturing and thus to support R&D decisions in manufacturing for the private sector (to guide industrial R&D strategies) and for the public one (to support policy making and funding strategies). Some of the most relevant work carried on in the last decade are:

- IMSS – International Manufacturing Strategy Survey (international – 1st round – 1992, 2nd round – 1996, 3rd round – 2000 and 4th round 2004);
- NGMP – Next Generation Manufacturing Project (USA - 1995);
- VISIONARY 2020 – Visionary Manufacturing Challenges for 2020 (USA – 1998);
- IMTI – Integrated Manufacturing Technology Initiative (USA – 1998);
- IMTR – Integrated Manufacturing Technology Roadmap Project (USA – 2000);
- NGMS – Next Generation Manufacturing Systems (European Union, Japan, USA – phase I – 1999 & phase II – 2000);
- Informan+ (European Union – 2000);
- FORESIGHT 2020 (UK – 2000);
- VISION 2025 (KOREA – 2000);
- FutMan – The Future of Manufacturing in Europe, 2015 – 2020 (European Union – 2003);
- *ManuFuture* (European Union – 2003);
- MANU 20/20 – Manufacturing in 2020 (CANADA - 2004);
- MANU INITIATIVE – The Manufacturing Initiative (USA – 2004);
- *ManVis* – Manufacturing Visions – Integrating Diverse Perspectives into Pan-European Foresight (European Union – 2004).

## 3    SURVEY OF TECHNOLOGY FORESIGHT INITIATIVES IN MANUFACTURING

Some of the above initiatives have been analysed to identify specific features and common elements. The *ManVis* and the *ManuFuture* initiatives are presented elsewhere in the book, so are not discussed further in this paper. In the following there are introductions to a selection of the analysed technology foresight projects. The projects, presented in chronological order, have been selected to highlight the evolution and the profound changes which manufacturing has been facing in industrialised countries in the last decade.

### 3.1    *Next-Generation Manufacturing Project – USA (1995)*

The *Next-Generation Manufacturing Project* (NGMP), started in 1995, aimed at developing a framework which US manufacturers could use as a guide to understand the future market and the necessary counteractions to remain competitive.

NGMP was funded by the National Science Foundation, the National Institute of Standards and Technology, the Department of Defence, the Department of Energy and 10 other associations. Close to 500 experts, from more than 100 companies, industry associations, government agencies and academic institutions participated in the initiative.

NGMP output is structured into Drivers, Attributes, Dilemmas, Imperatives and Action Plan Recommendations.

The *drivers* represent the forces that will shape the future competitive environment that manufacturing organisations will have to compete in. Among them, particular relevance is given to:

- The continuous development of ICT, which will allow information to be universally and instantaneously available.
- The worldwide spread of scientific education.
- More competitive markets.
- A general increase in customer expectations.
- A higher environmental responsibility and consciousness of resource limitations.

To remain competitive, and in response to the described drivers, manufacturing enterprises will have to develop specific *attributes*, such as:

- More intimate relationship with the customer, thus achieving customer responsiveness and even anticipating customers' requirements in terms of products and services.
- Flexible, adaptable and responsive processes, equipment, plants, human resources and strategies.
- Intra- and inter- enterprise team working and ability to face increasing problem complexity.

Many desired attributes of the future vision are in apparent conflict. This highlights the emergence of some *dilemmas* (or key barriers), that on the other side may also represent opportunities for achieving competitive advantages. NGMP identified three kinds of dilemmas.

- For the enterprise the dilemmas will be: How to guarantee employee security and loyalty within rapid skill shifts, turnover in the labour force and a flexible workforce? How to achieve collaborative knowledge sharing within knowledge-based competition? How to control core competencies without owning them? How to recover rising plant and equipment costs with shorter product and process lifetime? How to profit from long-term relationships when customers, suppliers and partners are becoming less loyal?
- For the nation the dilemmas will be: How to keep domestic jobs while developing global markets? How to deal with transnational corporations?
- For the individual the dilemmas will be: How to have *good jobs* with individual security while employed in flexible workplaces?

The dilemmas can be overcome through the identified *imperatives*, that is, practices and technology solutions, grouped into four categories: people-related, business process-related, technology-related and integration-related. All the imperatives are strongly interdependent; there is a need to simultaneously address all actions from an integrated people, business process and technology viewpoint across all elements of the enterprise to realise the desired objectives.

People-related imperatives stress the importance of: a flexible workforce, which includes not only the individual worker, but also the workplace and government policy; and knowledge supply chains, which must provide and spread knowledge among industry, university, school and associations.

Concerning business processes, NGMP underlined the need for rapid product/process realisation, innovation and change management; a necessity given rapidly changing environment and customer expectations. This search for flexibility, adaptability to customer requirements, and short time to market will be enabled by re-configurable, scalable and cost effective processes and equipment, pervasive modelling and simulation, and adaptive, responsive information systems.

The increasing range of knowledge and competencies needed to produce products or services or both will drive a growing integration; this has to be viewed from two perspectives: integration and collaboration among companies, which are grouped in extended enterprises, and integration within the company, among people, information, business practices and processes.

NGMP addressed all the stakeholders involved (enterprises, industry associations, academia, government and individuals) introducing several cross-cutting *action plan recommendations*, such as the need to: develop systematic processes for knowledge capture and knowledge-based manufacturing; establish a government partnership with industry and academia for a supportive manufacturing infrastructure; enable and promote pervasive use of modelling and simulation; and develop intelligent processes and flexible manufacturing systems.

## 3.2 *Visionary Manufacturing Challenges for 2020 – USA (1998)*

www.nap.edu/html/visionary

The *Visionary Manufacturing Challenges for 2020* initiative aimed to identify key R&D areas for investments to support US manufacturing. Funded by the National Research Council in 1998, it convened a committee of manufacturing and technology experts, with representatives from small, medium and large companies from a variety of industries.

Starting from the Next Generation Manufacturing Project and other roadmapping initiatives in manufacturing, Visionary 2020 relied on an international Delphi survey that helped the committee to prioritise future industry needs.

Above all, Visionary 2020 underlined the need for a strong collaboration among research centres, academia, industry, and government institutions as a critical success factor to keep US manufacturing competitive in the global market.

Visionary 2020 depicted a vision of the future competitive environment that will be characterised by:

- A growing development and diffusion of information and communication technologies which will allow the instantaneous availability of knowledge and information.
- In turn this will bring to a sophisticated, global and highly competitive market, where enterprises must base their success on continuous innovation, a skilled workforce and knowledge sharing.
- A higher customer demand for customised products and services, which will push the enterprise to be flexible and rapid in its responses.
- An increased sensitivity to environmental protection, which will be essential for the global ecosystem.

The changed market competition can be faced only by tackling the so called *Six Grand Challenges*, which are:

1. *Concurrent manufacturing*. Concurrent manufacturing reduces time to market, encourages innovation and improves quality. Concurrency in all operations needs new technologies in process management and in rapid prototyping, and more flexible machines and interactive computer networks that will allow skilled workers to share knowledge.
2. *Integration of human and technical resources*. Technologies will be adaptable to the changing needs of the market, and people will have the know-how to optimise and enhance them. Individual workers will continue to specialise and will share their knowledge with the other workers in networks that will include suppliers, partners and customers. Factory organisation will be less structured, allowing workers to reorganise themselves, the equipment and the processes to meet customer demands.
3. *Conversion of information to knowledge*. Enterprises will need to instantaneously transform information from a vast array of diverse sources into useful knowledge and effective decisions, and to make this knowledge available to users (human and machine) instantaneously.
4. *Environmental compatibility*. Production waste and product environmental impact will be reduced to *near zero*. This challenge can be won through developing cost-effective products and processes that do not harm the environment, using recycled materials and minimising wastes in terms of energy, material and human resources. A proactive approach is needed, with a strong collaboration among governments, academia and enterprises.
5. *Re-configurable enterprises*. In response to the changing customer demand for customised products and services, enterprises will need to reorganise themselves, rapidly forming and dissolving alliances with other organisations and teams within the enterprise.
6. *Innovative processes*. New concepts need to be applied to manufacturing operations, leading to dramatic changes in production capabilities. Great attention needs to be given to nanotechnologies and biotechnologies.

Visionary 2020 have identified some key technology research areas. These are: adaptable and re-configurable systems; technologies for the minimisation of waste and energy consumption; biotechnology and nano-technology; modelling and simulation; product and process design methods; human-machine interfaces; technologies to convert information into knowledge for effective decision making; new educational and training methods; and software for intelligent collaboration systems.

### 3.3 *IMTI – IMTR – USA (1998 – 2000)*

www.IMTI21.org

The Integrated Manufacturing Technology Roadmap (IMTR) project was launched in 1998 by the National Institute of Standards and Technology, the US Department of Energy, the National Science Foundation and the Defence Advanced Research Projects Agency with the intent to produce an integrated suite of plans to help and guide national manufacturing technology investments. *Six grand challenges*, which should be faced to stay competitive in the future manufacturing environment, and four technological areas, needed to win these challenges, were identified.

IMTR represented a *planning phase* and was followed in 2000 by the Integrated Manufacturing Technology Initiative (IMTI), the implementation phase, which aims at stimulating the implementation of high-priority R&D projects to support the goals outlined by IMTR. IMTI serves as a point of synergy for manufacturers, technology suppliers, research organisations, universities, associations and consortia, to promote co-operation and to facilitate collaborative development of high-priority technologies.

Based on existing roadmaps, for example NGMP, Visionary 2020, and Internet-based surveys, the entire project involved more than 300 individuals, representing over 150 organisations.

The developed vision shows a competitive environment characterised by dynamics and uncertainty, an accelerated pace of technological change, growing customer expectations, and competition among *extended enterprise*. The above mentioned *six grand challenges* are:

1. *Lean, Efficient Enterprises*: enterprises will integrate business process improvements and workforce performance enhancements with new technologies that raise manufacturing to a new level of efficiency.
2. *Customer-Responsive Enterprises*: future manufacturing enterprises will leverage a global communications infrastructure and customer-centric design, manufacturing, and lifecycle management systems to conceive, build, deliver, and support innovative products and services that directly satisfy diverse customers' needs.
3. *Connected Enterprises*: future manufacturing enterprises will be interconnected among all their internal functions and external partners, suppliers and customers; these extended enterprises will inter-operate as an integrated entity, irrespective of geographic boundaries.
4. *Environmental Sustainability*: manufacturers will face conflicts between the drive to industrialise and the need to protect the global environment; this problem must be addressed, developing innovative materials, zero net lifecycle waste management, and recycling products, with no negative impacts to the environment.
5. *Knowledge Management*: future manufacturing enterprises will be able to capture individual expertise and experience for efficient reuse and draw on a rich, openly accessible shared base of scientific, business, and process knowledge to make accurate decisions and ensure that the right people get the right information at the right time.
6. *Technology Exploitation*: it is important to identify and master the right technologies, which will give competitive advantages. Among them: nano-technology, Internet, computers and electronics, polymers, plasma physics.

Four technological areas were identified to achieve and win the six grand challenges: information systems for manufacturing, modelling and simulation, manufacturing processes and equipment, technologies for enterprise integration.

### 3.4 *Next Generation Manufacturing Systems Program – European Union, Japan, USA (Phase I – 2000; Phase II – 2003)*

http://cam-istandards.org/ngms.html

The Next Generation Manufacturing Systems Program is made up by an international consortium drawing together 22 companies and 11 research groups in the USA, Europe and Japan.

For NGMS the future environment, which manufacturing enterprises will have to compete in, is characterised by strategies based on global networks of self-organising, autonomous units. These units may be part of one company, or part of several companies, located globally, all co-operating to address customers' requirements.

These networks of companies need to rapidly adapt to changing requirements, new technologies and increased globalisation; shorter response times will not allow for experimentation and iteration with real artefacts, so all decisions will be made on the basis of modelling and simulation, rather than build-and-test methods. The entire supply chain (or virtual enterprise) will be modelled and simulated before actual operation allowing choices to be tried and evaluated quickly.

The Next Generation Manufacturing Enterprise will be characterised by:

- Re-configurability: the ability for fast adaptation to erratic and unpredictable environment changes;
- Capability of development: the ability to make evolutionary adaptations;
- Capability to manage turbulence: the ability to create and control turbulence in defined, demarcated markets;
- Capability to realise changes: the ability and the readiness of all employees to change internal structures;
- Evolutionary capability: the ability to continuously change through analysing and learning from the weaknesses and potencies of the past;
- Uniqueness: permanent differentiation compared to competitors;
- Focus on core competencies: competencies to produce unique core products, not only with knowledge, but also with wide practical experience.
(Source: *NGMS White Paper*).

## 3.5 *FutMan – European Union (2003)*

http://europa.eu.int/comm/research/industrial_technologies/articles/article_410_en.html
The FutMan project sought to assist the European Commission in examining what technological, knowledge and organisational capabilities might be required by European manufacturing to remain competitive and sustainable by the year 2020. Particular attention was paid to technological priority areas and to any policy changes required.

FutMan enabled the creation of a forum of manufacturing experts who participated, among other activities, in a scenario generation exercise with the intent to develop coherent long-term visions of European manufacturing in 2015–2020 which could be used as a basis for strategic planning. The scenarios represent imaginative, coherent views on potential socio-economic and technological developments in the future that are likely to shape the European manufacturing sector. Each scenario includes technological, economic, environmental and socio-political factors.

Four individual scenarios were developed, structured along two qualitative dimensions of change. The first dimension refers to policy making and specifically to the balance between central and decentralised decision-making in Europe, and to the co-ordination level between different policy areas. The second dimension refers to the extent to which social and environmental consciousness and public values will impact on future consumer behaviours and demand patterns. The four scenarios, synthesised in terms of *scenario features* and *implications for manufacturing, are:*

Global Economy

- Scenario features:
  - Global governance: World Trade Organisation (WTO) and the interests of large multinational companies shape international trade policies. Policy-making principally aims to strengthen market mechanisms and competition. Little co-ordination of policies among nations.
  - Consumer: consumers pursue personal utility without paying too much attention to environmental and social impacts of production and consumption.
- Manufacturing implications:
  - Manufacturers focus on customisation and individualisation of products.
  - The engineering processes are assumed to be quick and flexible.

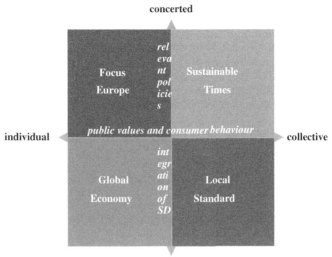

concerted

| Focus Europe | *rel eva nt pol icie s* | Sustainable Times |

individual — *public values and consumer behaviour* — collective

| Global Economy | *int egr ati on of SD* | Local Standard |

loose    SD: Sustainable Development

Figure 1.   FutMan scenarios (Source: *FutMan – The Future of Manufacturing in Europe 2015–2020 – The Challenge for Sustainability – Final Report*).

  – The scenario favours short-term industrial research activities.
  – Sustainability improvements are just a second-order effect owing to the search for energy consumption efficiency.

Local Standard

- Scenario features
  – Global governance: new global protectionism, local authorities have gained new powers. Regional governments determine policy priorities and drive regulation.
  – Consumer: consumers and citizen groups (organised in Non-Governmental Organisations (NGOs)) push their agendas on local environmental issues.
- Manufacturing implications
  – High innovation dynamic on a regional level but disparities among regions.
  – Regional peculiarities lead to a complex environment where centralisation and decentralisation of manufacturing operations will coexist depending on sectors, processes and products.
  – Regional demand structures require new solutions for flexible specialisation in manufacturing.

Sustainable Times

- Scenario features
  – Global governance: a global governance system has emerged that promotes sustainable development and environmental protection worldwide.
  – Consumer: citizens support government co-ordination to reconcile the economic, environmental, and social dimensions of sustainability.
- Manufacturing implications
  – Industry is a partner, closely collaborating with governments and the civil society. Emphasis is given to socially responsible technology development. The notion of competitiveness is broadened taking into account environmental and social aspects of production and consumption.
  – The manufacturing industry strongly pursues service-orientation in product design, and the product becomes less important within the value chain.
  – The industry strives for the optimisation of product lifecycles, introducing full lifetime control and management for their products.

Focus Europe

- Scenario features
  - Global governance: Europe emerges as powerful actor to guide societies toward sustainability. WTO has facilitated international trade but macro-regional disparities may prevail when it comes to sustainable development strategies.
  - Consumer: individualism prevails and citizens transfer their *responsibility* for sustainable development to their governments.
- Manufacturing implications
  - The priority given to strategically important sustainable technology development strengthens Europe's competitive advantage in advanced manufacturing technologies.
  - Industry works hard to attract and keep personnel experienced in using advanced manufacturing tools, managing virtual factories, using simulation methods, etc.

The developed scenarios are not exclusive and most likely a combination of them will occur. Some common developments for future manufacturing can nevertheless be identified:

- Human capital: human capital will replace physical capital at the core of competitive advantage.
- Value added and customisation through provision of services: product value and customisation will increase thorough incorporating a greater service element into the product, during design and during after-sales.
- Virtual enterprises: virtual enterprises via B2B will radically alter their organisational structures and the competitive positioning of firms. Co-opetition will be the rule.
- Flexibility: flexibility will be achieved at a low cost.
- New materials (nano- and biotechnologies) and improved conventional materials in new applications.
- Sustainability: the search for sustainability, either sought by enterprises as a source of competitive advantage or enforced by governments, will play a key role in future production systems.
- Closed-loop systems: waste outputs from one process are used as an input to other processes.

## 3.6  *Manufacturing 20/20 – Canada (2004)*

www.cme-mec.ca/mfg2020/index.asp
*Manufacturing 20/20*, an initiative led by Canadian Manufacturers and Exporters (CME), engaged business leaders, academics, policymakers and other interested parties.

The initiative aimed to develop a detailed roadmap for manufacturing, including a vision on the future of manufacturing in Canada, a list of gaps and challenges, and several recommendations to all the stakeholders.

Among the identified challenges and opportunities the project stressed the importance to:

- Compete on increasingly sophisticated products.
- Take advantage of new markets and business opportunities offered by the developing economies.
- Keep up with rapidly changing customer demands and the accelerating pace of innovation.
- Speed up the rate of new product introductions and process improvements.
- Secure a reliable and cost-competitive supply of energy.
- Obtain adequate financing for new investments, re-engineering and business growth.
- Overcome outdated perceptions of manufacturing on the part of young people, governments and the general public.

Looking to the next five to ten years, Canadian manufacturers highlighted that their future competitiveness and growth opportunities will depend on differentiation in competitive success determinants such as: time to market and time to delivery; service; product; innovation; global sourcing; global markets; continuous improvement; agility; highly skilled workforce; collaboration; productivity improvement; and cost competitiveness.

## 4 TOWARDS A NETWORKED, KNOWLEDGE-BASED, VALUE INTENSIVE MANUFACTURING

The analysed projects present specific features in their visions and, over coming years the introduction of new concepts and the refining of old ones. Nevertheless the most recent works share some basic ideas on the future of manufacturing.

Above all there is a common understanding that, while manufacturing will continue to move closer to emerging markets and to become outsourced into low-cost regions, including large parts of related design and engineering, industrialised countries can only remain competitive by being able to maintain manufacturing of high added value that is rapidly deployable.

The development, manufacturing, and continuous maintenance of new value intensive, intelligent and networked products will play a vital role in such a scenario. These products will be:

- Value intensive, since there will be hardly space for competition on a cost basis with enterprises from developing countries;
- Complex, since they will include a physical/hard part, and a virtual, digital, soft and intangible part;
- Multidisciplinary, since contributions from different scientific disciplines (nano-technology, biotechnology, infotechnology) will be brought together;
- Adaptable and customisable;
- Multi-stakeholder, since beside the *traditional* customer-product interaction, wider social and environmental considerations will drive customer purchasing behaviour and this has to be taken into consideration from the early concept generation phase;
- Long term / lifecycle oriented (lifetime responsibility; lifetime earning potential, …);
- Provided with embedded intelligence.

In parallel, production processes will face a profound shift that will bring them from a centralised, local paradigm to a distributed and global one. Isolated enterprises will not have anymore the know-how and the physical resources to compete in a highly aggressive global dynamic environment. Networks of enterprises (extended and virtual enterprises) will compete against other networks. In this framework the ability to manage co-opetition will be the key for success.

These networks will have to combine agility and leanness to achieve customer total satisfaction while being at the same time cost effective.

## REFERENCES

Agility Forum, Leaders for Manufacturing, and Technologies Enabling Agile Manufacturing (1997), "Next-Generation Manufacturing, A Framework for Action. Executive Overview".

R.E. Albright (2003), "Roadmapping Convergence", Albright Strategy Group, LLC. Available at http://www.albrightstrategy.com

C. Anderson & P. Bunce, CAM-I Next Generation Manufacturing Systems Program (2000), "Next Generation Manufacturing Systems NGMS White Paper". Available at http://cam-i.org/ngms.html

AUS Industry Science Council (2001), "Technology Planning for Business Competitiveness", Emerging Industries. Available at http://industry.gov.au/library/content_library/13_technology_road_mapping.pdf

CAM-I Next Generation Manufacturing Systems Program (2000), "NGMS-IMS project Phase II. Synergistic Integration of Distributed Manufacturing and Enterprise Information. Reference Manual". Available at http://cam-i.org/ngms.html

Canadian Manufacturers & Exporters (2004), "Manufacturing 20/20: Building Our Vision for the Future". Available at http://www.cme-mec.ca/mfg2020/index.asp

Canadian Manufacturers & Exporters (2004), "Manufacturing 20/20 – Update & Preliminary Report". Available at http://www.cme-mec.ca/mfg2020/index.asp

CIMRU (P. Haggins, S. van Dongen, A. Kavanagh, B. Wall, M. Smyth, etc.) (2002), "Roadmapping: An Overview", Intelligent Manufacturing Systems. Available at http://www.ims.org

Committee on Visionary Manufacturing Challenges, established by the National Research Council's Board on Manufacturing and Engineering Design (1998), "Visionary Manufacturing Challenges for 2020". Available at http://www.nap.edu/html/visionary

R.S. da Fonseca, "UNIDO Technology Foresight Programme" (2000), UNIDO. Available at http://www.unido.org/foresight

C. Dreher, H. Armbruster, E. Schlrrmeister, P. Jung-Erceg, "ManVis Report 2 – Preliminary Results from the first Round of the ManVis Delphi Survey" (2005), Fraunhofer Institute for System and Innovation Research.

European Commission (2003), "Working Document for The ManuFuture 2003 Conference". Available at http://www.manufuture.org

European Commission (2004), "ManuFuture: A Vision for 2020". Available at http://europa.eu.int/comm/research/industrial_technologies/manufuture/home_en.html

Foresight Manufacturing 2020 Panel Members (N. Scheele, A. Daly, M. Gregory, etc.) (2000), "UK Manufacturing: We can make it better. Final Report Manufacturing 2020 Panel". Available at http://www.foresight.gov.uk/manu2020

FutMan Project (2003), "The Future of Manufacturing in Europe 2015-2020, The Challenge for Sustainability", Available at http://www.cordis.lu

Industry Canada, (2000), "Technology Roadmapping: A Guide for Government Employees".

Integrated Manufacturing Technology Roadmapping (2000), "Integrated Manufacturing Technology Roadmapping Project: An Overview of the IMTR Roadmaps", Available at http://www.IMTI21.org

Integrated Manufacturing Technology Roadmapping (2000), "Manufacturing Success in the 21st Century: A Strategic View". Available at http://www.IMTI21.org

IMS-NoE website, www.ims-noe.org

ManuFuture, "ManuFuture: A Vision for 2020", 2004.

ManuFuture, "Working Document for The ManuFuture 2003 Conference", 2003.

ManVis, "Results from the *Manvis* first questionnaire", 2005.

Ministry of Education, Culture, Sport, Science and Technology, National Institute of Science and Technology Policy, "The Seventh Technology Foresight" (2001). Available at http://www.nistep.go.jp

Ministry of Education, Culture, Sport, Science and Technology, "White Paper on S&T Performance" (2002). Available at http://www.mext.go.jp

R. Phall (2002), "Technology Roadmapping", Centre for Technology Management, University of Cambridge, United Kingdom

Planning Committee organised by Korean Government (2000), "Vision 2025, Korea's Long-term Plan for Science and Technology Development". Available at http://www.most.go.kr

South Africa Manufacturing and Materials Working Group (2004), "Foresight Manufacturing Report". Available at http://www.sita.co.za

U.S. Department of Commerce, Washington, D.C. (2004), "Manufacturing in America: A Comprehensive Strategy to Address the Challenges to U.S. Manufacturers". Available at http://www.nam.org

Vision Consortium (2003), "Information Societies Technology Programme. Final Version of Strategic Roadmap", Contract Number IST-2002-38513.

*Advanced Manufacturing – An ICT and Systems Perspective – Taisch,*
*Thoben & Montorio (eds)*
*© 2007 Taylor & Francis Group, London, ISBN 978-0-415-42912-2*

# The IMS-NoE Delphi survey of ICT in manufacturing

Marco Montorio & Marco Taisch
*Politecnico di Milano, Department of Management, Economics and Industrial Engineering, Milano*

ABSTRACT: Within the framework of the IST project, IMS-NoE (*Network of Excellence on Intelligent Manufacturing Systems*), financed by the European Commission under the 5th Framework Programme, a Delphi study was undertaken with the intent of building a vision on the future of manufacturing with a special focus on the impact of Information and Communication Technologies (ICT) on the next generation of production systems. This paper summarises the results obtained from the Delphi study and proposes a critical analysis in terms of future competitive environment (external drivers), manufacturing enterprise attributes in response to the new market challenges (internal attributes) and key enabling ICT.

*Keywords*: ICT for manufacturing, next generation manufacturing, vision, Delphi study, intelligent manufacturing systems.

## 1 THE DELPHI SURVEY OF ICT IN MANUFACTURING

Within the IST project, IMS-NoE (*Network of Excellence on intelligent Manufacturing Systems*) (available at www.ims-noe.org), financed by the European Commission under the 5th Framework Programme, several activities have been undertaken aimed at developing a vision on the future of manufacturing: among these was a Delphi survey.

The survey, which aimed to develop a vision on the next generation manufacturing enterprises and the competitive environment they will have to face, had a special focus on the role which Information and Communication Technologies (ICT) will play in this context and how they will shape manufacturing in the decade to come. The Delphi was based on a two step survey with open- and closed- answer questions respectively on the first and second questionnaires.

The panel of the survey was made up of worldwide experts in manufacturing coming from industry, consulting, university and research, with a total of almost 1000 participants. The obtained response rate was equal to 16%. Figures 1 and 2 show respectively the composition and the geographical origin of the panel.

## 2 THE STRUCTURE OF THE OUTPUT: DRIVERS, ATTRIBUTES AND ICT

The Delphi questionnaire was structured into three sections. Within each section questions were asked to the interviewed experts with the intent to portray a vision on:

1. The future competitive environment (external drivers) which manufacturing enterprises will have to face.
2. The assets, features, capabilities and attributes that manufacturing enterprises will have to develop in response to the new market challenges (internal attributes).
3. The key enabling Information and Communication Technologies that will enable manufacturing enterprises to remain competitive and profitable in future markets.

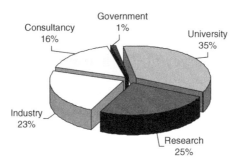

Figure 1.   Composition of the Delphi panel.

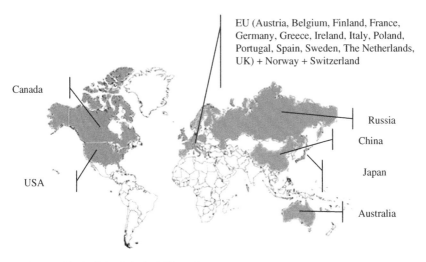

Figure 2.   Geographical origin of the Delphi panel.

For the first step of the Delphi study the experts were asked to answer open questions and were invited to use all their knowledge and creativity. The obtained answers were then clustered adopting an *ad hoc* statistical text-analysis technique to reduce them into a few statements. These statements were then sent back to the participants, who were asked to select the 5 most relevant among them according to their opinion.

In the following the main external drivers (D), the most important internal attributes (A) and the key Information and Communication Technologies (T) identified by the experts are listed, from the most voted for, down to the least one. Included in each element is the percentage of experts who selected the item.

### The competitive environment (external drivers):

D1.  (65%) Competition among manufacturing enterprises will be mainly based on customisation, on products and their related services. An ever more demanding customer in terms of price, quality, adherence to delivery times and customisation of product design, will require customised products and services to meet individual needs, in a total lifecycle oriented approach (customisation of design, production, after sales and end of life processes).

D2.  (58.4%) Every manufacturing enterprise will be highly focused on its core business. Centralised local production will be increasingly replaced by networked global production. Isolated enterprises will not have anymore the know-how and the physical resources to compete in a highly competitive global dynamic context. Competition will be among networks

14

of firms where every global, virtual extended enterprise will be made up by a large number of companies. Full transparency and sharing of information among the nodes will enable an efficient Supply Network management (e.g. real-time management).

D3. (45.1%) Co-opetition among enterprises will be the rule. Enterprises will compete and co-operate at the same time. This will be enabled also by the use of standard software interfaces. In particular pre- and post- competitive knowledge will be shared among co-opeting enterprises.

D4. (39.8%) The relationship between customer and supplier will become more complex (relying on articulated service contracts) and will last for the entire product/service lifecycle (from product/service conception, through its production, use and down to its disposal).

D5. (38.2%) Owing to B2B and B2C practices, markets will become increasingly global. The customer will get to know products and services independently of where these are produced and they will be able to purchase them from all over the world through the Internet.

D6. (36.7%) An efficient and effective Supply Chain will deliver high quality products and services to the final customer; in this view, rather than lowering prices, a continuous effort to improve quality and delivery times will be shared by all the Supply Chain actors.

D7. (35%) Academia and research, even in developing countries, will collaborate more deeply, creating consortia or networks of excellence, in which they will solve shared problems. Research centres will act as *expertise suppliers* to industry (e.g. technology adoption and improvements, continuous learning for employees…), in a real *knowledge supply chain*.

D8. (26.7%) Customers' purchasing behaviour will be led by *satisfaction* criteria (e.g. quality of products and services, delivery times…) as well as *environmental and social* criteria (e.g. environmental awareness, social influence, culture, fashion effect, …).

D9. (23.3%) Labour market will be global, leading to a high worldwide mobility of the workforce, which will be flexible, highly qualified, and oriented to high technology and services. The availability of skilled workforce will be a bottleneck for enterprises, while the unskilled workforce will decline.

D10. (21.7%) Government will have an active role in promoting public funded international long-term research programmes and in supporting the collaboration between R&D actors and enterprises (in particular SMEs).

D11. (20%) The market saturation, in developed countries, will increase, owing to the entrance of new enterprises from China, India and the Far East. The difficulty of finding new forms of competitive advantage will bring the market to a highly aggressive competition. As a consequence enterprises will find growing entry barriers to the market in the future. Niche manufacturing markets will tend to be saturated as well.

### *The enterprise answer to the foreseen external pressures (internal attributes):*

A1. (69.2%) The combination of emerging technologies such as bio-, nano- and new material technologies will enable the development of innovative products, which will be incredibly more complex, customised and provided with embedded intelligence. Mastering these technologies will be the key for success in developed countries for the next generation manufacturing enterprises.

A2. (48.4%) The enterprise will be highly integrated with the Supply Chain, through organisational and technological supports (e.g. automatic negotiation and contracting). Consequently the differences between managing in-sourced and outsourced activities will decrease.

A3. (45.7%) A total lifecycle approach will be implemented to manage the product/service from the concept, through the design, production, and use, down to the end of its life and its disposal.

A4. (41.8%) Process design and management will be automated, flexible, re-configurable and enhanced by ICT tools (e.g. online re-configuration).

A5. (39.9%) People will be at the heart of manufacturing process design. The production systems will be equipped with friendlier human-machine interfaces, aiming at deeper human-machine co-operation.

A6. (36.5%) There will be an enhancement of workforce flexibility, regarding contracts, time and mobility. Automation and service orientation will also lead to an increase of highly skilled support personnel. *Ad hoc* contracts will be used to retain key skills.

A7. (31.7%) Every enterprise product and process design activity will be customer driven and expressly carried on with the intent to satisfy a specific customer request. The enterprise will develop sophisticated internal procedures and solutions (CRM) to discover and formalise customer needs.

A8. (31%) Knowledge management solutions will be largely adopted and will be fundamental to translating enterprise information into knowledge and acquiring / managing the knowledge and experience developed by highly skilled workers.

A9. (30%) Production systems will have high environmental care. Products and production systems will be environmentally cleaner.

A10. (27.6%) The product and process management will implement lean approaches with the purpose of achieving continuous improvement and systematic innovation (based on TQM, BPR, . . . approaches).

A11. (25%) The organisation of the enterprise will be adhocratic: organised by processes (and not by functions), dynamic, flexible, lean, with a reduced number of hierarchical levels and team oriented.

A12. (Less than 20%) In developed countries there will be mainly space for just two kinds of production systems. The first solution will be a large, highly automated, high technology and almost unmanned plant. The second one will be a small, flexible, low-volume, re-configurable and labour intensive production plant close to the market (e.g. for final assembling, person-alisation...), aiming at mass customisation. There will be little space for other configurations in between.

### *Key enabling Information and Communication Technologies (ICTs):*

T1. (61.5%) Modelling, prototyping and simulation tools for plants, processes and products (e.g. enterprise modelling, online and collaborative prototyping, rapid and virtual prototyping, computer aided tools (CAx) ...).

T2. (56.8%) Intelligent production planning systems (e.g. expert systems, neural networks, genetic algorithms, vision systems, multi-agent systems ...).

T3. (54.8%) Tracking devices technologies for production and logistics processes (e.g. RFID, smart tags ...).

T4. (40.1%) Internet solutions (e.g. B2B and B2C solutions, semantic web, e-communications tools, Internet 2 ...).

T5. (34-3%) Software and hardware interfaces for the integration of different ICT systems (based on standards and protocols).

T6. (33-2%) Lean manufacturing solutions (e.g. design for manufacturing and assembly (DFMA), modularisation of components, Group Technology, Total Quality Management ...).

T7. (33.5%) Knowledge-based information systems to model, store and manage experts' knowl-edge (e.g. knowledge communities, virtual innovation centres, virtual partnerships among academia/research/enterprises ...).

T8. (32.5%) Adaptive and feed forward process control systems (e.g. relying on the correct Key Performance Indicators) and sensors (e.g. miniaturised wireless sensors, multi-sensors systems ...).

T9. (28.3%) Pervasive automation of processes and equipment highly integrated with humans (e.g. intelligent and friendly interfaces).

T10. (23.3%) Planning tools for the knowledge integration (e.g. ERP, SCM, CRM, MIS ...) and for data mining.

T11. (Less than 20%) Technologies for pervasive, always and everywhere available communication (e.g. wireless, GPRS, UMTS ...).

T12. (Less than 20%) Technologies for assuring information security (e.g. IPR defence, anti-hacking).

T13. (Less than 20%) Remote assistance for plant monitoring and maintenance (e.g. online diagnosis and repair of faults …).

## 3   DRIVER-ATTRIBUTE AND ATTRIBUTE-ICT RELATIONSHIPS

Specific questions within the Delphi survey were asked to the participants to identify the correlation among *internal attributes* and *external drivers* on one side, and *ICT* and *external attributes* on the other. In other words the questioned experts were invited to identify:

- Which enterprise attributes will turn out to be crucial to face the forthcoming competitive environment, or, in other words which internal attributes will be necessary to face each identified external drivers;
- Which Information and Communication Technologies will be vital for the enterprise to achieve profitability and sustainability, or, in other words which ICTs will be necessary to achieve each identified internal attribute.

Figures 3 and 4 represent the correlation identified by the experts respectively for *driver-attribute* and *attribute-ICT* couples. The table should be read according to the following legend:

| | | |
|---|---|---|
| ● | Dark spot | At least 75% of the interviewed experts have identified an existing relationship for the *driver-attribute* or *attribute-ICT* couple under evaluation. |
| ◓ | Intermediate spot | Between 50% and 75% of the interviewed experts have identified an existing relationship for the *driver-attribute* or *attribute-ICT* couple under evaluation. |
| ○ | Light spot | Between 25% and 50% of the interviewed experts have identified an existing relationship for the *driver-attribute* or *attribute-ICT* couple under evaluation. |
| | No spot | Less than 25% of the interviewed experts have identified an existing relationship for the *driver-attribute* or *attribute-ICT* couple under evaluation. |

## 4   INDUSTRY-RESEARCH BENCHMARKING

A benchmark between the answers given by the interviewed representatives of academia and research on one side and industry and consulting on the other has been carried on with the intent out of highlighting possible discrepancies. Identified drivers, attributes and ICTs have been compared, showing a general concordance, with a few exceptions.

### 4.1   *Comparison of the drivers*

Figure 5 shows a comparison between the drivers identified by representatives from university/research on one-side and industry/consulting firms on the other.

It is possible to notice a general correspondence between the answers given by the two groups. For example driver D1 (product and service customisation) and driver D2 (focus on core business and emergence of virtual and extended enterprises) are at the top of both lists.

Nevertheless it is valuable to point out how driver D11 (increased market saturation) is envisioned by researchers and academic people as a significant challenge for the future while experts from industry and consulting firms give less emphasis to this.

One more consideration is that D4 (complex customer-supplier relationships) and D5 (globalisation through B2B and B2C solutions) appear to be more important for researchers and academic people.

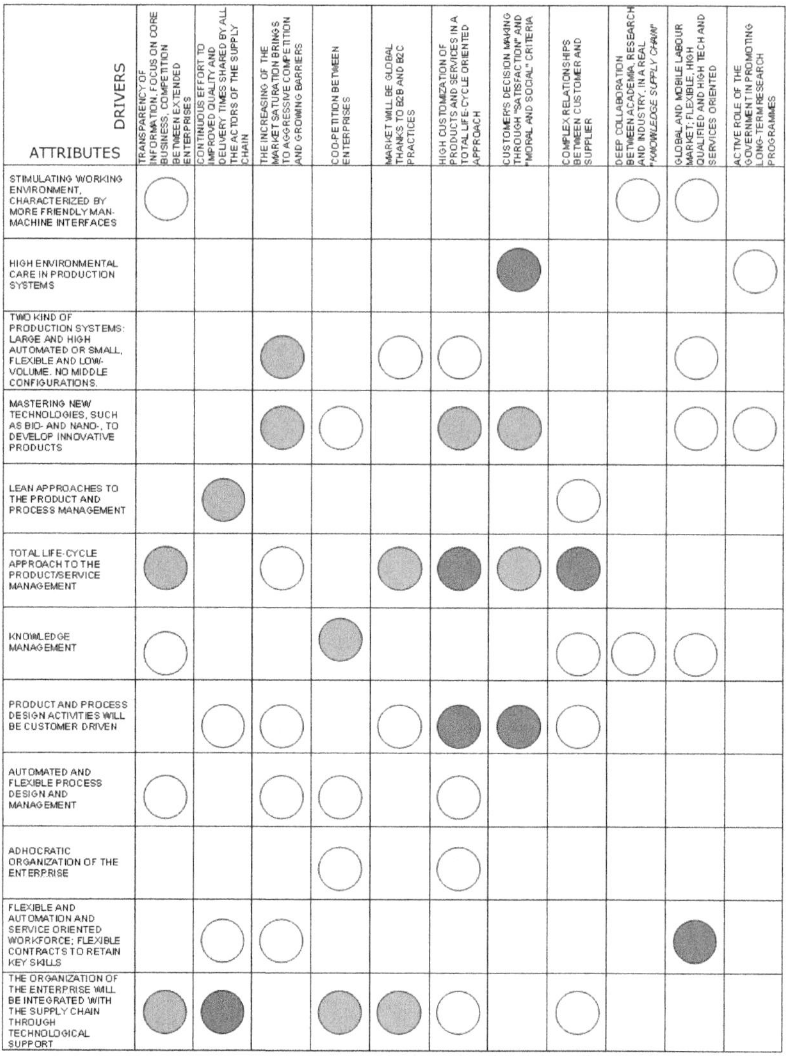

Figure 3. *Driver-attribute* relationships.

### 4.2 *Comparison of the attributes*

Figure 6 compares the attributes identified by representatives from university/research on one-side and industry/consulting firms on the other.

Even when it comes to attributes it is possible to notice a general homogeneity between the answers given by the two groups. It is worth highlighting for example how attribute A1 (combination of emerging technologies such as bio-, nano- and new material technologies will enable the development of innovative products) is considered to be of crucial importance for both.

As an exception it can be seen that:

- A9 (products and production systems will be environmentally cleaner) and A6 (importance of flexible and high-skilled workforce) are seen as key abilities of the future enterprise for the researchers while this is less the case for industrialists and consultants.

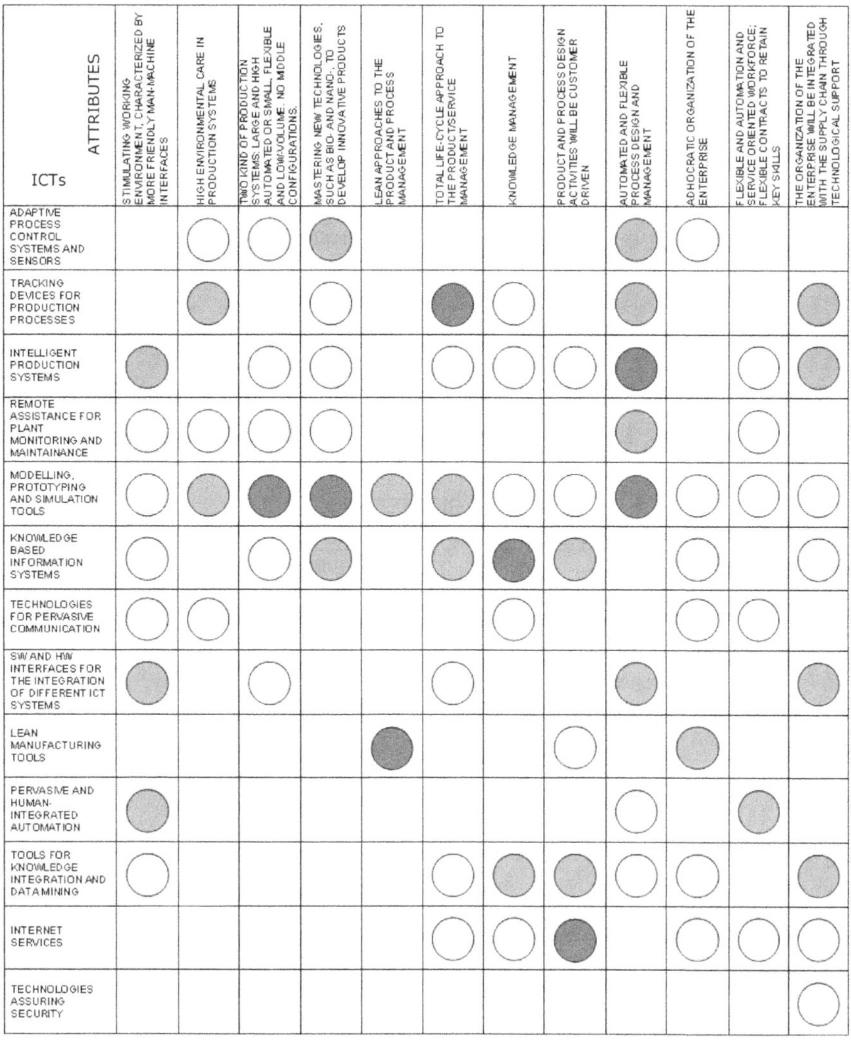

**Figure 4.** *Attribute-ICT* relationships.

- The opposite is true for A8 (efficient knowledge management solutions to translate information into knowledge) and A10 (lean approaches for product and process management).

### 4.3 *Comparison of the ICTs selected*

Figure 7 shows a comparison between the ICTs identified by representatives from university/research on one side and industry/consulting firms on the other.

One more time it can be seen a general coherence between the opinions of the two groups.

As an exception, T5 (software and hardware interfaces for the integration of different ICT systems) and T7 (knowledge-based information systems), are still seen as significant issues by industry and consulting, while this is less the case for academic people and researchers. However, T3 (tracking device technologies for production and logistics processes) is perceived as a *hotter* topic by researchers than by people from industry.

| UNIVERSITY & RESEARCH | | INDUSTRY & CONSULTING | |
|---|---|---|---|
| Driver | % | Driver | % |
| D1 | 61.6% | D1 | 70.2% |
| D2 | 54.8% | D2 | 63.8% |
| D4 | 49.3% | D3 | 51.1% |
| D5 | 46.6% | D6 | 38.3% |
| D3 | 41.1% | D7 | 34.0% |
| D6 | 35.6% | D8 | 31.9% |
| D7 | 35.6% | D9 | 29.8% |
| D11 | 30.1% | D5 | 25.5% |
| D8 | 23.3% | D4 | 25.5% |
| D10 | 21.9% | D10 | 21.3% |
| D9 | Less than 20% | D11 | Less than 20% |

Figure 5.   Comparison of the drivers selected by academia/research and industry/consulting.

| UNIVERSITY & RESEARCH | | INDUSTRY & CONSULTING | |
|---|---|---|---|
| Attribute | % | Attribute | % |
| A1 | 68.5% | A1 | 70.2% |
| A3 | 50.7% | A2 | 55.3% |
| A6 | 45.2% | A4 | 48.9% |
| A2 | 43.8% | A8 | 40.4% |
| A5 | 43.8% | A3 | 38.3% |
| A9 | 38.4% | A10 | 36.2% |
| A4 | 37.0% | A5 | 34.0% |
| A7 | 32.9% | A7 | 29.8% |
| A12 | 27.4% | A11 | 25.5% |
| A11 | 24.7% | A6 | 23.4% |
| A8 | 24.7% | A9 | Less than 20% |
| A10 | 21.9% | A12 | Less than 20% |

Figure 6.   Comparison of the attributes selected by academia/research and industry/consulting.

| UNIVERSITY & RESEARCH | | INDUSTRY & CONSULTING | |
|---|---|---|---|
| ICT | % | ICT | % |
| T1 | 68.5% | T2 | 63.8% |
| T3 | 61.6% | T1 | 51.1% |
| T2 | 52.1% | T5 | 46.8% |
| T4 | 38.4% | T3 | 44.7% |
| T6 | 38.4% | T7 | 42.6% |
| T8 | 30.1% | T4 | 42.6% |
| T9 | 30.1% | T8 | 36.2% |
| T7 | 27.4% | T6 | 25.5% |
| T10 | 27.4% | T9 | 25.5% |
| T5 | 26.0% | T12 | 21.3% |
| T11 | 20.5% | T11 | Less than 20% |
| T13 | Less than 20% | T10 | Less than 20% |
| T12 | Less than 20% | T13 | Less than 20% |

Figure 7.   Comparison of the ICTs selected by academia/research and industry/consulting.

## 5 FINDINGS

It is difficult to summarise in a few lines how ICT will shape the future competitive environment and how it will impact on the future of manufacturing. Nevertheless, a few conclusions can be drawn:

*From the customer perspective*

ICT solutions will enable customers to get information and to purchase products and services globally. Directly on the web through a PC, a pocket PC or a smart phone, customers will be able to set up a *unique* product. The physical artefact will be surrounded by a set of services (maintenance, assistance, loan plan, etc.) which will support customers along the entire lifecycle until the final disposal. Frequently customers will not buy the physical product itself; instead they will directly pay for the related service on a pay-per-use base.

*From the manufacturing company perspective*

Globalisation will lead to an aggressive competition coming from old and new actors and to a general increase of market saturation. Finding new forms of competitive advantage will be difficult for manufacturing enterprises.

Greater flexibility and re-configurability will be necessary to guarantee constant adherence to the always-unsteady customer requirements. In parallel, lean approaches will have to be implemented, aiming at a continuous improvement and innovation of direct and indirect processes.

Lifecycles of products and services of increased technological content and complexity will be difficult to manage by a single enterprise, which will not own the competencies and the physical resources needed to achieve competitiveness on a global scale. Manufacturing companies will therefore collaborate in the framework of supply networks or virtual enterprises with rapid creation-dissolution. The competition will then be on a network vs. network basis and on a global scale.

Different markets will require different industrial structures. Nevertheless, generally speaking, networks will see the presence of large scale, highly automated and low-manned industrial sites able to compete on costs, along side small dimension, highly flexible, re-configurable, low volume and close-to-the-market plants designed for the delivery of personalised high add-value products and services.

## REFERENCES

Agility Forum, Leaders for Manufacturing, and Technologies Enabling Agile Manufacturing (1997), "Next-Generation Manufacturing, A Framework for Action; Executive Overview". Available at http://www.agilityforum.com

R.E. Albright (2003), "Roadmapping Convergence", Albright Strategy Group, LLC. Available at http://www.albrightstrategy.com

C. Anderson & P. Bunce, CAM-I Next Generation Manufacturing Systems Program (2000), "Next Generation Manufacturing Systems NGMS White Paper". Available at http://cam-i.org/ngms.html

AUS Industry Science Council (2001), "Technology Planning for Business Competitiveness", Emerging Industries. Available at http://industry.gov.au/library/content_library/13_technology_road_mapping.pdf

CAM-I Next Generation Manufacturing Systems Program (2000), "NGMS-IMS project Phase II. Synergistic Integration of Distributed Manufacturing and Enterprise Information. Reference Manual". Available at http://cam-i.org/ngms.html

Canadian Manufacturers & Exporters (2004), "Manufacturing 20/20: Building Our Vision for the Future". Available at http://www.cme-mec.ca/mfg2020/index.asp

Canadian Manufacturers & Exporters (2004), "Manufacturing 20/20 – Update & Preliminary Report". Available at http://www.cme-mec.ca/mfg2020/index.asp

CIMRU (P. Higgins, van Dongen, S., Kavanagh, A., Wall, B., Smyth, M. etc.) (2002), "Roadmapping: An Overview", Intelligent Manufacturing Systems. Available at http://www.ims-noe.org

Committee on Visionary Manufacturing Challenges; established by the National Research Council's Board on Manufacturing and Engineering Design (1998), "Visionary Manufacturing Challenges for 2020". Available at http://www.nap.edu/html/visionary

da Fonseca, R.S.: "UNIDO Technology Foresight Programme" (2000), UNIDO. Available at http://www.unido.org/foresight

Dreher, C. Armbruster, H., Schlrrmeister, E., Jung-Erceg, P.: "ManVis Report 2 – Preliminary Results from the first Round of the ManVis Delphi Survey" (2005), Fraunhofer Institute for System and Innovation Research.

European Commission (2003), "Working Document for the ManuFuture 2003 Conference". Available at http://www.manufuture.org

European Commission (2004), "ManuFuture: A Vision for 2020". Available at http://europa.eu.int/comm/research/industrial_technologies/manufuture/home_en.html

Foresight Manufacturing 2020 Panel Members (N. Scheele, A. Daly, M. Gregory, etc.) (2000), "UK Manufacturing: We can make it better. Final Report Manufacturing 2020 Panel". Available at http://www.foresight.gov.uk/manu2020

FutMan Project (2003), "The Future of Manufacturing in Europe 2015-2020, The Challenge for Sustainability", Available at http://www.cordis.lu

Industry Canada, (2000) "Technology Roadmapping: A Guide for Government Employees".

Integrated Manufacturing Technology Roadmapping (2000), "Integrated Manufacturing Technology Roadmapping Project: An Overview of the IMTR Roadmaps", Available at http://www.IMTI21.org

Integrated Manufacturing Technology Roadmapping (2000), "Manufacturing Success in the 21st Century: A Strategic View". Available at http://www.IMTI21.org

IMS-NoE website, www.ims-noe.org

ManuFuture, "ManuFuture: A Vision for 2020", 2004.

ManVis, "Results from the ManVis first questionnaire", 2005.

Ministry of Education, Culture, Sport, Science and Technology, National Institute of Science and Technology Policy, "The Seventh Technology Foresight"(2001). Available at http://www.nistep.go.jp

Ministry of Education, Culture, Sport, Science and Technology: "White Paper on S&T Performance" (2002). Available at http://www.mext.go.jp

Phall R. (2002), "Technology Roadmapping", Centre for Technology Management, University of Cambridge, United Kingdom

Planning Committee organized by Korean Government (2000), "Vision 2025: Korea's Long-term Plan for Science and Technology Development". Available at http://www.most.go.kr

South Africa Manufacturing and Materials Working Group (2004), "Foresight Manufacturing Report". Available at http://www.sita.co.za

U.S. Department of Commerce, Washington, D.C. (2004), "Manufacturing in America: A Comprehensive Strategy to Address the Challenges to U.S. Manufacturers". Available at http://www.nam.org

Vision Consortium (2003), "Information Societies Technology Programme: Final Version of Strategic Roadmap", Contract Number IST-2002-38513.

*Advanced Manufacturing – An ICT and Systems Perspective – Taisch,*
*Thoben & Montorio (eds)*
*© 2007 Taylor & Francis Group, London, ISBN 978-0-415-42912-2*

# Manufacturing visions: A holistic view of the trends for European manufacturing

Carsten Dreher
*Fraunhofer Institute System and Innovation Research ISI, Department Industrial and Service Innovations,*
*Karlsruhe, Germany*

ABSTRACT: This paper presents the main findings of the *ManVis* project (*Manufacturing Visions – Integrating Diverse Perspectives into Pan-European Foresight*) in a very condensed way in the form of hypotheses (a detailed overview on reports and results can be found on www.manufacturing-visions.org). The paper investigates which technologies will be relevant for the future of manufacturing and which role, visions, challenges, and needed actions, will emerge for European manufacturing.

*Keywords*: *ManVis*, manufacturing visions, technology foresight, R&D policies.

## 1 INTRODUCTION

The specific support action *Manufacturing Visions – Integrating Diverse Perspectives into Pan-European Foresight (ManVis)* (Contract No NMP2-CT-2003-507139) started in early 2004. Its aim was to accompany the continuing policy process of enhancing European competitiveness in manufacturing industries and to include views of European manufacturing experts collected through a Delphi survey as well as views of stakeholders and overseas experts collected at workshops and in interviews. *ManVis* was an independently launched activity but has a supporting role in the policy process assembled under the catchword *ManuFuture* as well. In the meantime, *ManVis* has contributed to this process through presentations and inputs to *ManuFuture* and other conferences. This paper presents the main findings in a very condensed way in the form of hypotheses (a detailed overview on reports and results can be found on www.manufacturing-visions.org).

As a tool for initiating future-oriented thinking and to promote the linking of such diverse perspectives, a pan-European Delphi survey dealing with manufacturing issues was launched. In several workshops, manufacturing experts from all over Europe and overseas contributed to the shaping of the survey. To avoid an isolated view of Europe's manufacturing issues, experts from overseas were involved in the development of the statements of the Delphi questionnaire and commented on the results of the survey. Emphasising and elaborating the demand side perspective on manufacturing was an important aim of this project. Because of this, the views of users, consumers and other societal groups concerned with manufacturing discussed the findings of the Delphi survey. In parallel to the Delphi activities, scenarios on the development of the demand side of manufacturing were elaborated.

Because of the complex structure of the questionnaire, covering various areas of expertise, not all the 3112 experts completed it entirely, but chose to answer only those sections with which they felt most comfortable. Each statement was answered by more than 1200 experts, allowing a solid statistical analysis for all the statements. The median number of answers per statement is 1332. Since no systematic differences have been discovered after the first round (for instance with respect to expert origin, country etc.), it was considered risk-free to include all answers,

Figure 1. The *ManVis* Approach.

regardless of the number of statements each expert answered. For the second round, the team decided on a modification in the methodology. Instead of repeating the questionnaire of the first round, conflicting statements were regrouped and present jointly for re-consideration. In the second round, 1359 experts participated. Generally, the results of the second round only partly changed the results of the first round.

It is important to highlight the role of foresight exercises based on surveys and expectations like Delphi studies, workshops, and expert interviews as a *starting point* or *one of several inputs* to public debates on future developments. It does not replace other research or strategic planning activities as for instance scenario building, patent data analysis or other technology assessment methods nor interpretation of innovation indicators.

## 2   WHICH TECHNOLOGIES WILL BE RELEVANT FOR EUROPEAN MANUFACTURING?

From the presented findings on technology and their dynamics in manufacturing, some messages can be derived from *ManVis*:

- Micro electromechanical devices, smart materials, products using nano-coatings – in this timing order – are representing long-term developments in a new type of products with a disruptive character for markets. These product challenges offer an opportunity for strengthening competitiveness, which can only be exploited if appropriate manufacturing equipment is available and allowing the use of the technologies in new products. Hence, generic technology development needs complementary manufacturing technology research involvement.
- Such new manufacturing technology principles as bottom-up manufacturing technologies are only expected in the long run. Manufacturing technologies using biotechnologies for creating and manipulating inorganic material and products such as nano-manufacturing should also be on the long-term *radar* of RTD-policy.
- Micro electromechanical systems (here a European advantage in R&D is seen by the experts) as well as flexible organisation and automation strategies combined e.g. in re-configurable manufacturing systems supporting flexible business strategies are important for the short-term

research agenda. However, the people-less factory still receives a sceptical assessment by the experts. Humans working with flexible automation solutions in the near future will still play an important role in creating the flexibility. Experts expect therefore, people working with flexible automation technologies instead of a people-less factory.

- Only long-term automation visions comprise new ways of interactions between machines and humans such as human-machine interfaces, human-machine speech recognition, self-learning systems, and co-bots.

These issues underline the need for research on industrial adoption and innovation management practices in manufacturing industries and intensive communication and further debate of the *ManVis* results.

The development of new generic technologies and knowledge, challenges manufacturing research in two ways. First, it creates a need for manufacturing processes to produce the new products and provide the new services. Secondly, these new technologies and knowledge have to be integrated into the production processes themselves. *Basic manufacturing research* has to foresee and prepare for the new challenges, and *applied manufacturing research* has to adapt and transform existing technologies and organisational processes. Furthermore, manufacturing research plays a decisive role in combining the long-term horizon in technology trajectories with the short-term need of firms to innovate successfully. This requires a good *timing* of research activities to have solutions and tools ready for industrial adoption.

Considering these functions of manufacturing research, the *ManVis* messages on technology can be discussed using the presented concept of the combined science-technology cycle on innovation presented in *ManVis* Report No. 3. Four groups of technologies were discussed in several *ManVis*-statements:

- bottom-up manufacturing technologies (bio- or nano-processes)
- advanced materials
- micro systems technologies
- information and communication technologies.

For these technologies, the experts expressed different time horizons for realisation. Activities for basic and applied research have to be performed in advance (approximately 10–15 years for basic research, 5–10 years for applied research).

ICT will still play the decisive role in the short-term perspective in *manufacturing operations*, but only if the human–machine interaction is considered properly because the people-less factory is not foreseen in the future. Using the assessment and referring to existing diffusion studies, ICT use in manufacturing is in phase 6, where application oriented industrial research is predominant. Therefore, attacking dominant designs with new solutions, e. g. in simulation of processes and the product lifecycle, could be very costly and may have a lower success rate than developing solutions in more open fields. Hence, some mapping information on simulation solutions seems necessary for better-informed decisions.

As outlined by stakeholder assessments and the evaluation of the experts, ICT can play a *crucial role for customisation*. Software and ICT-components incorporated into products for customisation are important for product innovation. As said before, accompanying research on social and business impacts and standards (development and enforcement) should be important supporting elements in a public research strategy.

Micro systems (together with intelligent controls) are *key enablers* for plug-and-produce systems aiming at more flexible manufacturing systems as well as for process integration into multi-functional machinery. For this second technology, the necessary link from developing new machinery to creating new business models (although not emphasised by the experts) could be crucial as well as research bringing together equipment suppliers and users. This represents phase 4/5 as the most important stage of defining and setting dominant designs. The *ManVis* experts see Europe in an advantageous position (i.e. in some lead user industries such as automotive and medical equipment) and at the forefront. Industrial research is the main driver now. To be very

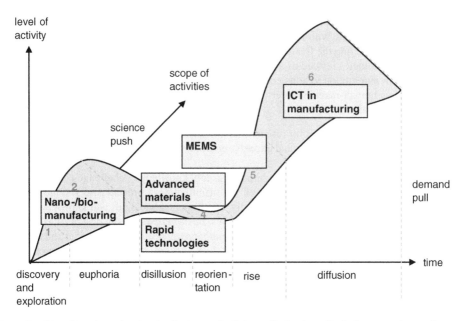

Figure 2.   Manufacturing related technologies on the Science-Technology Cycle for macro-innovations.

precise: Micro Electronic Mechanical Systems (MEMS) are not a basic research topic (and maybe because of that a little bit out of focus of the public attention) but are on the verge of a take-off in industrial use. It is important to maintain the existing advantages and exploit the commercialisation for the benefit of European manufacturing (Bierhals, R., Cuhls, C., et al. 2000).

For advanced materials, the problem of making the *processing and manipulation of new materials* feasible and (more importantly) competitive has already been identified as an important research topic by the FutMan study, i.e. smart materials and rapid technologies are in phases 3 to 4 representing a selection process and the search for breakthrough applications. These phases are characterised by search processes to assess and exclude non-viable options. It is a sobering phase of applied research in the concerned sciences and in engineering. Here, collective research efforts combining the related sciences, engineering and lead industries are helpful in bridging this period. The *ManVis* experts give a time horizon that leaves enough space for catching up in the R&D position that is considered lagging for the moment.

The new catchwords representing bottom-up manufacturing are in the middle of the first boom in the science cycle, close to euphoria (phase 2). The *ManVis* experts see the development as important for manufacturing but only on a very long-term horizon. Hence, basic research on nano- or biotechnology has to be carefully monitored for emerging manufacturing research topics. In addition, cross cutting manufacturing research issues like measurement, workplace safety of nano- or bio-based processes etc. may facilitate the basic research activities in other fields of nano- and biotechnology. A screening or roadmapping activity on nano-manufacturing connected to product roadmaps using nano-technology is useful to prevent an overlooking of possibilities. An additional action is the analysis of linking micro systems technologies with nano-based technology using existing advantages in micro systems to facilitate faster diffusion of nano-technologies.

The analysis in different sectors confirmed these views of the experts on the technologies, but is varying in the time horizon (cf. *ManVis* Report No. 3).

The *ManVis* experts see that e.g. ICT use in manufacturing and product development operations will be a driving element in traditional industries whereas in other sectors it will not play such an eminent role in operations and production technologies. More important is the role of ICT in customisation of products and product-service combinations. New materials may play a more

26

prominent role in the sector of fabricated metal products. In machinery, MEMS are predominantly seen for the next steps; i.e. in self-adapting systems as part of advanced machine tools. The car industry (mostly seen as a lead user for many technologies) is facing the challenge to master two more engine concepts (hybrid and fuel cells) parallel to the optimisation of the combustion engines. The *ManVis* experts do not prioritise one single option but the impacts on other industries could be dramatic.

This analysis is based on the *ManVis* experts' views and on secondary material. It could be useful to validate certain areas by using specialised innovation indicators (publications, patents, and diffusion data) in targeted foresight and forecast studies.

## 3 WHICH ROLE WILL EUROPEAN MANUFACTURING PLAY IN A MORE COMPETITIVE WORLD?

Probably the currently most debated international aspect of industry policy and a question on everybody's mind is the relocation of industry and jobs to other regions. During the last decade, upswings in the economy have not – as previously was the rule – been followed by expansions of employment in industry. Improved productivity seems to give the ability to respond to market expansion without any increase of staff. In the US, the economy is growing owing to the profits made on low price imports from China, new developments in the retail business sector, and because of a labour intensive upswing in the construction industry. The increases in productivity combined with the accessibility to very cheap labour in the expanding economies in Asia place industrial jobs in Europe under pressure. In most people's minds, outsourcing is making Europe lose lots of industrial jobs.

Relocation of industry, usually discussed as outsourcing, is in reality several structural changes coinciding in time, inside and across the borders of European Union. To make the discussion useful, it is necessary to consider distinctions between the different types of relocation. *Off-shoring* means the movement of industry (production) to *low wage regions* to reduce the pure labour cost per unit produced, i.e. for commodities. *Outsourcing* is understood as buying parts or services from suppliers, mainly to improve the economies of scale for the intellectual part of a product. Outsourcing is common within a country or a region, and is a normal way to achieve industrial improvement. When Europe starts to outsource to other regions, the movement is a bigger threat than off-shoring, because the creation of the intellectual capital is moved, and by that the more advanced jobs are lost.

Concerning relocation itself, there is on the one hand the relocation of production to locations near the consumers. With a rapid expansion of the product variants, it is increasingly efficient to locate production closer to the market to avoid long delivery times and a growing stock of finished goods in the market chain. On the other hand, there is relocation of innovative production and development close to lead-markets. This is a way to detect signals from the most demanding customers. This change of industry structure ought to be favourable for Europe, since the high living standard is creating demanding customers. Some of the Asian car manufacturers have recently announced openings of new production plants in Europe.

From the definitions above, it can be seen that off-shoring is driven by pure cost reductions, while the other forms of relocation are driven by a need to improve the speed of innovation (cf. Figure 3).

The current competitors for Europe in the field of cost are China and India, and in the area of innovation, the US and Japan, which is also verified by the *ManVis* Experts when studying the appreciation of relative research positions. With the ambitions shown by China and India to expand their research and education, these circumstances might change and in the foreseeable future competition for innovation might be seen also from these countries. The answers from the *ManVis* experts are not altogether conclusive in the timeframe for the effects of relocations and some contradictions in the answers on related statements can also be seen:

- The *ManVis* experts are convinced that the traditional sectors will be subject to high levels of off-shoring

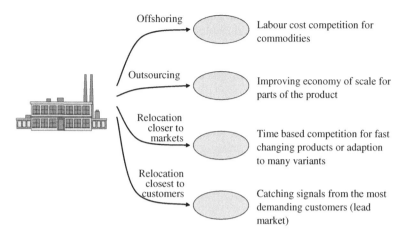

Figure 3.   Patterns of relocation.

- Despite a strong belief in the future development of successful automation in the traditional sectors, 80% of the industry is expected to relocate outside Europe in the foreseeable future.
- The *ManVis* experts are not consistent in their opinions, but the general impression is that more jobs in manufacturing will be relocated to other regions, as long as the relative differences in cost levels remain.

Given their fast rate of growth, it cannot be taken for granted that India and China will remain low wage regions for qualified work, since both show aggressive ambitions to grow into heavily industrialised regions on a much more sophisticated level, competing on innovation as well.

On balance, can low cost labour competition be expected to remain on the current level? The *ManVis* experts were asked for a number of possible reasons that could alter the balance between low cost, far away producers, and high cost producers near to market. Most experts believe in the statements pointing to a reduced difference. The possible conclusion is that in the foreseeable future all these factors in combination will ease off the worst levels of competition. Production will be more automated owing to a gradual development of more cost-effective automation equipment. Wages will go up in China and India, and since they are the most populous regions in the world, Morgan Stanley estimates that they will have a per capita income on par with the global average within two decades, so any new low cost regions entering the arena will not have a comparable impact. Product development will shrink the direct labour content in products even further. Maybe off-shoring for cost reasons is a temporary phase after all? Perhaps this development will continue at such a speed that the low wage profile for qualified personnel will come to an end in these parts of the world sooner than expected.

In looking outside of the European Union, it was decided to provide comments from three countries by additional expert interviews – the United States, Japan and China as obvious competitors and as key countries in the development of manufacturing (cf. *ManVis* Report No. 4). The common theme in each of the countries appears to be uncertainty. In the United States this is an uncertainty on whether manufacturing will remain a vibrant part of the economy, in Japan it is whether they can revitalise the manufacturing sector, and in China whether the rapid growth of manufacturing can be managed effectively. There are issues which are more individual, and these may turn out to be the most important, but the divergence of how manufacturing is perceived is crucial to understanding the probable future trajectories in each country. Uncertainty over China's global role is obviously a core issue across all three countries as China continues to expand its trade and moves into higher value activities.

Within each country there are distinct top-level issues in the debates:

- China's main concerns are managing growth and ensuring that power and infrastructure are in place to support this growth. Another issue is re-balancing the economy with services. China is encouraging the establishment of R&D activities, both Chinese public funding of R&D and international private R&D, and this is challenging the developed economies perceptions of what activities they will remain competitive in and retain.
- Japan's main concerns are revitalising manufacturing and combating high wage rates and changing demographics with increased emphasis on production technology and automation. At the same time, Japan is strengthening its position in key manufacturing technologies – particularly in emerging industries – creating further tension between the Japanese movement of production to China and their desire to lead in these new areas.
- In the US, rising costs of doing business (especially healthcare) and concerns about outsourcing are dominating, but with widely divergent views within the country about whether it matters. The continued pressure on US manufacturers to outsource and off-shore their manufacturing activities is leading to an emerging sense of protectionism. It is possible that barriers to such movements will emerge, depending on how threatening the developments are perceived to be by the American public and the political establishment.

*ManVis* results confirm the anxiety about the migration of European manufacturing. This anxiety is more often indicated by the experts from old Member States which is not surprising if the present structure and directions of intra-European Union foreign direct investment flows (New Member States and Candidate Countries migrate mainly low- and medium-technology industries and related sectors) are considered, and the nature of comparative advantages, which are possessed by new and prospective members of the European Union.

Considering these additional views, off-shoring and outsourcing are structural changes which are high on the importance ranking by the *ManVis* experts. Owing to the far gone development of well defined and documented procedures, in the views of the experts it will become easier to relocate industries and manufacturing plants in particular. Companies which are late in reducing the dependence of local knowledge on the production locations are instead more likely to outsource significant parts of the operation.

When analysing the answers the *ManVis* experts gave on regional leadership in different research and development areas, a clear pattern emerges. The fields that Europe is seen as the leader in are all issues concerning environmental protection and sustainability. Japan leads in most issues on production and the US in most issues on new technology. Although the estimated position in research, expressed by a relatively small selection of experts, should not be exaggerated in value, the question arises if Europe could do more to attract global industries to locate future core functions to Europe.

Production itself is becoming decreasingly a focal part of the industrial operation, while the connection to the customer and the interoperation with the customer to constitute an agile value chain, able to move closer to mass customisation, is growing in importance. The agile company must embrace many and frequent product changes as well as numerous models and options, thus leading to a need for very efficient changeover procedures. As representative evidence, the *ManVis* experts expect the development time for cars to be reduced to six months within ten years, e.g. Ford has recently reduced the development time to 18 months, and is aiming for 10 months. The whole innovation system of a firm, understood as product development, production, services and logistics, will be the core of a firm's global competitiveness.

## 4   WHICH VISIONS, CHALLENGES, AND NEEDED ACTIONS EMERGE FOR EUROPEAN MANUFACTURING?

*ManVis* had to explore the ambitions and expectations of manufacturing experts throughout Europe and reflect them with societal stakeholders and overseas views. During this exercise, diverse, and

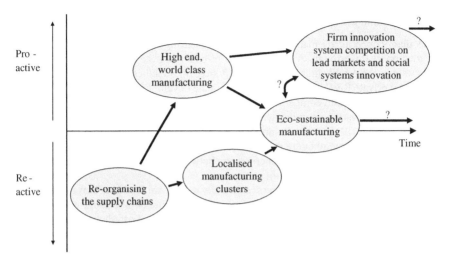

Figure 4.   The *ManVis* trajectories for Manufacturing of Tomorrow.

sometimes conflicting views emerged showing the scope and complexity of present and future manufacturing. The main difficulty of this policy paper is to bridge between the generalist views and the mass of details available in the various reports and databases. This part of the paper attempts to summarise the findings and views. It condenses them into several visions and discusses their impacts as well as the main challenges lying ahead. This should lay the ground for the recommended actions.

In particular, several trajectories for developments of the manufacturing of tomorrow emerge from the *ManVis* findings:

- The struggle on labour cost competition will prevail in the next years. There are two dimensions: the loss of operations to countries outside the European Union and the movement within the European Union. The strategies emerging from the *ManVis* expert consultation are mainly reactive, i.e. cost reduction through automation and enhanced labour productivity. The New Member States will exploit in the very near future an existing cost advantage but will lose it faster then competitors outside Europe. Without their own innovative capacity for absorption and enhancement, this foreign direct investment will just pass through these Member States in a decade. However, both developments are characterised by losses of employment in manufacturing and, of course, retreat from markets.
- Local manufacturing operations and local R&D excellence – as general options – are reactive patterns as well. Very often based on concepts originating from the sustainable development debate this vision is characterised by local operations and development based upon very close interaction with local users – who still have to have purchasing power. The consulted manufacturing experts were quite sceptical on the prospects of this option because of their assessments on the weak ties of modern manufacturing into its environment, contrary to the consulted stakeholders' representatives who value this concept as feasible and competitive.
- Eco-sustainable manufacturing based on new products, new materials, energy efficiency, and last but not least, on advanced product-service systems could be developed into a competitive advantage for Europe – in the view of experts and stakeholders. Regulations creating a technology pull, e.g. as outlined in the FutMan policy scenarios could be successfully mastered because of the excellent R&D position in this field.
- High-end manufacturing will be based on the efficient use of sophisticated manufacturing technologies, which will enable world class, highly automated operations for new products. This high ambition requires an exploitation of the expected potentials for micro electromechanical

systems, related nano-technologies, closing gaps in automation, and research on manufacturing with new materials. But this high efficiency approach will reduce or only maintain existing employment in European manufacturing.
- The most ambitious and far-reaching vision is the European best practice in competing all over the firms innovation system. This comprises user interaction, product development, production, supply chain, and logistics. The successful mastering of this *system* is considered the most promising way to ensure long-term competitiveness. To achieve that, two main crucial elements have to be realised:
  - innovative and adaptive lead markets have to give European companies the chance to be the first to learn, and
  - they have effective user/customer interaction mechanism to exploit this advantage.

These different visions or perceptions of trajectories are not independent from one another nor do they emerge in the same time period. Today, the struggle for labour cost competition dominates the debate, although several European big and small firms are successfully performing the high end manufacturing in their markets. The restructuring of the supply chains enables in the New Member States the establishment of new operations but does not yet ensure a real long-term impact on their innovation system. If the New Member States are not developing their own innovative and absorptive capacities very soon, the labour cost advantage will disappear – and with that the newly erected plants.

The scenario analysis by the IPTS furthermore indicates that in many respects advanced solutions are likely to emerge from a local level. For policy making, this implies that it could be useful to support such local *model* approaches to then be able to systematically foster their transfer and adaptation to other conditions. But contrary to that, global competition in the view of the *ManVis* experts requires global presence. Hence, the European Union's cluster oriented innovation policy could develop such needed islands in the New Member States. But this strategy will remain reactive considering global competition as long as it is not connected to newly created market opportunities.

High-end manufacturing will not – in the view of the consulted experts – create new employment but safeguard existing jobs. Further, it is a necessary condition for the more advanced and employment creating trajectories. For example, a successful and economically prevailing strategy on eco-sustainability requires high technology and professional organisation of product-service concepts.

The drivers for approaching competition based on the firm's system of innovation are lead markets. Innovation and adaptation take place where the demanding customers are. New lead markets attracting new companies have to emerge in Europe. To achieve this, besides new technologies and excellent products and operations, changes and social innovations are necessary to create the new demands and to allow firms to operate successfully.

To move along the different paths and create employment severe *challenges* can be concluded which have to be mastered:

- creating manufacturing based on sophisticated technology,
- developing knowledge-based and learning companies and industries,
- competing through the firms individual innovation systems,
- re-defining and innovating demand,
- keeping Europe economically united.

To master the challenges, *several types of actions* are recommended (cf. Figure 5).

- An advanced science and research base for manufacturing and engineering as basic conditions,
- excellent R&D projects (collective or academic) on manufacturing issues as core activities which bring forward manufacturing excellence,
- accompanying measures and research facilitating change and innovation,
- enhanced funding mechanisms to match the challenges in a more appropriate way,
- establishing diffusion policies for manufacturing technologies and best practices,

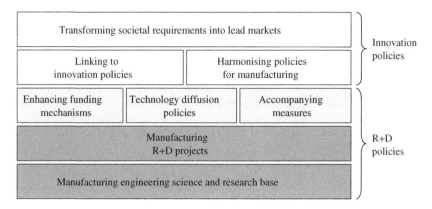

Figure 5.    Elements of policies for knowledge-based manufacturing.


Table 1.    Imminent technological research needs.

*Paving the way for new technologies in manufacturing*

- roadmapping and foresight on manufacturing relevance of nano- and (white) biotechnology
- measurement, workplace safety for nano-technology and biotechnology
- applied basic research for white biotechnology and nano-manufacturing

*Industrialising technologies*

- processing and manipulation of new materials
- incorporating smart materials into components for process technologies
- combining new materials with micro electrical mechanical systems (adaptronic)
- exploring new modelling knowledge and high power computing for simulation of product development, of material behaviour, and of virtual experiments

*Exploiting existing technology advantages*

- micro-systems in machine tools and products
- intelligent mechatronic systems for automation and robotics (e.g. self adapting components)
- new automation technologies considering advanced human-machine interaction by considering diverse workers capabilities
- ICT-tools for traditional sectors

*Technologies for customising products/services*

- Tagging technologies
- Approaches towards product customisation via software or electronic components that allow for maximum flexibility and user integration
- Technologies and concepts facilitating user integration into innovation processes
- Technologies and concepts facilitating personalisation and build to order concepts
- SME appropriate tools for networks and logistics


- linking manufacturing R&D policies to innovation policies,
- harmonise and re-directing various policies with impact on manufacturing towards innovation and competition of firm systems of innovation,
- transforming societal requirements into lead markets.

Through research on technologies for manufacturing and on organisation and management, manufacturing research provides manufacturing engineering with the necessary knowledge, tools, and solutions. Basic manufacturing research has to foresee and prepare for the new challenges and applied manufacturing research has to adapt and transform existing technologies and organisational processes. Furthermore, manufacturing research plays a decisive role in combining the long-term horizon in technology trajectories with the short-term need of firms to innovate successfully.

Even for practical and industrial manufacturing research the science base is of growing importance. The inclusion of manufacturing topics and issues into the funding mechanism of the planned European Science Council is of crucial importance. Other existing mechanisms on transfer and mobility of researchers have to be maintained as well as international co-operation.

Excellent research projects in manufacturing are needed. It is important to notice not to concentrate on technological developments only but the whole system of innovation in the firms has to be considered. This implies tools, strategies, methods, procedures etc. for product development, logistics, innovation management, business concepts, etc. have to be added to the technological research agenda. The main barrier towards more proactive strategies lies in the successfully implementation of learning companies which can adapt their innovation system fast.

Harmonising policies for manufacturing is something which has started already (cf. KOM (2005) 474). But as mentioned above, the relationship between research issues and topics for manufacturing engineering and has to be in two directions; e.g. developing policy support for new ways of biotechnology based production (*white* biotech), or what should workplace safety for nano-technology workers look like, or how to ensure product reliability through simulation tests. These cross cutting issues require parallel and co-ordinated activities to adapt sector-based innovation systems in the concerned fields as quickly as possible. In addition, education and training policies have to consider the needs for learning firms, i.e. foster in-house activities much more than external institutions. It will be in the firms where the struggle for a competitive corporate innovation system is decided.

This harmonised policy approach is necessary if societal requirements and existing competencies should converge into a lead market. First mover advantages can only be obtained if quick and decisive moves in demand shaping and competence building are made. To be successful, a thorough analysis of long-term demand and interactive participation of stakeholders and users is decisive for policymakers and industry. Hence, while closing the loop, exercising these practices in the R&D projects and efforts becomes of crucial importance.

ACKNOWLEDGEMENT

This work was funded by the European Commission through NMP Project *Manufacturing Visions – Integrating Diverse Perspectives into Pan-European Foresight (ManVis)* (Contract No NMP2-CT-2003-507139). The author wishes to acknowledge the contributions from Heidi Armbruster, Maurits Butter, Gerald Jan Ellen, Petra Jung-Erceg, Per Kilbo, Finbarr Livesey, Krsto Pandza, Bogdan Piasecki, Anna Rogut, Fabiana Scapolo, Jan Sjögren, and Philine Warnke.

REFERENCES

Bierhals, R., Cuhls, C., et al. (2000): Mikrosystemtechnik – Wann kommt der Marktdurchbruch? – Miniaturisierungsstrategien im Technologiewettbewerb zwischen USA, Japan und Deutschland, Heidelberg
KOM(2005)474: Umsetzung des Lissabon-Programms der Gemeinschaft: Ein politischer Rahmen zur Stärkung des Verarbeitenden Gewerbes in der EU – Auf dem Weg zu einem stärker integrierten Konzept für die Industriepolitik; (5.10.2005)
ManVis Report No.3 – Delphi Interpretation Report, October 2005 (www.manufacturing-visions.org)
ManVis Report No. 4 – Overseas Views: International Perspectives on the Future of Manufacturing, October 2005 (www.manufacturing-visions.org)

*Part II*
*Product lifecycle management*

Product Lifecycle Management (PLM) has become one of the most used acronyms in modern markets. Within IMS-NoE, SIG 1 activities have focused on this new emerging topic. Over the course of three years, SIG1 organised 7 international workshops and 2 special sessions in international conferences, with more than 50 international experts participating.

Considering the reference manufacturing scenario for the next decades (i.e. globally scaled and networked manufacturing, knowledge intensive manufacturing, innovation based manufacturing) which is being addressed by the European Commission's Framework Programme 7 and the second phase of IMS, the IMS-NoE SIG1 research activities investigated the role of PLM approach in the processes of design and manufacturing of new products. It is the SIG1 community's belief that PLM will be one of the key organisational and technological enablers for the sustainable and competitive manufacturing of developed countries in the near future.

PLM can be defined as an integrated ICT-supported approach to the co-operative management of all product-related data along the various phases of the product lifecycle. As such PLM involves:

- a strategic management perspective, where the *product* is the only source of enterprise value creation,
- the application of a collaborative approach for the empowerment of all the enterprise core-competencies distributed among different actors,
- the adoption of a large number of ICT solutions and tools to practically establish a co-ordinated, integrated and access-safe product information management environment.

In a few words, PLM can be seen as a complex *platform* consisting of organisational and technological tools that may be used as enablers by companies to get into the design and manufacture of new products using the PLM paradigm. It is SIG1's belief that this effort is synergistic with the development of an IMS visionary approach to manufacturing of the future.

Part II of the book is organised in four main contributions: an introductory contribution, written by IMS-NoE SIG1 members, and three contributions illustrating two IMS projects (SMART-fm and PROMISE), wherein PLM is addressed from complementary perspectives.

The IMS-NoE SIG1 contribution introduces PLM. In this contribution, a state-of-the-art of PLM is reported, while trends and gaps are identified. A definition of a roadmap for PLM application, defined by international experts participating in SIG1 activities is also reported.

The second contribution deals with the IMS project SMART-fm, which involved partners from USA, Europe, Canada and Australia. The SMART-FM contribution shows the main results obtained in the project in terms of interoperability, which is one of the most relevant topics related to PLM from an ICT perspective. PLM deals with managing information along the entire product lifecycle; interoperability of systems is needed for managing product information in a cost-effective way.

The third section deals with a promising project: PROMISE, an IMS and European Union Framework Programme 6 Integrated Project (IP). PROMISE aims to develop appropriate technology, including product lifecycle models, so called Product Embedded Information Devices (PEID) with associated firmware and software components and tools for decision making based on data gathered through the lifecycle of a product. The PROMISE contribution shows one of the most interesting perspectives for PLM in the IMS area: the management of the product data is fostered by new emerging technologies, providing new answers to product management.

The last contribution of Part II addresses closed-loop PLM. Nowadays the whole product lifecycle can be visible and controllable by tracking and tracing the product lifecycle information, thus creating closed-loop information flows across whole product lifecycle. As a result of these closed-loops, PLM can help to streamline several product lifecycle operations. In this study, the authors look into the concept of closed-loop PLM and describe the gaps between traditional PLM and closed-loop PLM.

*Advanced Manufacturing – An ICT and Systems Perspective – Taisch,*
*Thoben & Montorio (eds)*
© *2007 Taylor & Francis Group, London, ISBN 978-0-415-42912-2*

# Product lifecycle management: State-of-the-art, trends and challenges

Marco Garetti, Marco Macchi[1] & Sergio Terzi[2]
[1] *Politecnico di Milano, Department of Economics, Management and Industrial Engineering,*
 *Milano, Italy*
[2] *University of Bergamo, Department of Industrial Engineering, Dalmine (BG)*

ABSTRACT:   It is the IMS-NoE Special Interest Group (SIG) 1 community's belief that PLM will be one of the key organisational and technological enablers for the sustainable and competitive manufacturing of developed countries in the near future. The IMS-NoE SIG1 contribution introduces the PLM concept. In this contribution, the state-of-the-art of PLM is reported, while trends and gaps are identified. A definition of a roadmap for PLM application, defined by the SIG1 community, is also reported.

*Keywords*:   Product Lifecycle Management, state-of-the-art, gaps, trends and vision, roadmap

## 1  INTRODUCTION

The modern global context is characterised by the evolution of ICT tools and infrastructures, which are promising technological capabilities for supporting collaboration in extended enterprises. In such a context, Product Lifecycle Management (PLM) is representing one of the most promising management paradigms to answer the challenges of present and future industrial scenarios.

PLM indicates an integrated business approach for the management of product data, which are generated, managed and stored along the product lifecycle, from design to end-of-life, *from the cradle to the grave*. PLM entails:

- a strategic management perspective, wherein the *product* is the enterprise value creator;
- the application of a collaborative approach to better use the enterprise competencies distributed amongst diverse business actors in the extended enterprise context;
- the adoption of several ICT solutions to practically establish a co-ordinated, integrated and access-safe product information management environment in the extended enterprise.

Even if PLM has a so wide definition, it is addressed generally as an ICT question and market. The *PLM market* is one of the most increasing ICT markets, composed of a galaxy of solution-providers, coming from three main backgrounds: (i) vendors of the digital engineering world, which sell many kinds of Computer Aided (CAx) tools, (ii) vendors of the ERP world, which commercialise broad integrated suites, (iii) vendors which provide a series of web-based solutions for virtual collaborative environments.

As a result of these diverse ICT backgrounds and capabilities, the market PLM suites can enable diverse degrees of collaboration among many actors and users along the lifecycle of a product, from design to production, and from distribution to the end-of-life (Table 1). Collaboration is physically realised through the exchange of information related to the product; this exchange may be implemented in different ways and may regard different kinds of data (design specifications, drawings, customers' feedback, maintenance instructions etc.).

Table 1.  Product lifecycle phases, business actors and collaboration.

|                          | Customer | OEM | Supplier |
|--------------------------|----------|-----|----------|
| Concept                  | ←Collaboration→ | | |
| Design                   | | ←Collaboration→ | |
| Production & Distribution| | ←Collaboration→ | |
| Use                      | ←Collaboration→ | | |
| End of Life              | | ←Collaboration→ | |

This paper aims to introduce the diverse elements and perspectives that exist under this emerging acronym. In particular, section 2 illustrates the state-of-the-art of two main *elements* of PLM, while section 3 illustrates the trends and vision of PLM that are arising in the market. Section 4 defines the current gaps and summarises the main open issues of PLM. A tentative roadmap towards a large application of PLM is specified in section 5, while section 6 concludes the contribution.

## 2  STATE-OF-THE-ART OF PLM ELEMENTS

A complete definition of PLM considers at least three perspectives: (i) the collaborative business strategy of PLM, (ii) the collaborative business practices, and (iii) the role of ICT for collaboration. In respect to this paper, these three implicit dimensions of PLM will be the main perspectives of PLM under analysis. The actual position of the PLM approach, its methodologies and supporting technologies, will be positioned to understand the fulfilment of these three hidden perspectives, in particular investigating two main elements:

- Technology enablers, where the state-of-the-art and roadmap of ICT PLM-oriented technologies and standards will be studied.
- Enterprise business models, which will address interoperability in PLM at a business level.

### 2.1  *State-of-the-art of the technology enablers*

From the 1970s, enterprises have deployed several ICT systems, supporting increasingly complex activities and processes. The growth of ICT adoption by enterprises has passed through several phases: from the installation of minicomputers in the 1980s, to the revolution of workstations and Personal Computers in the 1990s, through to the revolution of the Internet era. All these phases have been supported by re-engineering of business processes; a clear example is the establishment of collaboration through organisational ideas like co-marketing, co-design, co-engineering, co-manufacturing, co-selling, which have been defined since the 1980s. These would have been only theoretical exercises without the evolution provided by Internet-based ICT.

Looking to the main process of New Product Development (NPD), the design activities are supported by diverse ICT tools, which are continuously developing and evolving. For example, in the area of product development, ICT tools supporting product engineers have been existing since more than 30 years and have now reached third generation: the first 2D Computer Aided Design (CAD) systems, introduced in the 1970s, were replaced in the 1980s by 3D CAD; in the 1990s, owing to hardware innovation, more functional features were introduced, such as assembly supporting definition, or design path recording. Nowadays, 3D technologies are assuming a relevant role: Digital Mock-up for product development provides to engineers the possibilities of a well-defined 3D simulation for stylistics, designing and also maintenance purposes. Other 3D approaches are now under development and diffusion in the market, such as the functional approach or the most advanced Knowledge-based Engineering systems (KBE), which automate sophisticated designing procedures. CAD systems can increasingly communicate with other CAx tools, such as Computer Aided Styling systems (CAS) and Computer Aided Manufacturing systems (CAM), which automates NC machine programme generation. This path to integration has been supported by the development of international standards, such as STEP and IGES.

Something similar has been happening in the area of manufacturing planning: since the 1970s, several ICT tools for Computer Aided Process Planning (CAPP) have appeared to support engineers in the definition of manufacturing plans. CAPP tools have evolved from simple approaches to more complicated ones. In recent years, CAPP tools have being developed in distributed and collaborative environments, evolving from standalone applications into more sophisticated CAPP platforms, where engineers, coming from diverse departments and enterprises, can co-operate to develop co-ordinated manufacturing planning solutions.

Also the world of factory design and planning has been subjected to a similar evolution; single and separated ICT tools adopted by engineers for plant layout designing, planning and simulation have been replaced by more integrated platforms and tools, connected also with other CAx systems.

Eventually, in the last years, many tools which enable information sharing among engineers in distributed environments appeared, under a lot of diverse names and acronyms: EDM (Engineering Data Management), PDM (Product Data Management), PIM (Product Information Management), TDM (Technical Data Management), e-BOP (Electronic Bill of Processes) to name a few. All these systems, generally defined as Document Management (DM) tools, are physically based on a central database, and provide central services (vault) for managing design data (product, plan, plant design), such as access rights control and design release management. These stored data are Bill of Materials (BOM), Bill of Resources (BOR), Bill of Processes (BOP), CAx files, manuals, guidelines, spread sheets files... Especially because of the evolution of these DM systems and also because of the evolution of diverse interoperability standards, a large integration among IT tools in the area of design process is under development; this integration is now defined as Digital Manufacturing and Engineering, which indicates how the whole Design Process, composed by Product Development, Manufacturing Planning and Factory Planning, could be realised using an integrated platform where engineers can co-operate, to reduce the development time. Internet-oriented technologies are the key-success factors, fostering integration of software and hardware platforms, in particular because of their independent protocols (e.g. XML, eXtensible Mark-up Language).

Something similar happened in the area of ICT tools supporting production and distribution management (generally Operation Management) and related activities. As is well known and accepted, the first operational activities supported by ICT tools have been the production activities, where, since the end of 1970s, many ICT systems have been developed such as MRP (Material Requirements Planning), which evolved into MRPII and CRP (Capacity Requirement Planning), and larger ERP tools (Enterprise Resource Planning), which integrate and support a lot of activities, such as finance, accounting, inventory management. The high capital costs of technological solutions available until the early 1990s (based on EDI – Electronic Data Interchange), often decelerated the adoption of these integrated ICT tools, in particular by SMEs (small and medium-size enterprise). An inverse route, with an improvement on the diffusion of integrated ICT tools for operation management has been started with the adoption of Internet-based resources (e.g. TCP/IP protocol, or platform-independent languages such as HTML). Moreover, with the evolution of the markets and outsourcing trends, new ICT tools appeared: Supply Chain Management (SCM) tools for improving relations with suppliers; Customer Relationship Management (CRM) tools for managing customers and their requests; Advanced Planning Systems (APS) tools for improving single and multi-sites production scheduling; IT tools for automating, controlling and integrating manufacturing processes with upper level systems (MES – Manufacturing Execution System). At the present, all these kinds of tools are under consolidation into larger distributed ICT platforms for the operation management processes of large international companies, constituting integrated expensive software suites. At the same time, and at lower cost, the Internet is providing a good way for all related actions of B2B and B2C.

The ICT developments mentioned derive intrinsically from the evolution of numerous basic tools and of their integration. Conversely, a new wave of technology is nowadays represented by BPA and WFM systems as well as PM systems. With the diffusion of process orientation into enterprises, lots of instruments and tools for Business Process Automation (BPA, also defined Work-flow Management WFM systems) have been developed in the last ten years. These tools automate business processes improving speediness and agility in offices with repetitive activities;

a WFM system is physically a tool for managing information and documents (DM) based on a common repository, where access-safe rights are defined for diverse users, and where repetitive *secretarial* activities are automated using standardised electronic communications. These systems are the core elements of all the DM tools. They can be applied to enable a controlled actuation of PDM, EDM and TDM within design processes; they also can be applied for document management in enterprise operations management such as in SCM and CRM within distributed systems. Another important evolution might be traced in the area of Project Management techniques (PM). Originally developed as standalone tools, PM tools are nowadays assuming a relevant role in distributed ICT platforms and are integrated as basic techniques for managing projects and processes and tools both of Digital Manufacturing/Engineering, and Operation Management. The Internet offered a relevant contribution to the development of such basic tools, providing cheap services such as electronic mail and platform-independent languages, but also video and phone streaming conference. WFM systems, at first developed into expensive EDI networks, are nowadays easily accessible at a cheaper cost using the Internet, and also integrate mobile platforms, such as PDA (Personal Digital Assistant, e.g. Palm, Pocket PC), GPRS and mobile phones. Also PM techniques are implemented at a low cost into Internet-based tools, providing new uses and users.

### 2.2   State-of-the-art of enterprise business models

Within the globally scaled economy and markets, the product is re-emerging as the *value creation* element to provide enterprise revenues, even if its role is becoming increasingly sophisticated. Markets are growing in a worldwide manner and customers are becoming increasingly demanding in terms of quality and delivery times. The product is turning into something more complex than just a physical good: it is evermore a conjunction of services and extra components. All the required processes related to the product are then ever growing themselves as a natural consequence, constituting a complicated cycle, which starts from understanding markets, passes through product and process design, operations and distribution management, until after sales and end-of-life management. This is even more complex since the required competencies usually exceed the boundaries of a single enterprise and integration is then generally required beyond these boundaries.

An enterprise is a set of activities connected to one another, which are oriented towards the same goal: creating value. This value derives from the maximisation of revenues and the minimisation of costs and all inefficiencies hidden in the organisations. During the 1980s, enterprises expended efforts in cost reductions and productivity improvements; in particular, installation of deep automation of the factory was one of the most significant leverages required and addressed to improve efficiency. After the 1990s, the worldwide scenario led to an greater complexity: customers were becoming increasingly demanding in terms of product quality and related services, while the market competitiveness was increasing worldwide. Enterprises then had to create their value adopting new strategies for leverage, looking for a continuous improving of innovation of products, processes, production systems and organisation structures, trying to reduce time-to-market for their products and time-to-the-right-solution for their projects. As one consequence, enterprises were dismissing the competencies that were considered as non-core, and they correspondingly needed to improve their collaboration with their partners outside, both suppliers as well as customers. In this way enterprises had to re-engineer their own structures and business modes, looking to a re-orientation of their basic business processes. A new relevant leverage, increasingly required, was a better co-ordination and integration scheme both in intra and inter enterprise context, and was obtained after some BPR efforts.

These industrial trends of the 1990s are not dead but still alive after 2000. Present requirements are for a better integration and B2B collaboration. A new focus after 2000 is the orientation to product lifecycle, according to which collaboration is engineered, having as a reference model, the perspective of the product lifecycle phases and their importance in the whole. Different requirements can be identified in the product lifecycle in different sectors:

- In the area of manufacturing (e.g. automotive, textile...), product lifecycle phases are typically separated into two enterprise processes: the New Product Development (NPD) process and the production and distribution process (Enterprise Operation Management). The first process involves all activities that deal with design and implementation of the productive capacity. The second process involves all needed activities for managing production, transportation, and distribution, until eventual after sales services.

- In enterprises defined as Engineering and Contracting (e.g. construction, naval industries), the main process responsible for the value creation starts with engineering and budgeting definition activities, passes thorough procurement of sub-components and contractors, until the physical construction and installation *in-field* (these companies are referred to as EPC, Engineering Procurement Construction, organisations). Beside, service design/management can be identified after construction and delivery of the product, these have however specific features, rather than those of the after sales service management of manufactured products.
- In service companies (including a variety of sectors, e.g., hospitals, insurance), the enterprise value is usually created along activities of service design, service provision and its maintenance.

Specific requirements from these three different enterprise types are leading to on-going business modelling activities. Modelling activities are extrinsically connected, on the one hand, with identification and empowerment of enterprise core competencies (and consequent outsourcing of non-core competencies) and, on the other hand, with establishment of a collaborative attitude between functions and departments, inside and outside the enterprise. All the activities performed along the product lifecycle must be co-ordinated and efficiently managed to gain revenues and reduce redundancies and costs. Enterprise modelling is the process of building models of whole or part of an enterprise (e.g. process models, data models, resource models, etc.) from knowledge about the enterprise, previous models, reference models, or combinations of the three, using model representation languages. Then, enterprise modelling is the way to represent and define in an explicit way the rules of co-ordination and integration declared in PLM.

In the context of enterprise modelling, diverse effort have been spent at the manufacturing level, like the well know initiatives such as CIMOSA, ARIS ToolSet, GRAI, IDEF (Doumeingts et al. 1998). More recently, enterprise models have been defined for specific topics; in particular, the supply chain area has been investigated and studied by relevant initiatives. The SCOR (Supply Chain Operations Reference) model has been developed and endorsed by the Supply Chain Council (SCC), an independent not-for-profit corporation, as the cross-industry standard for supply chain management. SCOR is a process reference model that provides a language for communicating among supply chain partners, spanning from the supplier's supplier to the customer's customer. By describing supply chains using process building blocks, the model can be used to describe supply chains that are very simple or very complex using a common set of definitions. As a result, disparate industries can be linked to describe the depth and breadth of virtually any supply chain. The model has been able to successfully describe and provide a basis for supply chain improvement for global projects as well as site-specific projects.

PLM has a wider scenario; in terms of processes, PLM encompasses diverse series of them, depending by the level of application/implementation. A larger initiative was the project ENAPS (European Network for Advanced Performance Studies). The objective of this project was to develop a generic set of processes and related performance measures to be used in enterprise benchmarking. This set of performance measures might allow enterprises to view performance measurement data from other enterprises all over Europe and to see their position on a league table of performance results.

Another important initiative, which is now at an early stage, is the VCOR (Value Chain Operations Reference model). VCOR aims to enlarge the well-known and accepted SCOR (Supply Chain Operations Reference model) initiative, providing an international reference model for business processes which take into account also the NPD mainstream. The VCOR model consists of three process levels. Level 1 consists of Plan, Market, Research, Develop, Sell, Source, Make, Deliver, Support and Return value chain process categories. The model is defined in successive levels of detail at Levels 2 and 3. Level 4, not defined in this project, are where company specific implementation occurs. At each appropriate level, VCOR aims to provide the following information: (i) Standard Process Descriptions, (ii) Best Practices, (iii) Metrics, (iv) Inputs/Outputs.

Companies collaborate with other companies (local or not local) through the various phases of the product lifecycle (making co-design, co-engineering, co-production, co-maintenance). Today competitive pressure pushes these companies to deal more efficiently with collaboration, reorganising themselves and adopting software technologies supporting it. A guideline for supporting the process modelling and re-organisation of the company is necessary before adopting a PLM software tools, but at the present this guideline is still missing in the market and in the research. Also the relative performance metrics for the business processes are not well-defined and diffused in such contexts. Some relevant initiatives are coming up, more at a consultant level than in terms of research contributions.

## 3 TRENDS AND VISION

Trends in ICT may be fostered starting from the actual state-of-the-art, synthesised in the next table for the main technological features emerged in the last decade. The table scope is centred on features supporting (i) engineering activities, (ii) work flow management and (iii) project management. Out of scope are other technological features mostly related to support of the operations chain of an extended enterprise. Operations chain is at the boundaries of the PLM paradigm, being managed accordingly to existing business management paradigms other than PLM.

The table overview facilitates two main considerations on the on-going trends:

- the trend is generalised toward a digital support of collaborative business practices
- the collaborative business practices are situated in distributed systems and may involve different business actors in such distributed systems.

One may then envision that this trend will continue in the next few years directed to achieve an effective interoperation and collaboration in inter enterprise systems. In the longer term and, most of all, looking over the product lifecycle orientation that is now emerging from industrial requirements of the modern context, expectations may also be fostered that other business processes may find consolidated support in the future from ICT. Now, they are either partially consolidated or not clearly included under a paradigmatic *umbrella*. PLM is going to be this paradigmatic umbrella, wherein ICT evolution may find place to lead to an effective technological offer to answer these other business needs.

Table 2. Technological evolution of PLM ICT tools toward collaboration.

| Supported business process | Technological Features |
| --- | --- |
| Product design | Enhanced integration of CAD with other CAx systems, with Knowledge based Engineering systems (for knowledge sharing of design criteria, rules, ...), within co-design platforms (for exchanges of design models and ideas in a CPD practice) |
| Process planning | Digital integration of CAPP tools in distributed environments for collaborative process planning |
| Factory planning | Digital integration of CAPE tools in integrated environments for manufacturing resource planning and for layout planning and simulation |
| Data and document management | Information and document sharing in distributed and collaborative systems (PDM for management of bill of materials, CAx files...; WFM to automate exchanges of product documentation along the product lifecycle) |
| Project management | Collaborative project management in distributed and collaborative project management teams |

Some relevant trends in interoperability along the product lifecycle might be quoted, even if it might be said that the standardisation trend is usually a long trip. This is, of course, a general consideration but it is well applicable to the object of this study. Moreover, standardisation involves many users and developers investing a large amount of effort and money and, sometimes, this involvement has heterogeneous modes, approaches and interests. For example, R&D teams coming from PLM vendors developed many works: e.g.: (i) some vendors (e.g. IBM, SAP) are already providing their integration features according to the first OAG specifications, while within the product development area, all the most important vendors are using the STEP standard to guarantee the most open interoperability to their customers; and (ii) PLM XML development, supported by UGS PLM Solutions. Also standardisation bodies are obviously working in this standardisation context, being partly overlapping with the vendor efforts: e.g., ISO 62264 is being used as the basis for many control and MES (Manufacturing Execution System). Eventually, many projects are also now ongoing in the area of standardisation; for example, in Europe a new community is active, structured as a network of excellence (INTEROP NoE), on interoperability efforts, which partly derives from the effort of the SMART-fm project. The world of interoperability standards is like the Tower of Babel, wherein many expressions exist. For example, in the quoted standards there are many overlapping definitions and redundancies (e.g. many definitions for the same concept of BOM/BOR). However, about the future trends, one might be in someway optimistic: the road for standard setting and using, supported by the diffusion of the PLM business paradigm as a common umbrella, seems to be promising to facilitate some more rationalisation in this Babel-like circumstance. Of course, it should not be forgotten that any standardisation activity is related to lots of interests of different parties (vendors, users, standardisation bodies, etc.) and the identification of PLM as a paradigmatic umbrella cannot be envisioned as the solution, but only as a potential facilitator for future development trends.

Looking at the PLM market, PLM has been referred to by many definitions, depending on PLM vendors and their marketing strategies. PLM has been defined by vendors in different ways, such as (i) a piece of technology which can inter-operate with other solutions, (ii) an additional module of a larger suite, (iii) other various definitions. These kinds of technological and marketing-oriented definitions give a reductive idea of PLM, which is a more complicated enterprise phenomenon. Effectively, the PLM acronym becomes more widely accepted when new business needs arise in the market and enterprises need to change their strategies and visions, by giving more attention to their products creating value. Into this new business scenario, ICT is playing a fundamental role, but, even if they are enabling elements, they are not sufficient to enable the PLM diffusion and evolution as a whole organisational matter. The ICT evolution is certainly a leverage to speed up the PLM diffusion but it is not the only one. The comprehensive definitions of the phenomenon now named PLM available in literature are symptomatic of this actual market perception of what is PLM: they are defined in a very wide and heterogeneous scope. Moreover, still missing is a full consideration and definition of PLM in the scientific community (even if many conferences and workshops have been organised in the last years). This contributes to the general confusion and variety of PLM perceptions in the market.

In particular, there are many positioning white papers coming out mainly from PLM vendors providing their own vendor-oriented definitions. Also some of the most important centres of business research have elaborated and proposed their definitions of such phenomenon, like AMR Research, CimData, Daratech, ARC Advisor Group, Gartner, QAD. All these definitions provide some interesting issues to be considered looking for a more comprehensive idea of PLM. In all these definitions, beside the general wide scope and variety, it is possible to identify some common elements, such as (i) business process strategy, (ii) collaborative approach and (iii) role of ICT systems. PLM is a multi-layered and multidisciplinary approach, and all different perspectives might be taken into account for its full characterisation:

- the PLM acronym deals, at first, with a strategic vision of the enterprise, and its processes might be product oriented to answer to market needs and requests (along the product lifecycle);

- the PLM approach deals with an innovative solution for creating/managing/maintaining all the information shared along enterprise processes; in particular, PLM deals with the digitisation of all such information, from design, to manufacturing, to after sales service activities;
- at the same time, this comprehensive approach to digital information management, which physically enables collaboration amongst people, is provided by the ICT evolution trends and interoperability.

PLM is not perceived in the market only as an ICT tool more or less integrated, neither it is only an organisational matter, nor only a technique. Considering its comprehensive dimensions, the PLM acronym is resolutely useful to indicate a complex phenomenon, so it was introduced and defined (at the beginning of this paper) as a business process management paradigm and approach as it is now understood (more or less) into the industrial context. As a business paradigm, first of all, it unifies different and heterogeneous issues such as organisational processes, economics issues, techniques and ICT. On the other hand, this same complexity of the definition could explain the difficulties of its full acceptance, especially given the nature of the market. It is possible to notice that PLM vendors, even if they come from three diverse backgrounds, are adopting the same strategies: (i) vendors coming from the digital engineering world (e.g. UGS, IBM-Dassault) are trying to connect enterprise Operation Management processes; (ii) vendors coming from the ERP world (e.g. SAP) are turning to connect Digital Manufacturing and Engineering tools and platforms; (iii) vendors coming from the ICT world are aiming to establish collaborative environments for PLM integration (e.g. Microsoft, MatrixOne) by means of using web technologies. This cannot be considered a trend that would change in the near future. In the PLM market vendors are evidently acting with similar merger and acquisition strategy, in the areas of Digital Engineering and Manufacturing, and in enterprise operation management.

On the other hand, a clear trend in the market is that the PLM approach is being developed and adopted in several and diverse industrial sectors. The most part of experiences and tools are nowadays in the world of mechanical products (e.g. automotive and aerospace), but also some interesting applications are coming out from the world of Architecture, Engineering and Construction (AEC). Vendors and business research centres estimate that this will be one of the most important sectors for PLM vendors in the next years and new acronyms are arising to this end (e.g. PLM/AECO, or ILM – Infrastructure Lifecycle Management). Also in the world of services there are many interesting uses of PLM; for example, in diverse hospitals the PLM approach is adopted for collecting and managing information about patients and their *lifecycles* and a new acronym has been proposed (Service Lifecycle Management – SLM). Also industrial sectors, such as textile, fashion and apparel, PLM has been adopted for managing, in shorter time frame, product information: information about daily sales are reported to production managers and to designers to improve production scheduling and change and modify seasonal catalogues.

Then, PLM acronym signifies something new since it merges a strategic *product-centric* vision to the adoption of advanced ICT distributed solutions and related business practices for collaboration amongst people and organisations along the product lifecycle. The PLM orientation and its mapping with related management issues can be considered a relevant novelty of focus of the business operation, not in itself but because it is now being supported by the technological evolutions (such as in ICT tools, methods, etc.). A vision for the next 5–10 years is that this *product-centric* vision will find a consolidation beside other existing paradigms well accepted in industry (such as the SCM). This will be mainly owing to the consolidation of the now on-going technological and methodological evolutions.

In particular, adopting a PLM approach will signify, at first, a complete understanding of the role of information and its sharing into the enterprise along the product lifecycle and related management activities. According to CimData, it is possible to identify a PLM approach in an enterprise when: (i) an universal, secure, managed access and use of product definition information is provided; (ii) the integrity of product definition and related information throughout the life of the product or plant is maintained; (iii) business processes for creating, managing, disseminating, sharing and using product information are managed and maintained.

## 4  GAPS AND OPEN ISSUES

The product-centric vision of PLM could be fully achieved if several open issues are either resolved or fixed. This section provides a gap analysis by means of identification of these open issues (Table 3).

The analysis takes into account two main PLM elements: Technology Enablers and Enterprise Business Model.

All the previously discussed perspectives are still open issues for industrial and academic R&D efforts. Looking at the next 5–10 years, it is still unclear in which ways, and when, PLM will fulfil its goals and will reach widespread adoption.

Table 3.  Gap analysis to achieve PLM vision.

| PLM elements | Description of the open issue |
|---|---|
| Technology enablers | From the ICT perspective, a *centralised* product management is no more than a *database* unification problem, physically enabling the previously envisioned business process practices in the product lifecycle phases. Information about products and processes are dispersed along a variety of information systems, which – until now – have been executed in *isolated islands* (e.g. PDM and ERP). The trends and issues now ongoing deal with integration of these *islands* into a larger integrated (distributed) repository, to provide a wider and more effective use of product information. In the first times, these integration trends had been performed in a closed way with instantiation of several proprietary *suites*. The recent evolution in *standardisation* along the product lifecycle (e.g. PLM XML, ISO/DIS 10303-239, ISO 62264) is a promising, though not yet complete, step toward *centralised* product management. From this perspective, then, there are still several open issues and further research to be developed to achieve effective integration of information in a *product-centric* and (quasi) standard approach. |
| | From an infrastructure perspective, the instantiation of a *product-centric* management approach signifies the product-centric design and management of several elements: (i) an information infrastructure, which is concerned with issues of ICT network establishment; (ii) a resource infrastructure, which is concerned with the design and the management of all physical elements involved along a product lifecycle (e.g. machines, plants, people, suppliers, warehouses…); (iii) a product *infrastructure* itself; the same product has become a resource to be managed directly, traced and controlled over its lifecycle. |
| Enterprise business model | From a strategic and organisation perspective, the adoption of a *product-centric* approach may signify a (re-)consideration of all the relations established amongst the resources (people and equipment) involved into relevant business processes of an extended enterprise. Re-consideration should be directed to put the resources into the context of product lifecycle phases. To this regard, in particular, the definition of some methods for strategic understanding of potentials of a PLM (re-)orientation of business operation (and correspondent achievable values) is still one of the main open issues that should be consolidated. Nowadays, the value creation is more *centred* and evaluated around the business processes themselves not the product features in its lifecycle. |
| | Interoperability at business analysis level has not been already achieved: a PLM business operations reference model (similar to the SCOR model for SCM processes) is still missing. The availability of such a reference model can be an aid to facilitate implementation and deployment of PLM integration projects. In this sense, the availability of such a reference model may facilitate consolidation, around it, of some reference PLM business process analysis and deployment methodologies. |

## 5 ROADMAP

Some steps required for roadmap implementation toward the PLM vision have been identified. These steps are summarised in Table 4.

These general principles subsumed in these roadmap steps can be grouped in terms of synthesis statements as follows:

- **PLM is not a piece of technology**, but more like a strategic approach/paradigm to be defined and decided at first at the top level of the enterprise.

Table 4. Prioritised steps to achieve PLM vision.

| Step | Description of the Step | Level of Priority |
|---|---|---|
| 0 | Provision of a clear definition of the PLM scope and objectives<br><br>Different perceptions of PLM are available in industry; the confusion arises from the PLM solutions in the market; confusion may also arise from trying to understand better relationships with other existing paradigms. An essential step (pre-requisite) is required to achieve provision of a clear definition of what is PLM. | This step is a pre-requisite. |
| 1 | Provision of methods for strategic analysis of PLM projects<br><br>This is still an open issue of the strategies/methods area. These methods would facilitate a better strategic analysis of a PLM implementation project and should be based, beside traditional business process analysis, upon product features and their implications on the asset of business processes along the product lifecycle. A product-centric view is a relevant concept for this analysis (what does PLM strategy aim at and support across the product lifecycle phases? What benefits can be achieved with this support? What costs?) | This step is a priority on a short-medium term basis.<br><br>It is necessary to boost penetration in the market of PLM by means of clear understanding of its strategic relevance for industrial users. |
| 2 | Provision of methods of interoperability along the product lifecycle<br><br>This is a relevant topic of PLM. This may have plenty of positive effects: e.g., (i) to enable a *best-of-breed* strategy (at implementation level); (ii) to facilitate business analysis and PLM set-up amongst people within an enterprise (at business level). | This step has ideally a high priority.<br><br>As with many other standardisation efforts, however, it may require a long term prospective and priority should be defined taking this into account; this consideration is mostly applicable at implementation level since lots of stakeholders are involved in the standardisation process at this level. |
| 3 | Provision of methodologies and frameworks for PLM control along its lifecycle<br><br>It is important to achieve a consolidated set of methodologies and frameworks to be applied for more clever and structured control PLM along its lifecycle: (i) to select it properly based on cost/benefit analysis; (ii) to deploy it properly (in a selected industrial field context); (iii) to control properly during its operation (as any other enterprise system, this has of course its own complexity being it typically positioned in an extended enterprise context) | This step has a priority on a short-medium term basis.<br><br>It is necessary to enable PLM integration project deployment and operation and it is subsequent, and, as such, related to strategic understanding (step 1); it is necessary then to enable PLM penetration in the market. |

- An important need of future manufacturing is to achieve a **product-centric view**. This means managing products as individuals where ever possible (track and trace them, control the user adoption and needs and so on) to improve customer satisfaction and response times.
- A **PLM Business Reference Model** could be an interesting research effort to be spent in the next years by the scientific community, also to facilitate diffusion of a PLM culture. This is expected to be very relevant for the adoption of a suitable and pragmatic PLM approach.


## 6  CONCLUSIONS

PLM acronym signifies something new, since it merges more complex aspects and phenomena, from a strategic product-centric vision, to the adoption of advanced ICT solutions, fostering collaboration among people and organisations. Adopting a PLM approach signifies, at first, understanding the role of information and its sharing into the enterprise along the value-creation activities and processes.

This contribution has presented a comprehensive definition of PLM and has summarised the most relevant trends and open issues. Considering the reference manufacturing scenario for the next decades (i.e. globally scaled and networked manufacturing) which is being addressed by the new research programs (i.e. European Union Framework Programme 7 and IMS phase 2), the IMS-NoE SIG1 research activities have investigated the role of the Product Lifecycle Management approach into the processes of design and manufacturing of new products. It is the IMS-NoE SIG1 community's belief that PLM will be one of the key enablers for the sustainable and competitive manufacturing of developed countries in the near future. New organisational paradigms like extended enterprise, collaboration, co-design, co-engineering, continuous product innovation, are already well known, but many difficulties arise for their practical implementation on a large scale. Moreover, such difficulties are greater when SMEs are involved. This is because greater complexity in the company processes, relationships and time-synchronisation requirements are connected with them. On the other hand, companies in developed countries (and SMEs especially), need to adopt these manufacturing paradigms if they want to survive in the long run. To solve the gap between this need and the present state-of-the-art, organisational and technology enablers are needed: PLM is one of these enablers.

PLM has already demonstrated its efficacy in big companies, while leading software vendors are offering expensive solutions to the market. Nevertheless, true integration (better, interoperability) is still a chimera. Standardisation problems to be solved are quite impressive. Business models for getting true competitive advantage from the adoption of the PLM approach are missing and detailed roadmaps to correctly implement a PLM strategy are needed. In the forthcoming years western manufacturing companies will mostly rely for their competitive advantage on the use of PLM technologies so to safely control, in a distributed way, the different phases of the lifecycles of their products. Relying on such technological and organisational enablers they will be allowed to select, in a *best-of-breed* fashion, the best opportunities in the global scenario. ICT infrastructures for PLM will support all product related data management, thus allowing control over involved businesses and engineering processes.

Summarising, the main gaps to be addressed by the research community are:

- a PLM reference model,
- a PLM adoption roadmap,
- standards to facilitate software integration/interoperability,

In a few word a system made of organisational and technological tools that may be used as enablers by companies to get into the design and manufacture of new products using the PLM paradigm. This effort is seen as synergistic with the development of an IMS visionary approach to the manufacturing of the future.

Tools acronyms

- CAD (Computer Aided Design): The term CAD refers to 2D and 3D design software, which can be specialised for example in mechanics design, electrical engineering, electronics design, hydraulics design, pipe system planning or ship and aircraft building.
- CAPP (Computer Aided Process Planning): Computer aided work and task design. This term refers to the use of IT in production-related work planning.
- CAM (Computer Aided Manufacturing): The term refers to the adaptation of information technology to the control of the production machines, warehouses, or transport systems used in production.
- CAE (Computer Aided Engineering): The term refers to the analysis of a design for basic error-checking, or to optimise manufacturability.
- DMU (Digital Mock-up): Extensions to 3D modelling capabilities that enable complete products to be built in electronic form. DMU can be used to check for problems such as interference and clashes among components.
- EDM (Engineering Document Management): Process of sharing data across the enterprise and making information available to decision makers, regardless of the source of the data. In the past, this technology was thought of as a data conversion. Data was converted among applications that needed to integrate data from other applications. After several iterations, the technology is now known by the general term of data migration. Data migration is used to build EDM applications such as e-commerce, Enterprise Resource Planning, Customer Relationship Management, Business Intelligence (BI), Data Warehousing, and Data Marts.
- PDM (Product Data Management): PDM is a systematic, directed method by which to manage and develop an industrially manufactured product. With the help of PDM, the product process can be well managed all the way from idea workshop to scrap yard, with all product data handled in a systematic manner. PDM can also be used in the supply chain and order-delivery process. Almost without exception, the term PDM also refers to an information processing system developed for the management of product data. Today the term PDM can also considered as an older name for PLM.
- WMS (Workflow Management System): Interaction of people working with product data according to the predefined business processes of an enterprise to achieve corporate objectives. Repetitive workflows and processes can be programmed as part of a collaborative system to route data and work packages automatically, to control and monitor processes, and to provide management reporting.
- ERP (Enterprise Resource Planning): An ERP system typically involves a very large set of functions that support the planning and management of company resources. It is an integrated information processing system, which covers several different functions and serves as the backbone of the daily business of the company. It combines such things as planning, production, sales, marketing, management, accounting, human resources management, and financing into a single entity. The task of the system is typically to direct the staff, material, money and data flows of the company.
- MRP (Material Requirements Planning): Methodology and system used to plan and manage manufacturing operations, starting from Bills of Materials.
- SCM (Supply Chain Management): The control of the supply chain as a process from supplier to manufacturer to wholesaler to retailer to consumer. SCM does not only involve the movement of a physical product through the chain, but also any data that goes along with the product and the actual entities that handle the product from stage to stage of the supply chain.
- CRM (Customer Relationship Management): CRM entails all aspects of interaction a company has with its customers, whether it be sales for service related.

Other acronyms

- ECR (Engineering Change Request): A request or suggestion to engineering for an improvement in a process or procedure.
- ECO (Engineering Change Order): Official document from an engineer department that indicates a design change for a product.
- DFx (Design For x): Design principle according to which attention must be paid in the design to viewpoints related to following processes (e.g. Design for manufacturing, design for assembly, design for supply chain etc.)
- LCA (Lifecycle Assessment): Method developed to evaluate the mass balance of inputs and outputs of systems and to organise and convert those inputs and outputs into environmental themes or categories related to resource use, human health and ecological areas.
- TRIZ (Theory of inventive problem solving): A knowledge-based, systematic approach to innovation. TRIZ involves a systematic analysis of the system to be improved and the application of a series of guidelines for problem definition. TRIZ analysis includes an integrated system approach, function analysis and function modelling.

ACKNOWLEDGEMENT

The authors wish to acknowledge the European Commission for their support. They also wish to acknowledge their gratitude and appreciation to all the members (Main, Associated or simply interested) that participated to the diverse IMS-NoE SIG1 events.

REFERENCES

Bussler, C.: B2B Integration: Concepts and Architecture, Springer 2003

Doumeingts, G., Vallespir, B., Chen, D.: Decision modelling GRAI Grid. In: Handbook on architecture for Information Systems, (P. Bernus, K. Mertins, G. Schmidt (Eds.)) Springer-Verlag, 1998.

Onkvisit S., Shaw, J.J.: Product Lifecycles and Product Management, Quorum Books, Greenwood Press, Westport, Connecticut, 1998

*Advanced Manufacturing – An ICT and Systems Perspective – Taisch,*
*Thoben & Montorio (eds)*
*© 2007 Taylor & Francis Group, London, ISBN 978-0-415-42912-2*

# SMART-fm: Setting interoperability in SME-based industrial environments

Ricardo Jardim-Goncalves[1], Hervé Panetto[2], Maria José Nuñez[3] &
Adolfo Steiger-Garcao[1]

[1] *Fac. Ciências e Tecnologia, Univ. Nova de Lisboa, UNINOVA, Portugal*
[2] *Université Henri Poincaré Nancy I, France, Herve*
[3] *AIDIMA – Asociación de Investigación y Desarrollo en la Industria del Mueble y Afines,*
*Valencia – Spain*

ABSTRACT: PLM (Product Lifecycle Management) is a set of capabilities that enable an enterprise to manage its products and services throughout the business lifecycle. A significant trend in the present global market is the increasing need for co-operation among enterprises, through which organisations can increase flexibility and reduce operational costs by focusing on their core competencies. However, enterprise applications need to be interoperable to achieve seamless interaction across organisations, leading to the need for adoption of data standards. This contribution to the book proposes a framework to support PLM, enhancing the interoperability in networked environments, and assisting the integration of reference models described following dissimilar methodologies. This framework assists in the automatic mapping among ISO10303 STEP, UML and XML models. The proposed work results from research developed and validated in the scope of the IMS SMART-fm project (www.smart-fm.funstep.org, www.ims.org), involving partners from USA, Europe, Canada and Australia, using emerging approaches for modelling and technology.

*Keywords*: Product Lifecycle Management (PLM), Open Integration, Standards, Information Technology, Interoperability

## 1 INTRODUCTION

According to Wikipedia, Product Lifecycle Management (PLM) is a term used for the process of managing the entire lifecycle of a product from its conception, through design and manufacture to service and disposal. PLM tackles a set of capabilities that enable an enterprise to effectively and efficiently innovate and manage its products and related services throughout the entire business lifecycle.

Many software solutions have been developed to organise and integrate the different phases of a product's lifecycle. PLM should not be seen as a single software product but a collection of software tools and working methods, integrated together to address either single stages of the lifecycle, to connect different tasks, or to manage the whole process. Indeed, nowadays enterprises have information technology that could fulfil their requirements in each operational phase and with external partners, e.g., suppliers. For instance, in industrial environments, many applications are available to support operating their Product Lifecycle stages. However, organisations typically acquire these systems aiming to solve focused needs, without a complete view of the global enterprise's system integration. Even when enterprise models are interoperable, very often difficulties arise with respect to data semantics when information has to be exchanged, since common semantic models are not in place (Goncalves et al. 2002).

To gain revenues and reduce redundancies, all the activities performed along the manufacturing process must be co-ordinated and efficiently managed (Ulrich 2000). For realising such co-ordination, manufacturing has to become evermore an integrated manufacturing process, enabling the communication among all methods, tools and environments dispersed along the manufacturing process itself.

The PLM paradigm aims to integrate all activities disposed along the product lifecycle line to improve co-ordination and co-operation among them. However, the existing PLM suites cannot be connected and, therefore, the adoption of the PLM paradigm is typically reduced to one IT choice, which becomes a *trust* choice: after takening it, the enterprise is *obligated* to live with it and by it.

In this scenario, seamless data exchange among internal and external production agents plays a key role, where an initial product data model must be instantiated during the early design and tooling phases, being updated along the product lifecycle. To have this data flow accurate, data models and processes need to be interoperable (Drejer et al. 2003). However, this circumstance has been identified as difficult to achieve, because typically there are many different software applications in use, each one adopting its own data structure and semantics. Additionally, when developing a product each participant team normally has its specific method of work and self-containing language, which does not result in a flawless interaction with others.

To help attain enterprise and systems interoperability, several dedicated reference models covering many industrial areas and related application activities, from design phase to production and commercialisation, have been developed. Most of these have, however, been developed using divergent methodologies, and without the concern of being interoperable among themselves (standard to standard).

New modelling methodologies have taken place in the market, which have recognised importance. Among them there is the Unified Modelling Language (UML), which has distinguished itself in industrial environments, allowing the specification, visualisation, and documentation of reference models. Additionally, Extensible Mark-up Language (XML) has also become prominent in the last years, playing an increasingly important role in the exchange of a wide variety of data. XML is open and freely available from the World Wide Web Consortium (W3C) having the support of the world's leading technology companies. These languages present the advantage of being widely supported and having a large variety of tools, which help in the design and management of such models.

Nevertheless, to support interoperability in heterogeneous networked manufacturing environments, it would be valuable to adopt recognised standard models for Product Data Exchange, like the ISO10303 STEP Application Protocols. If these STEP standards can be harmonised with UML and XML, the users can benefit from the advantages of all these technologies.

## 1.1 *Industrial and research motivations*

Based on the number of people it employs, the furniture industry is the largest manufacturing sector in the world. Most of the companies in the furniture manufacturing and related sectors are small and medium-size enterprises (SMEs). To keep its competitiveness, Europe needs to accomplish rapidly the new requirements in the e-global marketplace, and push promptly SME-based industry to adopt PLM services and extended enterprise practices. To maintain and increase the competitiveness of these companies in the advent of the emerging digital economy and smart organisations, the use of modern information technologies and standards among all agents involved in the furniture Product Lifecycle Management has to be considered.

The problem of data exchange to support the manufacturing phase of the furniture product lifecycle when doing business among manufacturers, retailers, providers, and customers is well understood. The furniture community considers nowadays this problem as a significant inhibitor for e-commerce, IT support and smart enterprises (see FSIG: http//www.funStep.org) and, although identified as a problem for the furniture industry, there is a global concern in the SME-based industrial sectors.

The large number of proprietary systems operating in the furniture industrial sector (as well as in other sectors) makes this problem bigger and more difficult to solve. Thus, this industrial community is eager to have an International Standard for product lifecycle information support in which software providers can have confidence on its worldwide adoption. Support from the international research community to create such a standard is being looked for too.

The main objective of the IMS SMART-fm project was to research, develop and demonstrate in industrial environments, an open standards-based framework that supports the management of the complete product lifecycle in the furniture manufacturing industry. The aim was to establish new concepts, methodologies and technology frameworks supporting all phases of the furniture product lifecycle.

SMART-fm brought together manufacturers, retailers, suppliers, designers, and industrial trade associations from the furniture sector as well as research and development centres, academic institutions, software vendors, and consultants. It provided a forum for research and development in the state-of-the-art, publishing roadmaps and guidelines that assist industry's short-term planning and implementation, and identifying needs for continued research and development. Although focused on the furniture industry, owing to its open and general characteristic, SMART-fm was concerned to enable the reuse of its results and deliverables to other industrial environments based on SMEs.

One of the principal SMART-fm motivations was to create and support an interdisciplinary network of actors involved in the study of smart manufacturing having as a basis the FSIG – funStep Interest Group (http://www.funStep.org). FSIG, a worldwide interest group, had as its main objective, to follow and support the development of an International Standard for furniture product and interior design data exchange. FSIG has now more than 600 members from industry (75%) and research/academia (25%), from 21 countries, and was used by SMART-fm as a source of requirements, a forum for industrial review, and a privileged medium for the dissemination and acquisition of the project results.

Communication among the different stakeholders is at the core of developing any valuable business. ICT, which is part of the communication process in the information flow for PLM, is becoming a higher priority as businesses seek to reduce costs, improve design time, and manage production and inventory systems. Information is seamlessly exchanged among parties, giving the customer better choices, by offering them a degree of power in managing and customising their own particular product choices. This drastically reduces the wait time, from the time the customer orders a particular product until it is received.

SMART-fm focused on interoperability issues to assist in PLM, aiming to fill the communication gap among the product lifecycle phases using ICT. It has been identified that SMEs in the furniture sector need new ways of working for better integration, as well as, encouragement to work in co-operation. Instead of proprietary solutions, the use of standards, with the help of the funStep services, will enable the sector to work faster, giving SMEs the ability to be more efficient and cost-effective.

Standardisation was one of the main aims of the SMART fm. Standards for data representation like ISO 10303 (STEP), ISO 13584 (PLIB), OMG/UML and the Extensible Mark-up Language (XML) are used. SMART-fm combined proven methods and standards for data specification with the XML language to demonstrate how furniture product lifecycle information can be exchanged and shared as well as inside any enterprise among common applications and among networked enterprises in an e-commerce environment using low-cost, scaleable software tools. Nevertheless, specification of e-business services and XML documents for furniture industry is a key issue to be standardised to contribute to the open interoperable platforms operating in this industry.

Recent developments in ISO TC184/SC4 have created an environment in which data exchange capabilities based on STEP application protocols can be created in a modular fashion, reusing components of existing validated standards and, therefore reducing the time and cost to develop, implement, and deploy standards-based solutions (Goncalves et al. 2003).

## 2 TECHNOLOGICAL BACKGROUND AND INTEROPERABILITY IN PRODUCT DATA EXCHANGE

### 2.1 STEP models and technologies

The International Organisation for Standardisation (ISO) has been pushing forward the development of standards and models to foster the exchange of information related to goods and services. Efforts like ISO10303 STEP – Standard for the Exchange of Product model data – have tried to deal with the issues of integration and interoperability.

STEP represents the standard for the computer-interpretable representation of product information and for the exchange of product data. It aims to provide a neutral mechanism capable of describing products throughout their lifecycle. Nowadays, STEP has been recognised as appropriate to help in the integration of manufacturing systems in industries such as automotive, aircraft, shipbuilding, furniture, building and construction, gas and oil.

Widely used in Computer Aided Design (CAD) and Product Data Management (PDM) systems, STEP is today adopted by some of the leading industrial companies in the world in the automotive, aircraft, shipbuilding, furniture, building and construction, gas and oil industries. All the above industries are using it to help in the integration of their manufacturing systems and applications. Recent studies show that the use of STEP technology, whether in-house or among organisations, could generate savings of about $1 billion per year in the U.S. automotive, aerospace, and shipbuilding industries. The same result could be foreseen also in Europe. Through these numbers, STEP has proven its value to some sectors of the world industry.

EXPRESS is the STEP modelling language. It combines ideas from the entity- relationship family of modelling languages, with object modelling concepts. It provides general and powerful mechanisms for representation of inheritance among the entities constituting the conceptual model, and it also encloses a full procedural language used to specify constraints on populations of instances.

The extent of standards required to support all the detailed characteristics of systems in the product lifecycle, may lead to highly complete models i.e., Application Protocols (APs). These are described in STEP using the EXPRESS language and represent reference models for a specific industrial scope of application, providing the basic mechanisms to reflect interoperability for complex engineering product models. STEP APs provide the basic mechanisms for interoperability among product data models. However, STEP translators have typically been implemented only in a selected number of systems, owing to their complexity and specialised data exchange format.

Conversely, XML has been widely used to interface many different systems, owing to its standardised format, easier understanding by common users and wide availability of supporting tools. Owing to the knowledge embedded in the models developed using the STEP methodology, it would be relevant to reuse all those models and implement them using a more popular representation. This way, a convenient solution would be to continue using the EXPRESS language for APs definition, and then translating them, e.g. to the XML Schema Definition (XSD) format.

To represent an EXPRESS schema in equivalent XML Schema definition, the emerging Part 28 of the STEP standard specifies the mapping of XSD type definitions and element declarations that are dependent on the EXPRESS schema. Part 28 also specifies the rules for encoding conforming data in XML to match the XML Schema and certain configuration directives.

### 2.2 UML methodology

Unified Modelling Language (UML) is a language for specifying, visualising, constructing, and documenting software system components. UML has become more popular than other modelling languages, mainly because it is a general purpose *language*, with broad application and a large variety of tools that assist in working with the models. Just like STEP, not only the structural part of the model is represented in UML, but the behavioural. The last, is represented through constraints using OCL (Object Constraint Language), the UML constraint language.

Also, the OMG (Object Management Group) has developed a neutral language with the intention of providing a common mechanism for interchange of UML models using XML representation, the XMI (XML Meta-data Interchange). XMI has been adopted by most of the popular tools for data modelling, and bindings for the main modelling languages, such as UML and EXPRESS, have also been defined for it. In this last case, STEP is developing mapping specifications described by Part 25 of STEP, *Implementation methods: EXPRESS to OMG XMI binding*.

### 2.3 *The XML paradigm*

The Extensible Mark-up Language (XML) was developed under the auspices of the World Wide Web Consortium (W3C). It is considered to be a simple, very flexible text format derived from SGML (ISO8879). Originally designed to meet the challenges of large-scale electronic publishing, XML has been playing an increasingly important role in the exchange of a wide variety of data, in particular, for web applications.

XML can be used in a wide variety of platforms and interpreted with a wide variety of tools. XML Schema Definitions (XSDs) provide the mechanisms for defining the structure, content and part of the semantics of XML documents, and it is through them that XML allows representation of data models. However, the current XSD specification does not yet support restrictions to the model, like UML does with the OCL, or like EXPRESS does with the rules.

Compared with EXPRESS, the availability of XML tools, resources, and related technologies have made the use of XML more practicable, cost effective, and widespread. Furthermore, web-services using the XML technology are extending these capabilities to provide a common, technology-independent infrastructure for remote implementation of open services.

### 2.4 *PD/MSE and EE integration initiatives*

A reference framework for PLM has been developed addressing the PD/MSE (Product Development and Manufacturing System Engineering) and the Enterprise Engineering (EE) fields. PD/MSE integration efforts include three main directions of investigation and development: integrated, collaborative environment for supporting the PD/MSE activities (A-I); international consensus among users concerning enterprise engineering and integration based on modelling technology (A-II) and creation of a reference standard for enabling product data representation (A-III) (Figure 1).

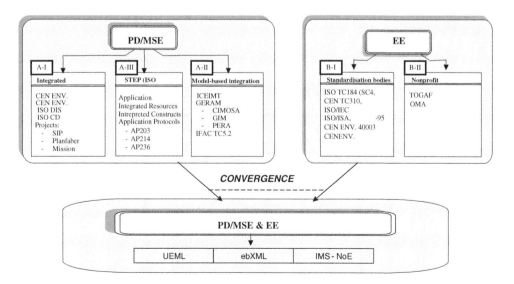

Figure 1.   PD/MSE and EE integration initiatives.

Within the EE area, the efforts in setting up a common data format for connecting different company processes can be divided in two groups. The first one (B-I) is strongly supported by international (European and worldwide) offices for standardisation, which are the leading actor in Enterprise Modelling and Integration (ISO TC184/SC5, CEN TC310/WG1), System Lifecycle Processes (ISO/IEC 15288) and Enterprise Control-System Integration (IEC 62264 series of standards).

The second group of relevant works (B-II) is done by non-profit organisations like OAG (Open Applications Group), developing TOGAF (The Open Group Architectural Framework) and Object Management Group (OMG) elaborating the Object Management Architecture (OMA) to develop integration for object-oriented applications.

In the PD/MSE area, the efforts spent during the last decade setting standard modelling methodologies (e.g. IFAC-IFIP Task Force GERAM, GRAI Integrated Methodology) are converging within a single project proposal for creating a Unified Enterprise Modelling Language (UEML) for describing and designing an enterprise system, composed of manufacturing, business and others processes. Something similar is presented within the development of ebXML language, which aims to identify a standard data format for managing an electronic market and IMS (Intelligent Manufacturing System) community.

The main features of all these efforts are that all the work is done by many different international projects, and there is close collaboration between science and industry. The most successfully efforts were accepted by some standardisation groups (at European and International level), and the most of the projects are supported by International organisations such as IFAC, IFIP, etc. (Garetti et al. 2003). As well, there is a close relation between integration activities and standardisation processes, and vendors of the PLM market participate in many activities. For example some vendors (IBM, i2, SAP, PeopleSoft) are already providing their integration features according to the first OAG specifications. As well, all the most important vendors within PDM area are using STEP standard to guarantee the most open interoperability to their customers. For instance, IEC 62264 standards are starting to be used by many control and MES (Manufacturing Execution System) vendors (such as Honeywell, Rockwell, Sequencia, Invensys-Baan, Siemens and Fisher-Rosemount). Moreover, some vendors (UGS-Tecnomatix) are collaborating to propose their PLM-XML standard integration, deriving their XML parser specifications according to SGML-ISO8879 strategy.

PLM aims to integrate three main enterprise *streams* (Figure 2):

- The stream of the PD&MSE activities, composed by such activities as product/process/factory design, distributed along the lifecycle of the generic product;
- The stream of the Production Planning and Control (PP&C) activities, such production planning, scheduling and control.

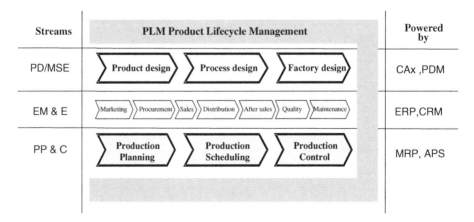

Figure 2.   Streams enabled within PLM.

- Between these two main streams, there exists an intermediate stream (Enterprise Management and Engineering), composed of activities that are connected with MSE and PP&C streams, such as Marketing, Procurement, Distribution, etc.

Several tools and environments, such as CAx and PDM in the PD/MSE stream, MRP and APS in Production Management, ERP and CRM in the Enterprise Management and Engineering, power each stream.

## 3 THE SMART-FM FRAMEWORK

Developed under the scope of SMART-fm, the ISO10303-236 (STEP AP236) standard is a foundation for data exchange in the furniture industry, so that all the software involved in designing, manufacturing, and selling a product understands the same vocabulary. Through the usage of this AP, seamless communication can be established among the several stakeholders of the furniture industry.

However, the acceptance of STEP technologies has been facing difficulties when applied to sectors primarily composed by SMEs, such as the furniture sector. The main problem is that this technology is unfamiliar to most application developers, and SMEs do not have the budget that larger companies have to hire or educate specialised personnel.

Moreover, STEP data (i.e., an instance population of an EXPRESS schema) is typically exchanged using an ASCII character-based syntax defined in ISO 10303-21 (also known as Part 21 of STEP). As well, the STEP Part 21 syntax, although adequate for the task at hand, lacks extensibility, is hard for humans to read, and is computer-interpretable only by software supporting STEP. ISO, predicting this circumstance, is developing standards for representation of EXPRESS schemas and data in XML and UML, which are technologies that are more popular and have better tool support.

For data exchange based on these models, the standard's recommendations should be followed, such as STEP's Neutral Format described in Part 21, or XML described in Part 28 of this standard. However, acting as explicit specifications of a set of concepts and relations among models, it seems relevant that the information contained at the meta-level is also addressed.

For the representation of data corresponding to an EXPRESS schema, the STEP Part 28 specifies the mapping of type definitions and element declarations to XML Schemas (XSD), the rules for encoding conforming data in XML, and also certain configuration directives. STEP Part 25 has similar purposes. It specifies a mapping of EXPRESS constructs into the UML Interchange Meta-model conforming to the XMI standard (Lubel et al. 2004).

AP236-XML is a standard-based framework developed by SMART-fm to assist in the integration of data models described following dissimilar methodologies, through a standardised meta-model harmonisation (Figure 3). In the framework used, Application Protocol models at the meta-data level are mapped from STEP, to UML and then to XML.

Using the directives defined in STEP part 25 (phase 1 of Figure 3) the model is mapped to XMI, which has been accepted as the UML standard for meta-model representation. Having the model represented in UML, another mapping must be made (phase 2) to reach XML. The combination of both phases results in an XML representation of the AP model through XML Schemas. These are equivalent as if the transformation was performed directly from EXPRESS using the Part 28 recommendations. In this way, a XML neutral data format, compliant with the XML Schemas, will be understood by every stakeholder using the same standard AP model.

When employing standards for meta-model representation associated to automatic code generators, the traditional effort required for the development of the translators can be significantly reduced. Thus, to help in the construction of the AP236-XML framework, automatic code generators are used. As a part of the framework, UNINOVA developed the STEP25 tool that translates an EXPRESS-based model to XMI, following the emerging STEP Part 25 directives. This tool implements and proves the concept of EXPRESS to XMI binding and, reporting to Figure 3, it also maps the STEP reference models to UML. The mapping from XMI to XSD is automatically

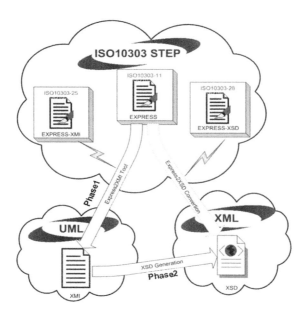

Figure 3.    Framework for mapping among STEP, UML and XML.

generated following a methodology based in the STEP Part 28 mapping specifications. The tool that enables to complete this last phase has been also developed.

However, the use of facilitators can be even more powerful. Owing to the capabilities associated to UML, especially after establishing the dynamic component of the model, it is possible to automatically deploy applications that conform to the specifications addressed in the model, using the functionality available in popular UML tools.

The SMART-fm framework includes the following set of tools:

- *STEP to UML* – STEP25 tool that translates an EXPRESS-based model to XMI, following the emerging STEP Part 25 directives;
- *STEP to XML* – UNISTEP28 tool, translating STEP model and data to XML according to Part 28 of STEP;
- *XML to RDBMS* –XSD2RDB is the result of the integration of the Apache Torque (*http://db.apache.org/torque/*) database resource generator with an XSD Part 28 compliant parser;
- *XML to OO* – the OO generator is the result of the integration of JAXB, (*http://java.sun.com/xml/jaxb/*) with a .NET extension.

After performing the first phase of the transformation sequence, the model is translated to XMI, UML and XSD formats. Figure 4 illustrates an example of these translation results. Since STEP Part 25 still does not define how to map the constraint mechanisms of EXPRESS, further processing on the mapping should be defined by SMART-fm thought it is desired to include these mechanisms in the transformation, e.g., *Product_Class* can only exist if the inherited attribute *target_market* is not assigned.

### 3.1    *The industrial demonstrators*

The implemented SMART-fm demonstrators cover the furniture's PLM, starting with designers creating a new furniture catalogue. Manufacturers publish such AP236 based catalogues in business portals, and automatically manufacture the requested products as soon as orders arrive through the Internet, either from customers or retailers. The manufacturer has as its central point ERP

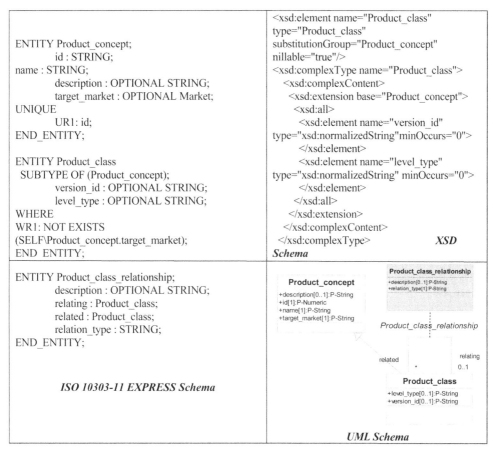

Figure 4.   Model and Meta-model harmonisation, among EXPRESS, XSD and UML.

(Enterprise Resource Planning) system that accepts requests according to the AP236 data representation and, later on, forwards those requests to production. The CNC (Computer Numerical Control) machines through computer programs that understand AP236 are responsible for automatic control, motion, tools and parts necessary to produce furniture objects.

The demonstrators show a scenario of complete manufacturing process integration in PLM. Manufacturers observe interoperability and integration, starting with item requests at ERP Level, passing through CAD systems and ending with CNC production machinery. The SMART-fm framework can be used with any STEP Application Protocol. However, it has only been tested in the area of product data exchange among manufacturers, distributors, suppliers, e-Marketplaces, and retailers, with information exchanged in XML format according to the AP236 model specification and Part 28 recommendations. The XML information that circulates among the several stakeholders must be compliant with the AP236 model, so that every stakeholder understands the data received. To achieve this, the XML data is validated against an XML Schema (XSD) that reflects the model.

The product catalogue exchange scenario illustrated by Figure 5 is one of the demonstrated PLMs scenarios that use the platform:

1. A furniture manufacturer has all data stored in an internal database, which has been used for years with good results. For this reason, the manufacturer does not want to change the way it works internally, but wants to publish its catalogues in an international marketplace, to expand its business;

59

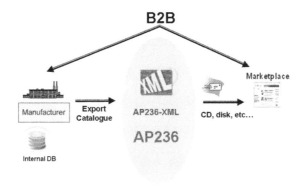

Figure 5. AP236-XML Catalogue Exchange Scenario.

2. Using AP236 would facilitate the execution of this task, so the manufacturer must map the catalogue to AP236 format;
3. The Manufacturer exports the AP236 catalogue to XML and sends it to the marketplace by CD, mail, etc;
4. The marketplace receives and understands the catalogue because it is compliant to AP236;
5. The catalogue is uploaded into the marketplace system;
6. Visitors to the marketplace can select some products from the specified manufacturer, configure them, and send an order directly to the manufacturer;
7. The manufacturer receives the order and imports the Purchase Order file to its ERP-system to proceed with the business.

The platform proves itself important for points 2 and 3 of the described scenario, because it provides the means to create an alternative storing system (e.g. database) where the manufacturer can map only the data he wants to use in the transaction. From that database, services exist to export data to AP236-XML format directly. In point 4, the marketplace can use a conformance checker service to determine if the data is AP236 valid. If it is valid, automatic generated OO interfaces can be used to read the information from the XML files, thus saving time in processing the data received.

Until now, the platform has only been validated among the SMART-fm partners. The comments received state that it is a very valuable work, reducing the implementation time of the AP236 standard. The platform is publicly available through standard web-based services publishing, providing a global open interface to access to the SMART-fm framework.

## 4 CONCLUSIONS

More than 95% of the companies working in the furniture industry are small sized ones. Based on the number of employees, the furniture industry is the largest manufacturing sector in the world and in Europe in particular. To keep its competitiveness, Europe needs to accomplish rapidly the new requirements for the furniture industrial sector in the e-global marketplace, and push manufacturers and their trade partners to promptly adopt PLM electronic services and extended enterprise practices, in the advent of intelligent agents, products and web-based services. The adoption of standard-based methods, models and tools to support interoperability among agents and applications is a key issue to achieve this objective.

The work developed by the IMS SMART-fm project contributes with a standard-based framework for PLM, to support data model integration and interoperability among applications through meta-model harmonisation (STEP, UML and XML). It resulted in a framework that enables the reuse and integration of knowledge and data described under heterogeneous methodologies, providing the mechanisms for a normalised framework, for dynamic collaborative business processes through and within the product lifecycles stages. The IMS SMART-fm project has been the test bed for assessment of this proposal. The results achieved clearly contribute to the revised European strategy i2010.

To show results to industry (SMEs and large companies), the SMART-fm partners developed 8 PLM scenarios of demonstration, based on industrial cases identifying the needs of the industry. A common understanding is that this integrated demonstrator is becoming a referential platform for PLM in advanced interoperability for SMEs.

These scenarios presented some solutions as examples on how to solve the PLM problems for interoperability, in this case between a manufacturer and an e-marketplace on catalogues and product information by exchanging XML files (being compliant with AP236 format) and order information. To solve the interoperability problems when exchanging catalogues between a manufacturer and a retailer, a catalogue was exchanged using an AP236-compliant database. It was also shown how manufacturers, architects and interior decoration companies could exchange 3D data through AP236-compliant XML files. Another group of scenarios showed how complex furniture could be configured by customers, including the use of 3D modelling. Now, the targeted user is mainly the retailer. Other groups of scenarios were developed to show solutions regarding e-marketplaces interoperability. Also, SMART-fm enables users to browse and place orders across multiple marketplaces in a seamless manner. The technology proposed for this is web services (including UDDI registries) in combination with AP236-compliant XML.

Significant results achieved by SMART-fm are within the standard-based platform to support data model integration and interoperability among applications, through meta-model harmonisation, normalised services, and standard data representation. This proposal enables the reuse and integration of knowledge and data described under heterogeneous methodologies, providing the mechanisms for a normalised framework to enhance PLM.

The validation performed in the scope of the SMART-fm piloting scenarios, has providing understanding on the requirements of each stakeholder and how they could be addressed by the platform. Its usage has resulted in a reduction of the AP236 standard implementation time. Nevertheless, the validation process also demonstrated that each technology binding might result in the loss of some expressiveness. This happens especially on the behavioural (knowledge) parts of the model, because neither STEP Part 25 nor Part 28 recommendations deal with this kind of information.

A significant challenge to face in future research is to contribute with enhanced proposals to cover this gap. Further processing on each transformation is required, and some complementary technologies specifically designed for knowledge representation can be used:

- Schematron, a language for making assertions about patterns in XML documents, is used along with XSD;
- UML Object Constraint Language (OCL);
- And PL/SQL, a trigger and procedural language, to specify rules and assure accuracy in RDBMS.

Also, the platform's framework has to consider integrating the work from the emerging Semantic Web technologies. Ontology Web Language (OWL) and Resource Description Framework (RDF) provide interesting new possibilities for the exploitation of STEP application protocols on the web. The W3C Semantic Web standards are based on work in ontology and logic languages providing means for moving beyond typical data exchange scenarios. Thus, challenging possibilities include integrated product knowledge-bases distributed on the Internet, and usage of technologies developed in the area of the Semantic Web such as reasoning, advanced visualisation, and comparison.

ACKNOWLEDGEMENT

The authors would like to thank all the organisations supporting the international projects that resulted in the development of the FSIG (funStep Interest Group) platform presented in this paper.

Namely, the European Commission that funded the IMS SMART-fm project; SMART-fm partners that somehow contributed for the presentation of this work; CEN/ISSS and ISO TC184/SC4 for the effort in developing industrial standards and binding guidelines. The SMART-fm inter-regional consortium includes companies from 4 IMS regions: USA, Europe, Switzerland and Australia. Also Brazil, although not an IMS region, is part of the consortium joining the European region and Canada is studying its participation. The Network is co-ordinated in Europe by the Portuguese Research and Development Institute: UNINOVA (Technical co-ordination) and by the Spanish RTD Association of furniture manufacturers: AIDIMA (Administrative and Inter Regional Co-ordination).

## REFERENCES

Bititci, U.S., Integrated Production Management in the 21st Century: A Vision and a Research Agenda, IFIP WG 5.7 Brussels Workshop, 2002.

Bourke M., PLM – new approach, www.cisco.com, 2003.

Clements, P. Standard support for the virtual enterprise, Int. Conf. on Enterprise Integration Modelling Technology – ICEIMT'97, 1997, Italy.

Drejer, A. and Gudmundsson, A.: Exploring the concept of multiple product development via an action research project, Integrated Manufacturing Systems, 2003, Vol. 14, Number 3, pp. 208–220.

Garetti M., Macchi, M. and Van De Berg, R., Digitally supported engineering of industrial systems in the globally scaled manufacturing, IMS-NoE SIG 1 White Paper, Milano, 2003.

International Organization for Standardization (ISO); www.iso.org/iso; 2005.

ISO TC184/SC4/WG11 N204; "Product data representation and exchange: Implementation methods: EXPRESS to XMI Binding"; 2003.

ISO TC184/SC4, "ISO10303 Part 236; Application Protocol: Furniture product and furniture project decoration data"; 2002.

ISO/IEC FDIS 19757-3; "Schematron: Final Committee Draft"; http://www.schematron.com/spec.html; October, 2004.

ISO TC184/SC4/WG11 N223; "Product data representation and exchange: Implementation methods: XML Schema governed representation of EXPRESS schema governed data"; 2004.

ISO10303-21; "Industrial automation systems and integration - Product data representation and exchange", www.tc184-sc4.org; 2005.

i2010, europa.eu.int/information_society/eeurope/i2010/index_en.htm, 2005.

Jardim-Goncalves, R.; Nunez, M. J.; Roca-Togores, A. and Steiger-Garcao, A.; "SMART-fm: to get ready industrial SMEs in the emerging digital economy"; The eBusiness and eWork Conference (E2002); Prague, Czech Republic, Oct 16–18, 2002.

Jardim-Gonçalves, R. and Steiger-Garção A., "Implicit Hierarchical Meta-Modelling. In Search of Flexible Inter-Operability for Manufacturing and Business Systems"; BASYS'02, Cancun, Mexico, September 25–27, 2002.

Jardim-Gonçalves, R. and Steiger-Garção, A. Implicit multi-level modelling for integration and interoperability in flexible business environments, submitted to Communications of ACM, special issue on Enterprise Components, 2002.

Jardim-Goncalves, R.; Malo, P.; Vieira, H. and Steiger-Garcao, A.; "Improving competitiveness through SMART Furniture Manufacturing in Extended Environments"; 10th ISPE International Conference on Concurrent Engineering: Research and Applications (CE 2003); Madeira, Portugal, Jul 26–30, 2003.

Jardim-Goncalves, R.; Olavo, R. and Steiger-Garcao, A.; "Modular Application Protocol for Advances in Interoperable Manufacturing Environments in SMEs"; 10th ISPE International Conference on Concurrent Engineering: Research and Applications (CE 2003), Madeira, Portugal, Jul 26–30, 2003.

Lazcano et al.; The wise approach to electronic commerce, International Journal of Computer Systems Science & Engineering, 2000, Vol. 15, No. 5.

Lubell, J.; Peak, R. S.; Srinivasan, V; and Waterbury S. C.; "STEP, XML, and UML: Complementary technologies"; ASME 2004 Design Engineering Technical Conferences and Computers and Information in Engineering Conference (DETC 2004); Salt Lake City, Utah USA, Sep. 2004.

Mello A., Better Products, Faster through Product Life-Cycle Management: Product life-cycle management applications spur innovation and cost savings. www.cisco.com, 2003.

Nagi, L.; Design and Implementation of a Virtual Information System for Agile Manufacturing, IIE Transactions on Design and Manufacturing, 1997, Vol. 29(10), pp. 839–857.

O'Marah K., PLM market, AMR research, www.amr.com, 2002.

OMG: "Unified Modelling Language (UML)"; www.uml.org; 2005.

Pahl, G. and Beitz, W. Engineering Design – A Systematic Approach, Springer-Verlag, 1996, London,

NIST, Manufacturing Engineering Lab, http://www.mel.nist.gov/proj/interop.htm, 2005.

Oasis/UN/CEFACT, Creating a single electronic market: ebXML Business Process Specification Schema, Version 1.01, 2003.

OMG: "XML Meta-Data Interchange (XMI)"; www.omg.org/technology/documents/formal/xmi.htm; 2003.

Panetto H., Pétin, J.F. and Méry, D. (2002). Formalisation of enterprise modelling standards using UML and the B method. In: Proceedings of the 8th International Conference on Concurrent Enterprising, ICE2002, 2002, Rome, Italy, pp. 93–101.

Panetto H., Berio G., Benali K., Boudjlida N. and Petit M. (2004). A Unified Enterprise Modelling Language for enhanced interoperability of Enterprise Models. Proceedings of the 11th IFAC INCOM2004 Symposium, April 5th-7th, Bahia, Brazil.

Price, D.: "A Brief Foray into Semantic Web Technology and STEP"; www.exff.org; October, 2003.

PDES, Inc., http://pdesinc.aticorp.org, 2005.

Poole, J. D.: Model-Driven Architecture: Vision, Standards and Emerging Technologies, ECOOP 2001, 2001, Hungary, University of California.

Pugh, Stuart, Total design: integrated methods for successful product engineering, Addison-Wesley, 1997, Wokingham.

Resource Description Framework (RDF); http://www.w3.org/RDF/; 2004.

Semantic Web; http://www.w3.org/2001/sw; 2005.

Seth, H.; "XML and Integration"; Industry Report, XML Journal; www.sys-con.com/xml/article.cfm?id=591; March, 2003.

Smart-fm Deliverable 1.3; "Definition of SMART-fm pilot demonstrators"; 2003.

UEML IST–2001–34229 Thematic Network: "Unified Enterprise Modelling Language", http://www.ueml.org, 2003.

Ulrich, K. T. and Eppinger, S. D.; Product Design and Development, McGraw-Hill, 2000, New York.

Web Ontology Language (OWL) Language Reference: http://www.w3.org/TR/owl-ref; 2004.

World Wide Web Consortium; www.w3.org; 2005.

W3C Consortium: "Extensible Mark-up Language (XML)"; www.w3.org/XML; 2004.

W3C Consortium: "Web Services Activity"; www.w3.org/2002/ws; 2005.

*Advanced Manufacturing – An ICT and Systems Perspective – Taisch,*
*Thoben & Montorio (eds)*
*© 2007 Taylor & Francis Group, London, ISBN 978-0-415-42912-2*

# Ubiquitous PLM using Product Embedded Information Devices

Dimitris Kiritsis[1] & Asbjørn Rolstadås[2]
[1] *EPFL, Lausanne, Switzerland*
[2] *NTNU, Trondheim, Norway*

ABSTRACT:  This paper reports on the European Commission Framework Programme 6 Integrated Project (No. 507100) PROMISE (IMS No. 01008). PROMISE will develop appropriate technology, including product lifecycle models, and Product Embedded Information Devices with associated firmware and software components and tools for decision making based on data gathered through a product lifecycle. This is done to enable and exploit the seamless flow, tracing and updating of information about a product, after its delivery to the customer and up to its final destiny (de-registration, decommissioning), with the information being sent back to the designer and producer. The breakthrough contribution of PROMISE, in the long term, is to allow information flow management to go beyond the customer, to close the product lifecycle information loops, and to enable the seamless e-transformation of Product Lifecycle Information into Knowledge. The PROMISE R&D implementation plan includes fundamental and applied research activities in the disciplines of information systems modelling, smart embedded systems, short and long distance wireless communication technologies, data management and modelling, Design for X and adaptive production management for Beginning-of-Life (BOL), statistical methods for preventive maintenance for Middle-of-Life (MOL) and planning and management of product End-of-Life (EOL).

*Keywords*:  Beginning-of-Life (BOL), Middle-of-Life (MOL), End-of-Life (EOL), PLM, Knowledge Transformation, Product Lifecycle Information Tracking, Product Embedded Information Devices, PEID.

## 1  INTRODUCTION

A product system's lifecycle is characterised by the three phases: Beginning-of-Life (BOL), including design and manufacturing, Middle-of-Life (MOL), including use, service and maintenance and End-of-Life (EOL), where the product ceases to exist. EOL has been the subject of much focus in connection with environmental issues and is characterised by various scenarios such as: reuse of the product with refurbishing, reuse of components with disassembly and refurbishing, material reclamation without disassembly, material reclamation with disassembly and, finally, disposal with or without incineration.

To properly manage the full product lifecycle, it is necessary to develop a seamless data flow through all three phases. This requires a Product Embedded Information Devices (PEID) to store information and communicate with local and centralised databases, as well as a product data model capable of structuring this information for easy access and updating.

During BOL (between design and manufacturing), and partly into MOL, the information flow is quite complete and supported by intelligent systems like CAD/CAM. Product Data Management (PDM), and Knowledge Management systems are effectively and efficiently used by the industry and, through their influence, by their suppliers.

The information flow becomes far less complete when moving from the MOL phase to the final EOL scenario. For most of today's technological products, and especially for consumer electronics, household machines, vehicles etc., it is fair to say that the information flow breaks down after the delivery of the product to the customer. The fact that the information flow in most cases is interrupted shortly after product sales prevents the feedback of data, information and knowledge, from service and maintenance and recycling experts back to the designers and producers.

The PROMISE project (FP6 507100 and IMS 01008) has been launched by the European Commission to develop technology to close this information gap. The development will be supported by demonstrators in selected product sectors such as railway, heavy load vehicles, automotive, brown goods and white goods. The project started in November 2004 and will finish in 2009. It has 22 partners covering 9 European countries. In addition it is part of an international (IMS) co-operation with additional partners in Japan and USA. The main objective of PROMISE is to develop a new generation of product information tracking and flow management system. This system will allow all actors that play a role during the lifecycle of a product (managers, designers, service and maintenance operators, recyclers, etc.) to track, manage and control product information at any phase of its lifecycle (design, manufacturing, MOL, EOL), at any time and any place in the world. The PROMISE concept is shown in Figure 1.

The main elements of the PROMISE concept and requirements are:

- Local (short distance) connection mode for product data and information exchange,
- Internet (long distance) product information and knowledge retrieval
- Data and information flows
- Envisaged decision support software
- PEID will amongst others be based on RFID technology

The above concepts and requirements compose what can be defined as seamless e-transformation of information to knowledge.

This paper presents the research aspects of the project and discusses some of the main issues of the proposed approach as well as about Closed-Loop Product Lifecycle Management (Closed-Loop PLM), which is the main research area at the beginning of the project and provides and most recent research results.

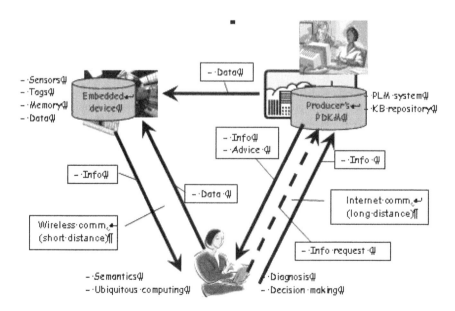

Figure 1. The PROMISE concept and requirements for seamless e-transformation of information to knowledge.

# 2  STATE-OF-THE-ART

## 2.1  *Business aspects*

(From Adhanda Enterprises, LLC, published in e-INSIDER, December 4, 2002, http://www.supplychainbrain.com/news/e12.04.02.newsletter.htm)

*Until recently, companies satisfied their needs for inventory and order fulfilment visibility down the supply chain with a series of manual processes, including physical counts and bar code scanning. The upshot: slow information flow, since bar codes are a line-of-sight technology that requires manual scanning – with trade-offs like labour cost associated with every read and the possibility of human error – and allows only one item to be read at a time. Consequently, bar codes are only read at a few control points in the supply chain. Thus arose the need for a technology such as RFID, which overcomes these limitations.*

*AMR Research expects this market to progress rapidly through 2003 and explode into a multibillion-dollar market in 2004 and beyond. This view is seconded by Forrester analysts, who believe that real-time data monitoring throughout the supply chain will expand greatly by 2006 when tiny microchips will be embedded into physical objects like pallets, trucks and machine tools to track the objects' identity, status, and location and provide other decision-enabling information.*

*According to research firm Venture Development Corp., the RFID tag market, which was $76 million in 2000, is projected to reach $330 million by the year 2005. Also, the hardware opportunity for this market was worth $660 million in 2000 and is forecast to be worth about $2 billion in 2005.*

*Despite the level of excitement generated by the new technology, considerable apprehension surrounds RFID. Edward Rerisi, a senior analyst at Allied Business Intelligence, attributes this to three main factors:*

*Confusion about low cost tags. "People are hesitant to invest in RFID because they are hearing about a five-cent tag. It may be possible one day to create a five-cent tag, but you can still do a lot of things with RFID and make a compelling business case for it today."*

*The general decline in spending on IT systems.*

*A lack of skilled integrators who can install RFID systems. "You need people who know how to handle RFID technology, but you also need people on the backend. Especially for larger installations, you need people who can take the RFID component and integrate it neatly into a large scale enterprise resource planning system or a local database."*

*Additionally, many companies are waiting until standards emerge so they don't get stuck with a proprietary technology that may become obsolete in a few years. However, by enabling co-ordination across an organisation's entire supply web of partners, customers, suppliers, and distributors, it would not be hyperbole to call RFID B2B's new killer application.*

Additionally, in a recent report of the AUTO-ID centre made by the well-known firm *Accenture*, published February 1, 2003, it is stated:

*"While it is clear that Auto-ID applications will unlock unprecedented value in the future, the path to adoption and implementation will be a challenging one. Although many companies have implemented point solutions using Auto-ID technology, the scope and scale of the EPC™ revolution coupled with item level tagging are unprecedented. All the obstacles along the path to widespread Auto-ID adoption are unknown, but some of the known challenges are data storage requirements, network bandwidth and the accuracy of read rates. These issues will likely force adoption of case-level tagging applications, with a gradual transition to item-level tagging as issues are resolved. Managing this implementation process will require detailed planning and a deep familiarity with the industry."*

In today's challenging global market PLM still plays an important role, and its market is still increasing above average (Source: AMR Research, 2002, http://www.amrresearch.com). PLM can be described an a strategic business approach that applies a consistent set of business solutions in support of the collaborative creation, management, dissemination, and use of product definition information across the extended enterprise from concept to End-of-Life – integrating people, processes, business systems, and information (Source: CIMdata, 2002, http://www.cimdata.com/). PLM is the next significant wave of opportunity for manufacturers looking to improve business performance. Just as Enterprise Resource Planning (ERP) consolidated disparate back-office activities into a cohesive environment for running business operations, PLM consolidates diverse business activities that create, modify and use data to support all phases of a product's lifecycle (Source: Marc Halpern, 2PLM e-zine, 2003, http://www.johnstark.com/2PLM131.html).

## 2.2 Scientific and technical aspects

The scientific and technical approach for the realisation of PROMISE at the system level as described above will be based on the application of the following technologies:

- Product Embedded Information Devices (PEID) based on Transponders or RFID IC, integrating sensors and Wireless Communication technologies,
- Data, Information and Knowledge Modelling and web-based Programming,
- Distributed Decision Making Logistics (e. g. by using multi-agent technologies),
- Innovative web-enabled and Embedded Predictive e-Service technologies to deal with MOL (use, service and maintenance) issues,
- Innovative Product Lifecycle Modelling and Simulation technologies that allow the evaluation and validation of a product system through its whole lifecycle.

There are two main aspects of the enabling technologies that are at the background of the PROMISE concepts: RFID (which is at the origin of the motivation of the PEID concept of PROMISE) and Information Processing.

### 2.2.1 RFID

Radio Frequency Identification (RFID) technology is over 15 years old. The present labels were developed with respect to the demands of baggage tag and ticketing applications. This means that the proper range of operation is approximately one metre. Future labels will have short range (and, therefore, less anti-collision problems owing to the smaller reading volume). On the other side, we will see long-range labels, working in a semi-passive or semi-active mode. The short and medium range systems work with electromagnetic field reading at 125 kHz or 13.56 MHz. Long range transponders work at higher frequencies (900 MHz, 2.45 GHz, 5.8 GHz, etc). Most of them contain battery for power in operation.

Most RFID labels work in a passive mode, i.e. they are woken up by the read-write base station, so called Reader Talk First (RTF) systems. This functionality prevents interference with other systems. One of the biggest handicaps (besides cost of transponder labels) is the problem of sufficiently proper operation of stickers or labels mounted directly on the surface of metals, as it will occur widely in machinery, but also in food labelling (tin cans, aluminium foils in paper containers). There are some approaches to overcome this *metal surface labelling* issue by some improved shielding and tuning of the labels. Transponders or RFID devices for labelling objects have been discussed already for a long period of time. Recently the activities have concentrated around the so called smart labels, low cost RFID tags for applications in logistics, brand protection and other areas. Standardisation has been done and is in the process for vicinity cards (ISO 14443), proximity cards (ISO 15693) and item management (ISO 18000). The work with the 18000 standards is not finished.

### 2.2.2 Information processing

The following describes state-of-the-art developments at the former Auto-ID Centre (actually Auto-ID Labs, http://www.autoidlabs.org/) which had laboratories at MIT, Cambridge University,

University of Adelaide, Fudan University (China) and Keio University. This general approach can be used also for the PROMISE project.

- **EPC Code**

The Electronic Product Code (EPC) is a numbering scheme that can provide unique identification for physical objects, assemblies and systems. Information is not stored directly within the code rather; the code serves as a reference for networked (or Internet-based) information. In other words, the code is an *address* – it tells a computer where it should go to find information on the Internet. The EPC requires few parameters to determine its design: *Number of bits* – i.e.: How much information is needed to provide a unique identity to every single product manufactured, sold and consumed in the global supply chain? *Bit partitions* – i.e.: What is the best way to organise – or *break up* – the numbers/figures so that as many unique combinations as possible is achieved, while also expediting Internet searches?

Consider this an exercise in determining the best *search hierarchy* – like a postal address – which goes from country, to city, to zip code, to street, to house and individual.

As the detail or levels of the hierarchy increase, the speed and accuracy of the search will likewise increase, but the possible combinations of unique numbers will decrease.

- **Product Mark-up Language**

The Product Mark-up Language (PML) is a standard *language* for describing physical objects. It will be based on the eXtensible Mark-up Language (XML). Today, HyperText Mark-up Language (HTML) is the common language on which most web sites are based, allowing individuals to surf the Internet from their desktops. Where HTML tells a computer how information should be displayed (e.g., what colour and size it should be) – XML goes a step further, telling the computer what information it is viewing (e.g., an address or a telephone number). The PML will go even further, building in layers of increasingly specific data to describe physical objects, their configuration and state. In the end, PML should translate or contain static data, such as: dosage, shipping, expiration, advertising and recycling information. It should provide instructions for machines that *process* or alter a product, such as: microwaves, laundry appliances, machine tools and industrial equipment. Moreover, it may need to communicate dynamic data: information that changes as a product ages or as it is consumed, such as: volume, temperature, moisture and pressure. In addition, it may need to include software, or programs, which describe how an object behaves, for instance: a PML file may contain the program which describes how fast the tyres on a car will wear before they need to be replaced, or how fast an object may burn in case of a fire.

- **Object Naming Service**

The Object Naming Service (ONS) tells computer systems where to locate information on the Internet about any object that carries an EPC (Electronic Product Code). ONS was developed at the Massachusetts Institute of Technology by Dr. David Brock, Professor Sanjay Sarma and Joseph Foley. ONS is similar to – and (in part) based on – the Internet's existing DNS (Domain Name System), which allows Internet routing computers to identify where the pages associated with a particular web site are stored. The DNS is used every time a web site is accessed. The ONS will be used every time information is needed about a physical object. It is likely that the ONS will be many times larger than today's DNS. Although conceptually simple, designing ONS was a challenge. The system must be capable of quickly locating data for every single one of the trillions of objects that could potentially carry an EPC code in the future. The ONS must serve as a lightning fast post office that, on a daily basis, receives and delivers millions (if not billions) of letters.

- **ID@URI link between physical objects and their information**

As an alternative to the Auto-ID system, Helsinki University of Technology (HUT), will test another technology. In the Dialog platform developed at HUT, an email address-like notation *ID@URI* is used for identifying physical product items and linking to the *product agents* that handle their information. The ID@URI notation fulfils the same functionality as the combination of EPC and ONS. The URI part is a domain name of a computer, whose uniqueness is guaranteed by the DNS. The ID part then only needs to be a locally unique identifier inside the address space of the URI.

Just like email addresses are globally unique by the fact that they use DNS-unique URIs and locally unique user names, the same is true for product item identifiers.

- *World Wide Article Information (WWAI)*

WWAI is an application level protocol for distributed article information, peer-to-peer networking. With WWAI, users can find other systems providing information on the same ID attached to an object. WWAI enables companies to share real-time product information, regardless of the RFID tag (or other AIDC method) used, over the Internet. The WWAI protocol lets that information be distributed on the computers of the companies that have participated in the manufacturing, assembling or transporting of the product. Only the ones who are interested will be affected by the information flow. Every participant has control of its own product information and whether the information is public or private. Because there is no central point or routing information attached on the identifier (RFID or barcode), WWAI makes every participant independent and able to operate even if some of the participants are not available. Distribution of information makes WWAI networks scalable and able to grow as the number of products grow. The WWAI network is self organising, meaning it does not require name server infrastructure, and it can work with existing EAN coding standards and Auto ID EPC code.

## 3  THE PROMISE VIEW OF WHOLE PRODUCT LIFECYCLE

The whole product lifecycle implies all phases related to the product generation, usage, and disposal. It consists of the following lifecycle activities: design (product lifecycle planning and product design), production (procurement, manufacturing, and assembly), logistics (distribution), usage, maintenance (service), collecting, re-manufacturing (disassembly, refurbishment, re-assembly, etc.), reuse, recycling, and disposal. Figure 2 shows the whole product lifecycle in PROMISE.

In general, relevant activities and information flows have complex interactions during the lifecycle. Therefore, it is important to control and steer the process and information flow of the product lifecycle. For this purpose, modelling issues related to product lifecycle activities and lifecycle information are considered.

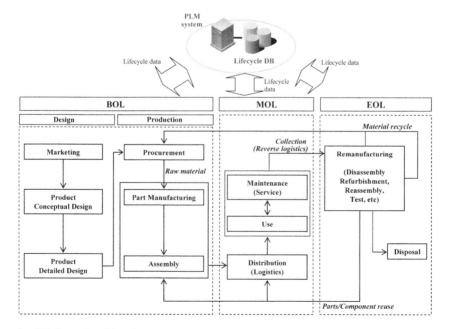

Figure 2.   Whole product lifecycle.

70

## 4 PRODUCT LIFECYCLE DATA

Product lifecycle data indicate all objects that are created, updated, deleted, and stored during whole product lifecycle, such as conceptual ideas, CAD/CAM/CAE models, technical documents, manufacturing data, logistics data, usage information, disposal information, and so on. Depending on their characteristics, they can be classified into some generic types: *product definition*, *product history*, and *best practice*; *content* and *meta-data*, *internal* and *external*, and *stationary* and *dynamic*. In this study, the product lifecycle data has been classified by four dimensions. First, the lifecycle data can be classified into three groups depending on its characteristics: product, process, and organisation (resource). Second, the lifecycle data can be grouped into static data and dynamic data by *degree of variation*. Third, depending on the *chronological order of lifecycle phase*, these can be divided into three groups: BOL data, MOL data, and EOL data. Finally, depending on *degree of abstraction*, these can be divided into content data and meta-data.

There are many product, process, and organisation related data that are generated during product life span. These are called product, process, and organisation data, respectively. Static data indicate the data that do not change during product lifecycle. Most are determined and fixed at BOL phase: e.g. specifications of products and End-of-Life information such as disassembly sequences. However, dynamic data implies the data that can change during product lifecycle: e.g. service/maintenance history information. BOL is the phase where product concept is generated and its physical model is realised. Therefore, there are many BOL data related to product generation such as CAD/CAM data, technical documents, and so on. MOL is the phase where products are distributed, used, maintained, and serviced by customers or engineers. For example, usage conditions, failure, and maintenance or service events are MOL data. EOL is the phase where products that that have use their use value are collected, disassembled, refurbished, re-assembled, recycled, reused, and disposed. Hence, there are many EOL data with regard to conditions of retirement and disposal of products. Meta-data implies the data to describe the content data while content data means tangible data occurring over whole product lifecycle. Table 1 shows examples of meta-data of product data.

## 5 THE PROMISE PLM SYSTEM

To co-ordinate and efficiently manage PLM activities, it is necessary to develop the PLM system framework which addresses how the PLM system and its context will be represented in viewpoints of system components and their relations, and what are the key principles of the PLM system. This framework is very important because it is a basic sketch for developing PLM system. In this framework, it should be defined how to integrate and co-ordinate components with respect to

Table 1. An example of meta-data of product data.

| Meta-data description | Meta-data |
| --- | --- |
| Who created you? | *Creator* |
| Who owns you now? | *Owner* |
| What kind of a product are you? | *Type* |
| Do you contain hazardous materials? | *Hazardous_materials* |
| What has been happening to you? | *History* |
| Where are you going? | *Next_destination* |
| What is your destination? | *Final_destination* |
| When should you arrive at your destination? | *Due_date* |
| To which order do you belong? | *Order_number* |
| To what shipment do you belong? | *Shipment_number* |

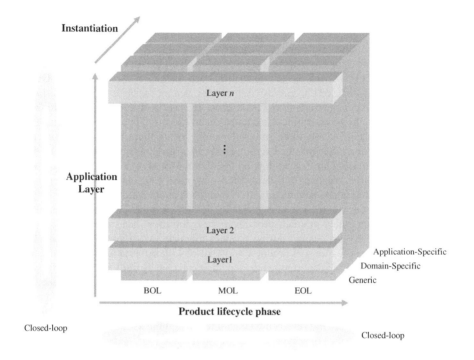

Figure 3.   Generic PROMISE PLM model.

business model, hardware, and software. Moreover, it is necessary to define system architecture in a standard and flexible way for adapting it to various application domains.

To satisfy generic requirements, objectives, and principles, a generic PROMISE PLM system model has been proposed as shown in Figure 3. It has three perspectives: application layer, instantiation, and product lifecycle phase.

### 5.1  *Application layer*

As shown in Figure 4, the generic PROMISE PLM model consists of 9 applications layers. This shows the logical and hierarchical view of application layers needed to build up a closed-loop PLM model. The right side of Figure 4 is the instances of PROMISE project for each application layer. PEID layer represents information devices built into the product itself such as RFID tag or on-board computer that takes a role of gathering data. The firmware is located at the embedded software layer, which is installed in PEID and takes a role of managing and processing data of PEID. Middleware layer handles data transferred between PEID layer and PDKM layer.

Network layer takes a role of specifying communication ways among each application layer, in particular, between PEID and data management system layer. Data management system layer contains the applications that can store and manage gathered data. Information/knowledge transformation and decision support layer plays an important role in PLM because it generates the core of knowledge needed to implement several PLM applications from gathered data.

The PDKM is located at knowledge management system layer for managing the knowledge and sharing them with other lifecycle actors during whole product lifecycle. Back-end system layer indicates the area of legacy systems of a company such as Enterprise Resource Planning (ERP) and Supply Chain Management (SCM) systems. Finally, PLM business application layer contains several business applications to streamline product lifecycle operations such as predictive maintenance and EOL decision making.

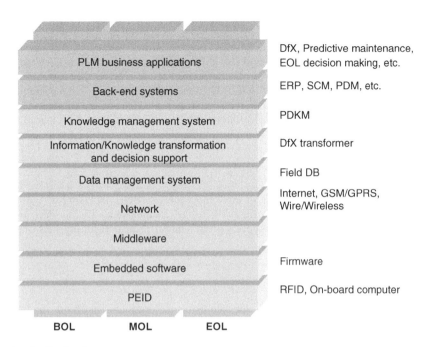

| | |
|---|---|
| PLM business applications | DfX, Predictive maintenance, EOL decision making, etc. |
| Back-end systems | ERP, SCM, PDM, etc. |
| Knowledge management system | PDKM |
| Information/Knowledge transformation and decision support | DfX transformer |
| Data management system | Field DB |
| Network | Internet, GSM/GPRS, Wire/Wireless |
| Middleware | |
| Embedded software | Firmware |
| PEID | RFID, On-board computer |

**BOL      MOL      EOL**

Figure 4.   Application layers.


Table 2.   Details of instantiation.

| Generic | Domain-specific | Application-specific | |
|---|---|---|---|
| Generic model for closed-loop PLM | Lifecycle phase-specific<br>7.  BOL-DfX<br>8.  BOL-Re-configuration<br>7.  of production system<br>9.  BOL-Production logistics and warehouse management<br>10. MOL-Predictive maintenance<br>11. EOL-Decision making | Target product<br>Locomotive<br>Car bumper<br>(Plastics material)<br><br>Heavy vehicle<br>Passenger vehicle<br>Tractor | Main issue<br>DfX<br>Production logistics and warehouse management<br>Predictive maintenance<br>EOL decision making<br>MOL decision making |
| | Target product-specific<br>12. Automotive<br>13. Machinery product<br>14. Network device<br>15. White goods<br>16. Locomotive | Tractor<br><br>Milling machine<br>Gas boiler<br><br>Passenger vehicle<br>Broad band access system<br>Refrigerator | EOL decision making<br>Predictive maintenance<br>Predictive maintenance<br>Re-configuration of Production system<br>MOL maintenance/service<br>Preventive maintenance |

73

## 5.2 *Instantiation*

In an instantiation viewpoint (see Figure 3), there are three layers: generic, domain-specific, and application-specific. In the application-specific layer, there are specific system architectures for 11 applications in the PROMISE project. In the domain-specific layer, there are semi-generic architectures that can be grouped into BOL, MOL, and EOL according to lifecycle phase. In addition, they can be grouped into automotive, machinery products, heavy vehicle, electric and electronic equipment (EEE), and so on, according to the characteristics of products that are targeted in 11 applications. In the generic layer, common and flexible architecture will be designed, which is the scope of this study.

## 5.3 *Product lifecycle phase*

In general, a product lifecycle consists of three main phases: BOL, including design and production; MOL, including logistics (distribution), usage, maintenance, and service; and EOL, including reverse logistics (collecting), recovery (disassembly), re-manufacturing, reuse, recycling, and disposal as shown in Figure 2 above.

## 6 PROMISE PLM ARCHITECTURE

PLM architecture can be defined with three components: business, hardware, and software architecture.

## 6.1 *PROMISE PLM business architecture*

In PLM, all activities performed along the product lifecycle must be co-ordinated and efficiently managed. Figure 5 shows the business architecture of a closed-loop PLM that exactly reflects the PROMISE concept shown in Figure 1. The operations in the closed-loop PLM are based on the interactions among three organisations (PLM agent, PLM system, and Product). The PLM agent can gather product lifecycle information from each product at a fast speed with a mobile device such as a personal digital assistant (PDA) or a fixed reader built in antenna. It sends information gathered at each site (e.g. retailers, distribution sites, and disposal plants) to a PLM system. It enables a PLM system to manage lifecycle information reported at each site (e.g., retailers, distribution sites, and disposal plants) by reading the RFID tag via an information network. The PLM system provides lifecycle information or knowledge made by PLM knowledge agents through an information network whenever requested by related persons and organisations. The following figures show BOL, MOL, and EOL business architecture which describe business applications and information flows.

## 6.2 *PROMISE PLM hardware architecture*

In general, hardware architecture addresses the infrastructure that supports business applications. Here, the physical components that constitute a PLM system are defined. For this, the PEID and network architecture is described.

### 6.2.1 *PEID architecture*

PEID stands for product embedded information device. In PROMISE, since there are various kinds of information devices to gather and manage product information, which can be grouped into two parts – RFID tag and on-board computer – the term, *PEID*, covers all things. *Product embedded* means that product lifecycle information can be tracked and traced in a real-time way over the whole product lifecycle by attaching PEID to the product itself. For this, PEID should possess a unique identity and its work should not be dependent on availability of its power. It requires product identification function and power management function.

The *information device* indicates that the PEID should have the data processing and storing functions with sensor reading function that enables PEID to gather signals from several sensors,

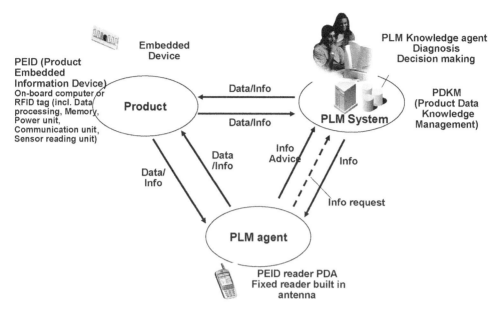

Figure 5.   Generic business architecture.

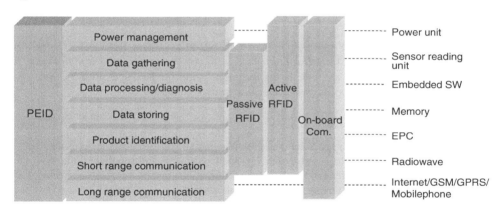

Figure 6.   PEID architecture.

and to retain or store them, and if necessary, to analyse or make decisions. In addition, it should have a communication function with external environments for transmitting data. For this, the PEID should have a processing unit, communication unit, sensor reader, data processor, and memory. In PROMISE, as the PEID device, RFID tag or on-board computer that has the above functions will be used. The architecture of PEID is depicted in Figure 6.

### 6.2.2   *PLM network architecture*

PLM aims to create a framework that will allow efficient sharing of product data across the extended enterprise and throughout the product lifecycle. In this regard, it is important to describe how each component of the PLM system communicates with one another for sharing the data. The related network architecture provides a guide for the technical design of network. In the PLM system, generic network architecture is required to standardise network protocols. Over the last decade, a rapid development of Internet, wireless mobile telecommunication technologies such as the Zigbee standard, wireless sensors, machine-to-machine communication, and RFID have changed

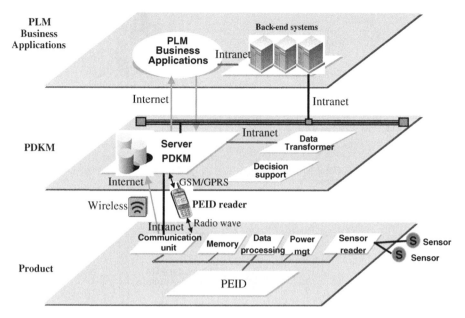

Figure 7.   Generic network architecture.

the traditional environment for networks. In the PROMISE-PLM system, these technologies will be used to build up a network infrastructure for a closed-loop PLM. As can be seen in Figure 7, in the PROMISE-PLM, there are three layers for the network: product layer, PDKM layer, and application layer. Each layer has its own technical network architecture.

### 6.3   PLM software architecture

PLM has emerged as an enterprise solution. It implies that all software tools/systems/databases, such as CAD, PDM, CRM, etc., used by the various departments and suppliers throughout the product lifecycle have to be integrated such that the information contained in these systems can be shared promptly and correctly among people and applications.

Hence, it is important to understand how application software in PLM fit with others that manage product information and operations. For this, software architecture is required. Software architecture is the high level structure of a software system, and is concerned with how to design software components and make them work together. Software architecture is commonly defined in terms of components and connectors. Here, the focus is on structural views of software architecture in terms of components and their relationships. Also described are behavioural views that describe how the component interacts to accomplish their assigned responsibilities. Execution views are not considered.

Figure 8 shows the software architecture in the closed-loop PLM. It has a vertical viewpoint in the sense that its structure represents a hierarchy of software from gathering raw data to connection to the legacy systems.

Embedded software (called firmware) has a role of controlling and managing data of PEID. Database (DB) software also is required to store sensor data and manage them efficiently. Middleware can be considered in general as an intermediate software layer between applications and the operating system. Particularly for distributed communication, co-ordination, and data management, enterprise applications typically rely on functions of the underlying middleware. It is used most often to support complex, distributed applications, for example, applications between RFID tags and business information systems to manage gathered data from RFID tags.

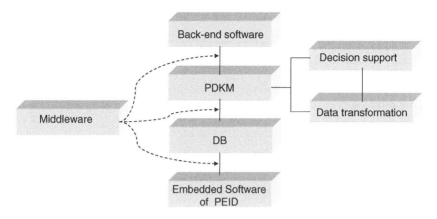

Figure 8.   Generic software architecture.

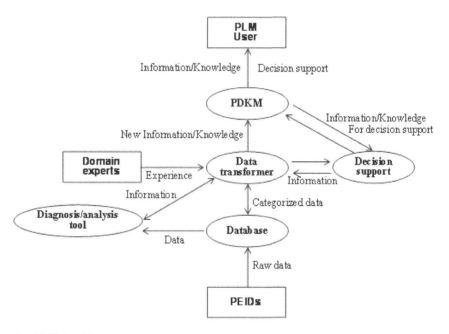

Figure 9.   PDKM architecture.

PDKM should link not only product design and development such as CAD/CAM but also other legacy systems. e.g. Enterprise Resource Planning (ERP), Supply Chain Management (SCM), Customer Relationship Management (CRM) to achieve interoperability of all activities that affect a product and its lifecycle.

PDKM manages information and knowledge generated during product lifecycle. It is generally linked with decision support system and data transformation software. PDKM is a process and technologies to acquire, store, share and secure understandings, insights and core distinctions. PDKM is a very important in PLM because it generates and manages core properties of product lifecycle and is used for improving competence. Back-end software can be defined as the part of a software system that processes the input from the front-end system dealing with the user.

Figure 9 shows the detail of PDKM architecture.

77

## 7 CONCLUSIONS

The international project PROMISE (IMS 01008 and FP6-IP-507100) and its main modules, the European and Swiss ones, started in November 2004. The first reported research results deal with Closed-Loop PLM issues.

New environments such as Closed-Loop PLM provide opportunities to reduce inefficiencies and improve competitiveness.

Resolving these issues will provide PLM actors with the ability that efficiently make decisions on several operational problems from design to end-of-life based on analysing the product lifecycle activities and information model.

Because of the short history of PLM, research on PLM operational level is still in an early stage especially in what concerns closing of product lifecycle information loops. It is expected that further results of the research activities in PROMISE will demonstrate the usefulness and applicability of Closed-Loop PLM.

## ACKNOWLEDGEMENTS

The work reported in this paper was based on the PROMISE project (FP6 IP 507100 and IMS 01008, (http://www.promise.no). The authors wish to express their deep gratitude to all PROMISE partners and particularly to the EPFL team for their contributions on PLM modelling.

## REFERENCES

PROMISE project FP6-IP-507100, 2004, Annex I-Description of work.

PROMISE project FP6-IP-507100, 2005, DR. 1.1 PROMISE system requirements, specifications and system architecture.

Foley J. T.: An infrastructure for electromechanical appliances on the Internet. BE and ME Thesis, Massachusetts Institute of Technology, Cambridge, MA; May 1999.

Huvio E, Grönvall J, Främling K.: Tracking and tracing parcels using a distributed computing approach. NOFOMA'2002 Conference Proceedings, Trondheim, Norway; 12–14 June 2002.

Kärkkäinen M, Främling K, Ala-Risku T.: Integrating material and information flows using a distributed peer-to-peer information system. In: Jagdev H, Wortmann H, Pels H-J, Hirnscall A, (eds.): Proceedings of APMS 2002, Eindhoven, Netherlands, 8–13, September, 2002. pp. 463–73.

Worldwide article information (WWAI), http://www.stockway.fi/wwai.html.

Kiritsis, D.: 2004, Ubiquitous Product Lifecycle Management using Product Embedded Information Devices, Proceedings of the International Conference on Intelligent Maintenance Systems.

Kiritsis, D., Bufardi, A. and Xirouchakis, P.: 2003, Research issues on product lifecycle management and information tracking using smart embedded systems, Advanced Engineering Informatics, 17, 189–202.

Datamation limited (2002) 'Understanding Product Lifecycle Management', Draft, Ver. 1.0.

Jun, H. B., Kiritsis, D., Xirouchakis, P.: 2005, Product lifecycle information modeling with RDF, July 13–15, 2005, Proceeding of the International Conference on Product Lifecycle Management, Lyon, France.

Zhu, J., Yassine, A. A., and Sreenivas, R. S. (2004): 'Information Incorporation Policies in Product Development', Proceedings of the ASME 2004 International Design Engineering Technical Conferences and Computers and Information in Engineering Conference, Utah, USA, DETC2004-57351.

*Advanced Manufacturing – An ICT and Systems Perspective – Taisch,*
*Thoben & Montorio (eds)*
*© 2007 Taylor & Francis Group, London, ISBN 978-0-415-42912-2*

# Closed-loop PLM

Hong-Bae Jun, Dimitris Kiritsis & Paul Xirouchakis
*Swiss Federal Institute of Technology in Lausanne, Lausanne, Switzerland*

ABSTRACT:  This paper reports on work developed under PROMISE, an IMS (01008) and European Commission Framework Programme 6 Integrated Project (507100). PROMISE will develop appropriate technology, including product lifecycle models, Product Embedded Information Devices with associated firmware and software components and tools for decision making based on data gathered through a product lifecycle.

Recently with emerging technologies such as wireless sensors, telecommunication, and product identification technologies, product lifecycle management (PLM) has been in the spotlight. PLM is a new strategic approach to manage product-related information efficiently over the whole product lifecycle. Now, the whole product lifecycle can be visible and controllable by tracking and tracing the product lifecycle information with product embedded information device (PEID). These information flows during whole product lifecycle now have closed-loops, with the information being transmitted back to designers. Thanks to the closed-loops, PLM can streamline several product lifecycle operations. This environment is called *closed-loop PLM*. In this study, the concept of *closed-loop PLM* is examined and the gaps between traditional PLM and *closed-loop PLM* are described.

*Keywords*:  Closed-loop PLM, Product lifecycle, Product identification.

## 1  INTRODUCTION

In general, the product lifecycle consists of three main phases: beginning-of-life (BOL), including design and production; middle-of-life (MOL), including logistics (distribution), use, service, and maintenance; and end-of-life (EOL), including reverse logistics (collecting), re-manufacturing (disassembly, refurbishment, re-assembly, etc.), reuse, recycle, and disposal as shown in Figure 1.

During BOL, the information flow is quite complete because it is supported by several systems like CAD/CAM, product data management (PDM), and knowledge management systems. However, the information flow becomes vague or unrecognised after BOL. In other words, the information flow after BOL phase is interrupted. This prevents the feedback of data, information, and knowledge from service and maintenance and recycling experts back to the designers and producers [PROMISE 2004]. Therefore, several lifecycle phases after BOL such as maintenance, service, recycling, reuse, and disposal have limited visibility of information flow. Actors involved in each lifecycle phase make decisions based on incomplete and inaccurate data, which gives rise to the operational inefficiency (IMTI 2002).

However, over the last decade, a rapid development of Internet, wireless technologies with mobile telecommunications, and the introduction of smart tags that can be embedded in the product, have changed the stereotype view of the product lifecycle.

In particular, thanks to recent product identification technologies (please refer to table 1) such as RFID (Schneider 2003) and AUTO ID (Parlikad *et al.* 2003), the whole product lifecycle can be visible and controllable using these technologies, in the form of product embedded information device (PEID). These new technologies become an important driving force that is

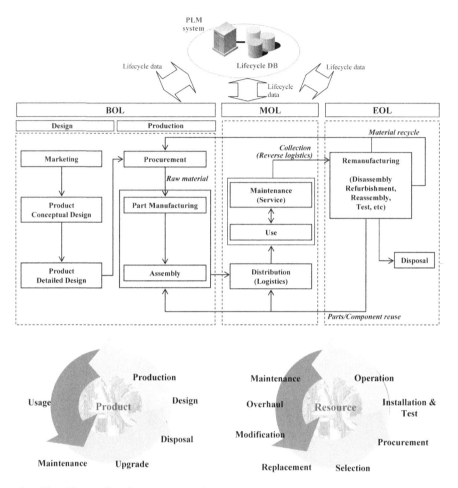

Figure 1.  Closed-loops of product, process, and resource.

needed in the propagation of product lifecycle management (PLM) because they can tackle the problems of PLM that have previously been obstacles to progress. The PLM under this new environment allows all actors across the whole product lifecycle to access, manage, and control product-related information, especially, the information after the delivery of a product to a customer and up to its final destiny, without temporal and spatial constraints. This information can be used to streamline operations of MOL and EOL. This information also goes back to the designer and producer at BOL so that the information flow can be horizontally closed over whole product lifecycle. In addition, the control of information flow is vertically closed. This means that based on gathered data, product related information can be analysed and decisions taken on behaviour of products, which will affect data gathering again. This concept and relevant systems is called *closed-loop PLM* in this study.

To co-ordinate and efficiently manage complex PLM implementation activities, it is necessary to clarify the concept for *closed-loop PLM*. This concept is very important because it is a basic sketch for developing a *closed-loop PLM* system. This can help achieve consensus about how PLM systems should be built.

For this purpose, in this study, the focus is on clarifying the concept of *closed-loop PLM*. The rest of the paper is organised as follows: In section II, the state-of-the-art regarding enterprise and PLM modelling framework is discussed. Section III introduces the concept of *closed-loop PLM*. In Section IV, the main components of *closed-loop PLM* are introduced. In section V, the gaps

Table 1. Product identification technologies.

| Technologies | Definition | Product lifecycle | | |
|---|---|---|---|---|
| | | BOL | MOL | EOL |
| Auto-ID: EPC (Parlikad *et al.* 2003) | Electronic Product Code: Product unique code | • | • | • |
| Auto-ID: PML (Brock *et al.* 2001) | Physical Mark-up Language: Mark-up language for product information | • | • | • |
| Auto-ID: ONS (Foley 1999) | Object Naming Service: Telling computer systems location information on the Internet about any object that carries an EPC. | • | • | • |
| ID@URI (Huvio *et al.* 2002) | Identifying physical product items and linking to the product agents that handle their information | • | • | • |
| RFID (Schneider 2003) | Radio Frequency Identification: Communication technology for collecting and transferring information via radio waves | • | • | • |
| GPS (Evers and Kasties 1994) | Global Positioning Systems: Satellite navigation system used for determining precise locations and providing a highly accurate time reference | – | • | • |
| GIS (Evers and Kasties 1994) | Geographical Information System: Information system capable of assembling, storing, manipulating, and displaying geographically-referenced information | – | • | • |

between traditional PLM and *closed-loop PLM* are described. Finally, the paper concludes with discussion and identification of further research topics.

## 2 STATE-OF-THE-ART

There have been much previous work dealing with enterprise architecture or enterprise model, as shown in Table 2. Some of these have been developed for special purpose. For example, CIMOSA was provided for modelling computer integrated manufacturing (CIM) systems. Architecture for integrated Information System (ARIS) has been developed to model and integrate the information system of enterprise level that is more business-oriented. Recently, for general purpose, the GERAM methodology (Generalised Enterprise Reference Architecture and Methodology) has been developed by IFAC/IFIP. GERAM essentially was built on results from CIMOSA, GRAI Integrated Methodology (GIM), and Purdue Enterprise Reference Architecture (PERA). The purpose of GERAM is to serve as a reference for the whole community concerned with the area of enterprise integration providing definitions of the terminology, a consistent modelling environment, a detailed methodology, promoting good engineering practice for building reusable, tested, and standard models, and providing a unifying perspective for products, processes, management, enterprise development, and strategic management (Vernadat 1996). For more detailed information of enterprise model or architecture, please refer to Vernadat (1996) and the work of Lillehagen and Karlsen (2004).

On the other hand, some have dealt with PLM related models. For example, CIMdata (2002) addressed a high level PLM definition, describing its core components, and clarifying what

Table 2. Previous research.

| Classification | Previous research |
| --- | --- |
| Enterprise architecture or model | • IDEF (Integrated computer aided manufacturing DEFinitions methodology) (Mayer 1994)<br>• IEM (Integrated Enterprise Modelling) (Vernadat 1996)<br>• PERA (Purdue Enterprise Reference Architecture) (Vernadat 1996)<br>• CIMOSA (Open System Architecture for CIM) (Bruno and Agarwal 1997)<br>• ARIS (Architecture for integrated Information System) (Scheer 1998a, 1998b)<br>• GERAM (Generalised Enterprise Reference Architecture and Methodology) (Vernadat 1996)<br>• UEML (Unified Enterprise Modelling Language) (Vernadat 2002) |
| PLM model | • High level PLM definition (CIMdata 2002)<br>• New business model in virtual enterprise (PLM) (Ming and Lu 2003)<br>• Conceptual lifecycle modelling framework with IDEF (Tipnis 1995)<br>• Conceptual architecture and the key components for total product lifecycle design supporting system (Kimura and Suzuki 1995)<br>• Conceptual product lifecycle model which consists of a product model and a process network for lifecycle simulation (Nonomura *et al.* 1999)<br>• IPPD (Integrated Product and Process Development) methodology: conceptual mathematical model for describing product lifecycle with a simple graph description technique (Yan *et al.* 1999) |

is and is not included in a PLM business approach. CIMdata mentioned three core concepts of PLM:

1) Universal, secure, managed access and use of product definition information;
2) Maintaining the integrity of product definition and related information throughout the life of the product or plant;
3) Managing and maintaining business processes used to create, manage, disseminate, share and use the information.

Ming and Lu (2003) proposed the new business model in virtual enterprise to tackle issues of product development in the scope of PLM. They proposed the framework of product lifecycle process management for collaborative product-services. The framework consists of industry specific product lifecycle process template, product lifecycle process application, abstract process lifecycle management, supporting process technology, supporting standards, and enabling infrastructure.

Although there has been much previous work on enterprise or PLM related modelling, few have addressed the concept for *closed-loop PLM* that focuses on the integration between PEID and PLM. Most did not consider PEID in PLM. Although some of enterprise modelling methodologies contained the concept of lifecycle in their methodologies, e.g. CIMOSA and PERA, however, their lifecycle concepts are focused on the development lifecycle of manufacturing systems or enterprise so that they are not suitable to represent the concept of *product* lifecycle. The GERAM framework gives a very good overview of enterprise modelling, but it is too general to design the characteristics of the *closed-loop PLM*. In addition, the previous PLM modelling methods are also not suitable to describe the *closed-loop PLM*. In summary, previous enterprise modelling frameworks are too sizeable to describe the *closed-loop PLM*. On the other hand, the previous PLM modelling works are too conceptual and implicit.

3  THE CONCEPT OF CLOSED-LOOP PLM

The concept of *closed-loop PLM* can be defined as follows:

> *A strategic business approach for the effective management of product lifecycle activities by using product data/information/knowledge which are accumulated in the closed-loops of product lifecycle*

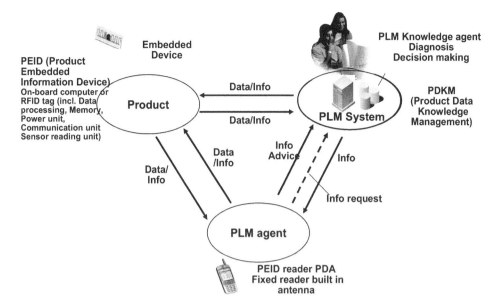

Figure 2.   Framework of closed-loop PLM

*with the support of product embedded information device (PEID) and product data and knowledge management (PDKM).*

The objective is to streamline product lifecycle operations, based on the seamless flow of product information, through a local wireless connection to PEIDs and through remote Internet connection to knowledge repositories in PDKM, after the delivery of the product to the customer and up to its final destiny (disassembly, re-manufacturing, re-use, recycling, disposal, etc.) and back to the designer and producer.

Figure 2 shows the framework of *closed-loop PLM*. The operations in the *closed-loop PLM* are based on the interactions among three organisations (PLM agent, PLM system, and product). PLM agent can gather the product lifecycle data from each product at a fast speed with a mobile reading device, or a fixed reader with built-in antenna. It also sends information gathered at each site (e.g., retailers, distribution sites, and disposal plants) to a PLM system. The PLM system provides lifecycle information or knowledge built by design-for-X (DFX) agent whenever requested by related persons and organisations. To implement the concept of *closed-loop PLM*, the following are necessary conditions.

- Every product has a PEID to manage its lifecycle data. If necessary, sensors can be built in products and linked to the PEID for gathering its status data.
- Each lifecycle actor accesses to PEIDs with its reader or accesses to a remote PLM system for getting necessary information.
- *Closed-loop PLM* should have decision support systems, and PDKM systems for providing lifecycle actors with suitable advice at any time.

In the *closed-loop PLM*, information flow is horizontally closed, which means that information flow is closed over product lifecycle phases: BOL, MOL, and EOL.

- Designers will be able to exploit expertise and know-how of the other players in the product lifecycle such as the modes of use, conditions of retirement, and disposal of their products and thus improve product designs.
- Producers will be provided in a real-time way with not only operation data form the shopfloor but also the usage status of product until its disposal phase.

- Service and maintenance experts will be assisted in their work by having not only product design information but also an up-to-date report about the status of the product during product usage.
- Recyclers and re-users will be able to obtain accurate information about *value materials* arriving through end-of-life (EOL) routes by the analysis of modes of use and conditions of product.

Moreover, the information control flow is vertically closed, which means that information are gathered and controlled in the vertical loops of hardware, software, and business process.

- PEID gathers product related data under specific conditions or periodically or in a real-time way over the whole product lifecycle.
- PEID sends gathered data to database under specific conditions or periodically or in a real-time way.
- Based on gathered data, information and knowledge are generated and stored at knowledge repository in PDKM system. They are based on decision making of lifecycle actors.
- Based on analysis and decision making, if there is any need to update product information, PLM server sends updated information to PEID directly or via PLM agents.

The core of *closed-loop PLM* is the information management of lifecycle objects such as product related data, processes, and resources over the whole lifecycle since it can support the ability to analyse data and make decisions in fast and consistent ways. For this, *closed-loop PLM* should support the following.

- Management of whole product lifecycle activities,
- Management of product related data and resources,
- Collaboration among customers, partners, and suppliers, and
- Enterprise's ability to analyse challenges and bottlenecks, and make decisions on them.

## 4 MAIN COMPONENTS FOR CLOSED-LOOP PLM

- **PEID:** This stands for product embedded information device. It is defined as a device embedded in (or attached to) a product, which contains product related information (e.g. product identification), and which is able to provide the information whenever requested by external agents during the product lifecycle.
- **PDKM:** This manages information and knowledge generated during the product lifecycle. It is generally linked with decision support systems and data transformation. PDKM is a process and technologies to acquire, store, share and secure understandings, insights and core distinctions. PDKM should link not only product design and development such as CAD/CAM but also other backend software (legacy systems), e.g., enterprise resource planning (ERP), supply chain management (SCM), and customer relationship management (CRM) to achieve the interoperability of all activities that affect a product and its lifecycle.
- **Decision making/support:** This streamlines the lifecycle operations by providing suitable information and knowledge through analysis of gathered lifecycle data. There can exist several types of decision making/support: automatic, semi-automatic, and manual. Moreover, according to the application area, there can exist many applications, e.g. for predictive maintenance, for design for X-ability, and for maximising the value of EOL product.
- **Data transformer:** This takes a role of converting raw data gathered by PEID to necessary information and knowledge.

Figure 3 shows the use case diagram of *closed-loop PLM* system. In the *closed-loop PLM*, PEID (Use case 2.1) is embedded into each product. PEID periodically sends data using communication network (Use case 2.8) to field DBs (Use case 2.3) directly or via PEID reader (Use case 2.2). The data transformer (Use case 2.5) transforms gathered data from field DBs to information to knowledge. The information and knowledge are generated with the help of several analysis tools (Use case 2.4) and decision making/supporting tools (Use case 2.6). All information and knowledge generated during product lifecycle will be managed by PDKM (Use case 2.7). Product (Actor 1.0) uses

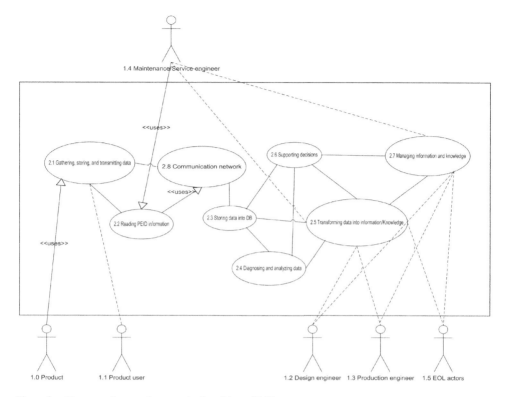

Figure 3.   Use case diagram for generic closed-loop PLM

PEID and product user/operator (Actor 1.1) access to PEID directly. Maintenance/service engineer (Actor 1.4) can assess PEID with the PEID reader. Product engineer (Actor 1.2), production engineer (Actor 1.3), and EOL actors (Actor 1.5) access to PDKM for getting necessary information.

## 5   GAPS BETWEEN TRADITIONAL PLM AND CLOSED-LOOP PLM

In this section, the difference between traditional PLM and *closed-loop PLM* are examined. In general, PLM is a new strategic approach to manage product-related information efficiently over the whole product lifecycle. The concept appeared in the late 1990s targeting with moving beyond engineering aspects of product and providing a shared platform for creation, organisation, and dissemination of product related knowledge across the extended enterprise (Ameri and Dutta 2004). PLM is defined in various ways as follows:

> *A strategic business approach that applies a consistent set of business solutions in support of the collaborative creation, management, dissemination, and use of product definition information across the extended enterprise from concept to end-of-life integrating people, processes, business systems, and information (CIMdata 2002).*

> *A strategic business approach for the effective management and use of corporate intellectual capital, which is the sum of retained knowledge that an organisation accumulates in the course of delivering its objectives (Datamation, 2002).*

PLM facilitates innovation by integrating people, processes and information throughout the product lifecycle and across the extended enterprise. It aims to derive the advantages of horizontally connecting functional silos in the organisation, enhancing information sharing, efficient change management, use of past knowledge, and so on (Macchi *et al.* 2004).

85

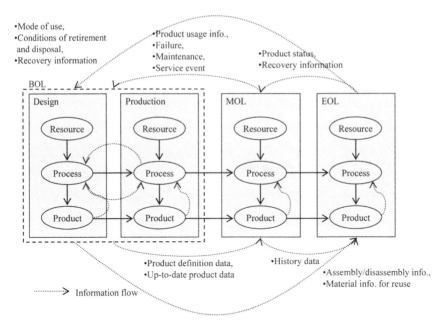

Figure 4. Product information flow.

Compared to the generic concept of traditional PLM, *closed-loop PLM* uses PEID, and the information flow and control flow are horizontally and vertically closed, respectively. The *closed-loop PLM* focuses on the complete lifecycle of a product with more emphasis on tracking and managing of information across the whole product lifecycle, and possible feedback of information to each product lifecycle phase. There are many lifecycle information flows among BOL, MOL, and EOL, as shown in Figure 4. Product lifecycle data, such as usage conditions, failure, and maintenance or service events, etc., can be gathered by the PEID that is embedded in each product over the whole product lifecycle. These data play an important role in analysing and making decisions of several operational issues in the product lifecycle. Based on the feedback information, *closed-loop PLM* can support the decision making of several operational problems over the whole product lifecycle. This provides opportunities to improve several operations over whole product lifecycle.

## 6  CONCLUSION AND FURTHER RESEARCH

Total management of the product lifecycle is critical to creatively meet customer needs throughout the entire lifecycle without driving up costs, sacrificing quality, or delaying product delivery (Kiritsis 2004). For this, it is necessary to develop a PLM system. Recently, with emerging technologies, it is possible to develop a *closed-loop PLM*. The *closed-loop PLM* system provides opportunities to reduce the inefficiency of lifecycle operations and gain competitiveness. To seize the opportunities, first and foremost, it is necessary to clarify the concept of *closed-loop PLM*. For this purpose, in this study, a concept for *closed-loop PLM* has been proposed and distinguished from traditional PLM. Implementation of this concept provides an opportunity to create leverage and synergies, and to avoid duplication and inconsistency in the product lifecycle operations.

Because of the short history of PLM and PEID technologies, the research on *closed-loop PLM* are still in an early stage. Hence, there exist many research challenges an open issues, for example, the development of the general modelling and development framework for *closed-loop PLM*. Moreover, resolving the detailed design problems regarding PEID and PDKM also needs to be addressed. Although the work discussed may not provide the exhaustive result for the *closed-loop PLM*, it should lay the foundations for further study of *closed-loop PLM*.

ACKNOWLEDGEMENT

The work reported in this paper is based on the PROMISE project (FP6 IP 507100 and IMS 01008, www.promise.no). The authors express their deep gratitude to all PROMISE partners for their assistance.

REFERENCES

Ameri, F. and Dutta, D.: Product lifecycle management: needs, concepts and components. Technical report, Product lifecycle management development consortium PLMDC-TR3-2004, 2004.
Brock, D. L., Milne, T. P., Kang, Y. Y., and Lewis, B.: The physical mark-up language-core components: time and place. White paper, Auto-ID center, 2001.
Bruno, G. and Agarwal, R.: Modeling the enterprise engineering environment. IEEE Transactions on Engineering Management, Vol 44, 1997, p. 20–30.
CIMdata: Product lifecycle management-empowering the future of business. Technical report, CIMdata 2002.
Datamation: Understanding product lifecycle management. Technical report, Datamation limited PLM-11, 2002.
Evers, H. and Kasties, G.: Differential GPS in a real-time land vehicle environment-satellite based van carrier location system. IEEE Aerospace and Electronic Systems Magazine, Vol 9, No 8, 1994, p. 26–32.
Foley, J.: An infrastructure for electromechanical appliances on the internet. BE and ME thesis, Massachusetts Institute of Technology, May, 1999.
Huvio, E., Grönvall, J., and Främling, K.: Tracking and tracing parcels using a distributed computing approach. Proceedings of NOFORMA' 2002 conference, Trondheim, Norway, 12–14 June, 2002.
IMTI Inc.: Modeling and Simulation for Product Life-Cycle Integration and Management. White Paper, 2002.
Kimura, F. and Suzuki, H.: Product lifecycle modelling for inverse manufacturing. Proceedings of IFIP WG5.3 International Conference on Life-cycle Modelling for Innovative Products and Processes, Berlin, Germany, 1995.
Kiritsis, D.: Ubiquitous product lifecycle management using product embedded information devices. Proceedings of International Conference on Intelligent Maintenance Systems (IMS 2004), 2004.
Lillehagen, F. and Karlsen, D.: Enterprise Architectures-Survey of Practices and Initiatives. Proceedings of INTEROP-ESA 2005, Industry track report, 2005.
Macchi, M.; Garetti, M.; and Terzi, S.: Using the PLM approach in the implementation of globally scaled manufacturing. International IMS forum 2004: Global challenges in manufacturing, 2004.
Mayer, R. J., IDEF0 function modelling. Knowledge-based systems. Inc., 1994.
Ming, X. G. and Lu, W. F.: A framework of implementation of collaborative product service in virtual enterprise. Proceedings of Innovation in Manufacturing Systems and Technology (IMST) 2003.
Nonomura, A., Tomiyama, T., and Umeda, Y.: Lifecycle simulation for inverse manufacturing. Proceedings of 6th International seminar on lifecycle engineering, 1999.
Parlikad, A. K., McFarlane, D., Fleische E., and Gross, S.: The role of product identity in end-of-life decision making. Technical report, Auto-ID center, Institute of manufacturing, Cambridge, 2003.
PROMISE: PROMISE-Integrated project: Annex I-Description of Work. Project proposal 2004.
Scheer, A.-W.: ARIS Business process framework. Springer, 1998a.
Scheer, A.-W.: ARIS-Business process modelling. Springer, 1998b.
Schneider, M.: Radio Frequency Identification (RFID) technology and its application in the commercial construction industry. Technical report, University of Kentucky 2003.
Tipnis, V. A.: Toward a comprehensive lifecycle modeling for innovative strategy, systems, processes and products/services. Proceedings of IFIP WG5.3 International Conference on Lifecycle Modelling for Innovative Products and Processes, 1995.
Vernadat, F. B.: Enterprise modeling and integration: principles and applications. Chapman and Hall, 1996.
Vernadat, F. B.: UEML: Toward a Unified Enterprise Modelling Language. International Journal of Production Research, Vol 40, 2002, p. 4309–4321.
Yan, P., Zhou, M., and Sebastian, D.: An integrated product and process development methodology: concept formulation. Robotics and Computer Integrated Manufacturing, Vol 15, 1999, p. 201–210.

*Part III*
*Sustainable products and processes*

Research into the environmental impacts of industrial activity shows that in the next decades citizens of the industrialised parts of the world will have to learn to live by relying on far less environmental resources than is the case today. Given this background, the role of business in society is about to be transformed. There are expectations that business should contribute at a societal level in the three areas: economy, social, and environment. The popularity of and attention to concepts like extended producer responsibility (EPR) lifecycle analysis (LCA), design for environment, industrial ecology, value chain management, corporate citizenship and CSR all reflect the shift from looking at companies as isolated profit-maximisers to treating them as integrated subsystems of larger industrial systems and positive value chain in society. Different legislative actions have recently been introduced by the European Commission as measures which, individually and jointly, will lead e.g. to sustainable electronic product-services with an optimal eco-efficient lifecycle.

Product-service system strategies where products and services are seen as interrelated systems, may contribute to economic growth without significant growth in material and energy throughput, and thereby contribute to the required de-materialisation goals. There is a growing understanding among industrialists and policymakers that environmental initiatives need to be seen in relation to one another rather than in isolation. However, sustainability has become a topic that is important for the European Union's research and for international research. Sustainability can be a competitive factor for manufacturing industry, but can also reduce competitiveness through environmental legislation that is only adopted in a single region e.g. in the European Union only, and not in other regions.

In the project period of IMS-NoE a main work item of SIG5, *Sustainable Product and Processes* has been the Sustainable Industrial System (SIS). Most of the approaches for sustainability claim to have a holistic view, but few of them considered this in their solutions. The Sustainable Industrial Systems (SIS) approach brings together the different challenges and dimensions of sustainability into a common platform comprising a system oriented approach for industry. As a result, this will lead to improved methodologies, as well easier and clearer standards and legislation. One important feature that has been almost absent in research projects is the consumer-market relationship, which has been called the consumption paradigm. Research challenges related to the SIS approach will be discussed in the first contribution of Part III.

In the two following contributions the results of the AEOLOS (An End-of-Life of Product Systems) project and the ECOLIFE II project is presented. A short introduction to these projects is given below.

The AEOLOS (An End-of-Life of Product Systems) project has developed an integrated methodology and tool-set that examine the issues relating to the sustainability of the end-of-life (EOL) treatment of waste products, and in particular, waste electrical and electronic equipment. The AEOLOS methodology enables users to systematically define EOL products and processes to a level of detail that will allow useful economic, environmental and societal costs and values of different EOL scenarios to be calculated, analysed and compared, and in so doing provide users with the means to assess and minimise the impacts of EOL scenarios. Targeted users include resource recovery and recycling companies, original equipment manufacturers (OEM), and policymakers. The paper presents the AEOLOS methodology and software, with illustration of the scope of the tool using the latest prototype version of the software.

ECOLIFE II focuses on the product-service lifecycle of electrical and electronic products, and involves key players in the electronics and automotive industries among the various stages of the product-service lifecycle – from component suppliers and product manufacturers, to service and logistic suppliers and the End-of-Life processors. The main activities of the Network focus on the environmental and economic aspects of product design, functional innovation and service-system innovation. The goal is to decrease the environmental load of electrical and electronic products by better design and pave the way to the provision of sustainable service systems. Further challenging work will be done in information/knowledge management within the lifecycle, product environmental communication, and socio-economic aspects. Finally the project provides a one stop shop for the state-of-the-art including aspects of design, production, operation/use and re-use until the end of the operative life, at technical and organisational levels.

*Advanced Manufacturing – An ICT and Systems Perspective – Taisch,*
*Thoben & Montorio (eds)*
*© 2007 Taylor & Francis Group, London, ISBN 978-0-415-42912-2*

# Sustainable products and processes: Challenges for future research

Odd Myklebust[1], Dimitris Kiritsis[2] & Trond Lamvik[1]
[1] *SINTEF, Trondheim, Norway*
[2] *EPFL, Lausanne, Switzerland*

ABSTRACT: Economics, environment and social issues are the three themes that cover sustainability. However, is sustainability a vision, an ultimate goal or a utopian society? Sustainability can also be seen as an operational strategy. There is a request for product systems and end-of-life management. Landfill sites are becoming full, new lifecycle legislation will be implemented to limit the material flows to landfill, and emission of toxic materials to the air, land and sea will be terminated or significantly reduced. There are many important actors in this area; manufacturing and re-manufacturing companies, recyclers and policymakers and there are also several end-of-life options: reuse (parts and products); recycling (materials); energy production from incineration; energy production without incineration; and landfill. This paper tries to set this scope into a systematic industrial approach.

*Keywords:* Sustainability, product lifecycle, recycling, landfill, industrial systems, environment.

## 1 INTRODUCTION

In the project period of IMS-NoE the focus and main work item of SIG5 *Sustainable Product and Processes* has been the Sustainable Industrial System (SIS).

Most of the approaches for sustainability claim to have a holistic view, but few of them considered this in their solutions. The Sustainable Industrial Systems (SIS) approach brings together the different challenges and dimensions of sustainability into a common platform comprising a system oriented approach for industry. As a result, this will lead to improved methodologies, and easier and clearer standards and legislation. One important feature that has been almost absent in research projects is the consumer- market relationship, which has been called the consumption paradigm.

The goal of SIS is to obtain a holistic view of product cycles in the manufacturing industry and optimise the lifecycle of industrial systems, products and services. Methodologies and tools to support the manufacturing of products and production must be increasingly lifecycle and service oriented, in addition to the requirements of intelligence, cost-effectiveness, safety and cleanliness. The objectives are to:

- Combine the environmental dimension of product development and manufacturing and plant operations;
- Develop viable and sustainable industrial systems that support customer satisfaction in a sustainable consumption paradigm;
- Develop corporate social responsibility (CSR) principles;
- Develop a framework for business organisations to measure progress of commitment and results regarding environmental, economic and social criteria.

The modules in SIS, as they are shown in Figure 1 below, can shortly be described as follows:

- 1st pillar: Product and Process Lifecycle Management;
- 2nd pillar: Sustainable Network creation (sustainable extended enterprise);

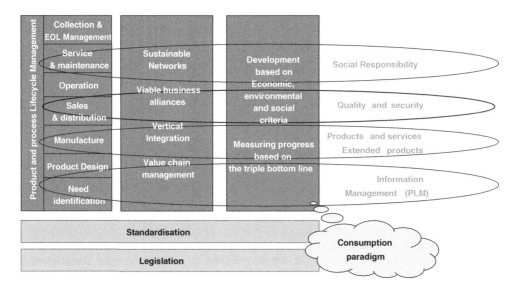

Figure 1.    Illustration of the Sustainable Industrial System framework.

- 3rd pillar: Social- economic- ecological combined assessment and performance measurement.

This will lead to a Sustainable Industrial Development Model where:

- Non-renewable resources can be used only at the rate at which others are developed or loss of opportunity compensated;
- Waste and emissions should not be generated at rates that exceed the ecosystem's capacity to assimilate them;
- Renewable resources should not be used at rates that exceed the systems ability to replenish itself.

The above three pillars are based on international/worldwide standardisation and legislation regulations that act as guidelines and frameworks on one hand and, on the other hand, serve also the purpose of direct normative value settings.

Examples of such standardisation and legislation frameworks are the WEEE and ELV directives in the European Union, the ISO 14000 family of standards etc.

## 2    SUSTAINABILITY AS A RESEARCH TOPIC WITHIN THE EUROPEAN UNION

Sustainability is a topic that is important for European Union research and for international research. Sustainability can be a competitive factor for manufacturing industry, but can also reduce competitiveness through environmental legislation that is only adopted in a single region e.g. in the European Union only, and not in other regions. Below are some examples for resource topics within sustainability that can help productivity for the industry, even if only adopted in one region. These topics can of course also have an international dimension.

### 2.1   *Product-service*

Product-service system strategies where products and services are seen as interrelated systems, may contribute to separate economic growth from growth in material and energy throughput, and thereby contribute to the required dematerialization goals. There is a growing understanding among industrialists and policymakers that environmental initiatives need to be seen in relation to one another rather than in isolation. One important approach within Product-service system is to

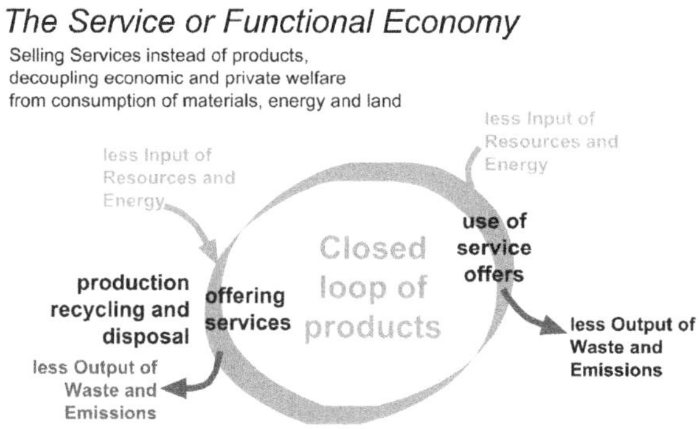

*The Service or Functional Economy*

Selling Services instead of products, decoupling economic and private welfare from consumption of materials, energy and land

Figure 2.    Closed-loop material flow in a service economy.

move industry away from the linear thinking that dominates present industrial practice, towards a cyclic rationale, as illustrated in Figure 2.

In this systems view, the service element is a vital component of the connection between the product and the customer. This strategic innovation of PSS is based on a new interpretation of the concept of *product*. In a PSS strategy a company offers *utility*, which means functions or end-results, instead of a tangible product.

Technological challenges involve understanding and improving the effectiveness of the material and energy flows. Social challenges are linked to how to redirect society at large, companies and individuals towards a more cyclic- and systems-thinking on production and consumption. The relationships among the different disciplines are vital to understand the action-effect mechanisms and interpret incentives to obtain desired results.

## 2.2    *Product lifecycle information management*

PLM can be described an a strategic business approach that applies a consistent set of business solutions in support of the collaborative creation, management, dissemination, and use of product definition information across the extended enterprise from concept to End-of-Life.

PLM is for the manufacturing industry looking to improve business performance. PLM should consolidate business activities that create, modify and use data to support all phases of a product's lifecycle. PLM technology with seamless access to product model structures with data and information on the one hand and distributed knowledge repositories.

Functionality for handling product information of Middle-of-life (operational phases) and End-of-Life (recycling phases) has only been conceived in a limited way up to now. In an End-of-Life perspective the product model supported by the PLM system also needs to *reincarnate* product or material data for reuse of material or parts or both (closing the information cycle).

State-of-the-art in this area are the projects FP6 IP 507100 and IMS 01008 PROMISE, initiated by the core IMS-NoE SIG5 actors. According to PROMISE, the whole product lifecycle implies all phases related to the product generation, usage, and disposal. It consists of the following lifecycle activities: design (product lifecycle planning and product design), production (procurement, manufacturing, and assembly), logistics (distribution), usage, maintenance (service), collecting, re-manufacturing (disassembly, refurbishment, re-assembly, etc.), reuse, recycle, and disposal. Figure 3 shows the whole product lifecycle in PROMISE.

This offers the following business proposition to the product lifecycle stakeholders: to create value by transforming information to knowledge during all phases of the product lifecycle and thus improve product and service quality, efficiency and sustainability. The product and service value may be created at various levels, with respect to the above statement, as follows:

- Technical: optimal accomplishment of the expected functions and user expressed and unexpressed needs, after exploiting *field* knowledge gathered through the product lifecycle.

93

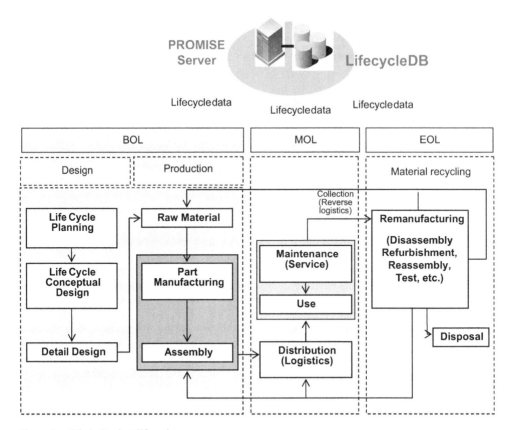

Figure 3.   Whole Product lifecycle.

- Economic: creation of value for the producer (better products, better CRM), for the service provider (new business opportunities, better CRM), and for the product owner (extended product life).
- Environment: minimisation of pollution, of resource use and of energy consumption by applying optimal BOL (Beginning-of-Life), MOL (Middle-of-Life) and EOL (End-of-Life) planning.

It is expected that the above concept will be further elaborated and used as a basis for further research. One of the main issues in product lifecycle management seems to be knowledge management, more particularly, knowledge sharing.

### 2.3   *Knowledge sharing*

*Knowledge sharing* is an approach of modern manufacturing that provides elements of the answer to the following two questions relevant to modern manufacturing: *how work is undertaken* and *how value is added*.

The main elements of *knowledge sharing* may be identified as follows:

- Knowledge sharing is already a challenge even within one single organisation;
- Knowledge owners (particularly SMEs) want to protect their know-how in their particular domain of activity, particular everything that concerns *design* aspects;
- Knowledge sharing has to be:
  - Clear
  - Fair

- In every case of knowledge sharing it has to be defined:
  - What is to be shared
  - How does the sharing happen
    (i) What are the necessary organisational structures?
    (ii) How any created value is to be shared?
  - When the sharing will happen
    (i) Implications with the product and business lifecycles
- New methods and tools to enable and facilitate *clear* and *fair* knowledge sharing have to be developed.
- Knowledge is created by the ability to transform information and already own knowledge.
  - The concept of *knowledge transformation* needs to be further investigated and formalised.
- Collaborating organisations should clarify and define their concerns about confidential information. This is an essential element in trust development. Some relevant questions are following:
  - How trust is developed among individuals?
  - How trust is developed among organisations?
  - What are the main elements of trust development in collaborative business?
  - How trust elements can be formalised?
  - Is it possible to develop and implement methods and tools for trust development in collaborative business?

### 2.4 *Combination of the environmental dimensions in product development, manufacturing and plant operations*

The goals for this topic are to develop methodologies and tools to enable the dynamic and integrated design and implementation of product/process/plant (3P) to support a sustainable lifecycle of plant, product and production. More particularly, the following two objectives have been identified:

- to realise a global holistic view of 3P (product/process/plant) integration in the manufacturing industry
- to develop a flexible and sustainable ECO-ECO-SOC (economic-ecological-social) 3P lifecycle management tool set.

Industries are no longer home-based, but are operating in a global market. Digital business has become a strategy to survive. The extended enterprise is being implemented. Parts are made where conditions are most favourable. Future products need to use less energy over its lifecycle and the materials should be available for reuse in other industrial settings.

The expected results will contribute to the systematic transition of product based manufacturing industry into innovative global business operations based on a sustainable *closed-loop* value chain.

- A systematic methodology and integrated models for sustainable design of product/process/plant;
- New business models for the global sustainable manufacturing industry.

The integration of the *plant* dimension in the product/process integration is important and it may provide useful answers to the questions faced by the 3P concept, because the environmental and social dimensions are mainly realised at the (plant) levels of integration.

## 3 INDUSTRIAL ECOLOGY CHALLENGES

### 3.1 *Improvements within an organisation*

Within a company the challenge is to make environmental improvements in accordance with industrial ecology principles along the internal value chain. In product development central aspects are design and material choices that secure good quality, long life for the product, and reuse and recycling. In production processes the challenges are to optimise the use of energy and materials. In the final link, the distribution processes, the challenges are to distribute and redistribute the products for use and reuse/end-of-life treatment with a minimum of environmental impact.

### 3.2 *Value chain management*

Industrial ecology goes beyond conventional environmental work in industry by expanding the system borders of a company. According to Graedel and Allenby (1995) all industrial activities are interconnected through thousands of transactions and actions. These transactions and actions all have environmental impacts, and by organising them together and looking for synergies, significant environmental improvements at a system level can be achieved. The customer relationship between manufacturers gives opportunities for demanding environmental performance to be the preferred supplier. A manufacturer has a lot of suppliers and dealers, and by putting forward environmental demands one single company can create large environmental chain reactions far beyond its own organisational borders. The challenge is to understand how green chain reactions are started and managed.

Based on experience from industrial eco-design projects, research is now in progress at several research institutions in Europe (e.g. TU Delft, IIIEE, CfSD, SERI, and 5th Framework Programme), to investigate how functions can be performed more eco-efficiently with the help of dematerialised services as an extension of the environmental design of physical products.

Several research projects indicate that in principal, by responsible design and realisation of such services, significant environmental improvement can be possible, which offer a better perspective than the redesign of products only (eco-design).

Based on these observations in industry, there are indications that there are at least three issues, which form the common denominator for developing product-service combinations. The three observations emphasise the need for:

- **Modularised design**
  A modularised design where a product platform and product structure is developed to be prepared for disassembly and re-assembly which makes the product easy to take apart completely, but also prepared for part disassembly for replacement of worn out subsystems or parts. This is the principle of utilisation of the separation in structure and among parts.
- **Organisation and infrastructure**
  This is an infrastructure that optimises the material flow, and an infrastructure that optimises the information flow. Strategic alliances through vertical integration align the two-way material flows to render supply and take back of the technical means or product possible. The information flow is needed to track utilisation of the product's functionality. The company supplying customers with the function, can track and compare the actual utilisation with the service and maintenance plan, and replace worn subsystems and parts when needed. This is the principle of utilisation of the separation in space.
- **Control**
  The control and ownership of the product is the key element for introducing functions to the customer. Ownership of the product is transferred from the customer to the producer or function supplier. In these circumstances, the customer no longer has full ownership of the product, but procures access to the product's functionality when needed. This is the principle of utilisation of the separation in time.

A further investigation of all these three dimensions of a product-service combination will contribute to an understanding of the framework that embraces the concept of eco-efficient services. Research questions includes

- How and under what conditions could eco-efficient services systematically and successfully be developed?
- What contribution could be delivered by the existing eco-design methodology? and;
- How could new electronics based technologies contribute to the development of eco-efficient services?

Hypothesis to be investigated are:

- Only under specific conditions can eco-efficient services be developed in such a way that they are beneficial to business and the environment. How does product life (i.e. the technical, economic and social life) affect this benefit? One hypothesis is that the social life of a product to a large degree affects the success of eco-efficient services. How do deviations between the different product life dimensions affect eco-efficient services?
- The methodology of eco-design of products is suitable for the development of eco-efficient services, but needs additional concepts, procedures and information.
- Electronics industries' driven, new ICT-based services have, via their dematerialising character, a positive contribution towards the development of eco-efficient services.

## 4 RESEARCH TOPIC WITHIN SUSTAINABILITY EUROPEAN UNION IMS – INTERNATIONAL DIMENSION

Sustainability has an important global dimension. Europe alone cannot solve most of the significant challenges. There are several international activities, committees and agreements that try to include the whole world (Rio Agreement, Kyoto etc.). The IMS program can from the European Union perspective be made a driver to focus on important international topics such as sustainability. IMS can then have the industrial dimension and define important global research topics within sustainability for industry (industrial ecology?). This is important when the growth in the global industrialisation and the total growth in consumption of materials and energy is considered. Draft ideas for such research topics are discussed in the following.

### 4.1 Methods for End-of-Life treatment

In Europe the regulations for recycling products has been developed for quite a while. There is for example the WEEE Directive and the automotive recycling Directive. This is probably just a start of the treatment of products in middle-of-life and end-of-life. This is a typical topic that can be made global. The global waste streams could be made optimal and the methods and regulations for recycling of the products could be standardised.

Reuse – re-manufacture – recycle

The concept of Sustainable Consumption and Production is at the focus of relevant research efforts in European Union and some discussion has already been undertaken in the *ManuFuture* workshops so far. Research in this domain may provide elements of answers in the question *how do we live*.

Some elements of the concept of Sustainable Consumption and Production, particularly in the area of re-manufacturing, may be identified as follows:

- This issue has strong societal aspects;
- There are strong multi-cultural characteristics, also within Europe, even among regions in Europe;
- There are also strong political implications; in a relevant discussion it will be difficult to avoid comparisons between *liberal* and *socialist* views. The terms *liberal* and *socialist* have to be seen under their general meaning. Other terms such as *conservative* and *progressive* may be more appropriate.
- A strong driver for sustainable consumption is the *environment* in general or even the need to sustain the *eco-system* in the longer term. But, it may not be the only one. The economic driver has to be taken into account as well.
  - An example: What will be the consequences in the European Union manufacturing industry if the European Union (or the world) citizens *consume* less cars?
  - Another example: What will be the consequences in the automotive supply sector if the use of *re-manufactured* components is allowed in new cars?

## 4.2 *Green logistics*

Industry is global and they work in networks and chains. Products and parts are distributed all over the world. For every km of transport $CO_2$ gas is emitted into the atmosphere. A more balanced assessment model among transport, cost and environmental factors should be developed. This is important for logistics in manufacturing and distribution and the return logistics after end-of-life treatment.

## 4.3 *Standardisation*

At present, national and international standardisation and legislation act as guidelines and frameworks for direct regulations and normative value settings. A significant contribution to further development and enhancement of national and international standards and legislation within environment and sustainability is an important activity that must continue.

Improved contribution towards international standards for global sustainable product designs and manufacturing, including End-of-Life treatment, is a large global challenge. One part of the world should try to start the research work to speed up this process. Regions like Europe and Japan could initiate a drive in such activities.

## 5 CONCLUSION

The main conclusion from the work within IMS-NoE and other project work is that focus on sustainable industry is important. This should concern future European Union framework programmes and national R&D programmes. However improved sustainability in industry is a global issue. International R&D programmes or a platform like IMS are important arenas to address sustainability. Unless research within environment and sustainability does not obtain a global perspective the main challenges will never be met.

*Advanced Manufacturing – An ICT and Systems Perspective – Taisch,*
*Thoben & Montorio (eds)*
*© 2007 Taylor & Francis Group, London, ISBN 978-0-415-42912-2*

# The end-of-life stage of product systems

Odd Myklebust & Trond Lamvik
*SINTEF, Trondheim, Norway*

ABSTRACT: The AEOLOS (An End-of-Life of Product Systems) project has developed an integrated methodology and tool set that examine the issues relating to the sustainability of the end-of-life (EOL) treatment of waste products, and in particular, waste electrical and electronic equipment. The AEOLOS methodology enables users to systematically define EOL products and processes to a level of detail that will allow useful economic, environmental and societal costs and values of different EOL scenarios to be calculated, analysed and compared, and in so doing provide users with the means to assess and minimise the impacts of EOL scenarios. Targeted users include resource recovery and recycling companies, original equipment manufacturers (OEM), and policymakers. This paper presents a current version of the AEOLOS methodology and software, with illustration of the scope of the tool using the latest prototype version of the software.

*Keywords*: End-of-Life, EOL, sustainability, product lifecycle, product systems.

## 1 BACKGROUND AND PROJECT DEFINITION

There has been a worldwide recognition that the current consumption rate of electrical and electronic equipment (EEE) is not sustainable. This realisation has led to the re-assessment of the design of these products, consumption patterns and also the end-of-life options for waste EEE (WEEE). An added incentive to European EEE designers, manufacturers, consumers and disposal agents, is the introduction of new legislation, which enforces the concept of producer responsibility. Examples of this new legislation are the WEEE Directive and the Restriction of Hazardous Substances (ROHS) Directive.

The European Union has invested in several research projects that focus on the above issues. One of these research projects is the AEOLOS (An End-Of-Life Of product Systems) project (http://www.sintef.no/aeolos), which has as its focus the development of a methodology and software tool for informing the decision-making process around the end-of-life (EOL) treatment of WEEE. The AEOLOS methodology is designed to enable users to define product and process to a level of detail that will allow useful economic, environmental and societal costs and values of different EOL scenarios to be calculated analysed and compared. In so doing, the methodology will assist the user to determine the most sustainable treatment of EOL products.

The AEOLOS methodology (Figure 1) comprises four main steps, preceded by user and facility/ product set-up steps (Adda et al. 2002).

## 2 FACILITY AND PRODUCT SET-UP

Initially the EOL product and EOL treatment option has to be defined. This initial set-up of information is to be completed by the user before starting on the AEOLOS *loop*, and can be updated when necessary. The user is able to enter data on their facility (e.g. a recycling plant) or product.

Figure 1.    The 4 steps of the AEOLOS Methodology.

The information entered here includes:

- Facility details;
- Facility processes;
- Item definition;
- Supply and demand;
- Product, functional component, component building.

## 2.1    *Item manager*

Before a product, functional component, or component can be built each individual item (or relevant items) must be defined. The product builder then allows the user to define the item's structure. The functionality to allow both operations to be performed simultaneously is also available. A product can be structured like in Figure 2. This figure shows the different levels of detail that can be defined for any product.

The levels of detail are as follows:

- Detail 1: the product only has material fractions associated to it. This level of detail is mandatory for recyclers, and any user who wishes to undertake an environmental analysis of their EOL scenario;
- Detail 2: the product has component data associated to it. The component or components can then have materials associated with them;
- Detail 3: the product has functional components associated to it. Materials or components or both can then be associated with the functional components.

The Item Manager contains the product, functional component and component definition, including information such as:

- Item type: product, functional component, or component;
- Weight: the total weight of the item;

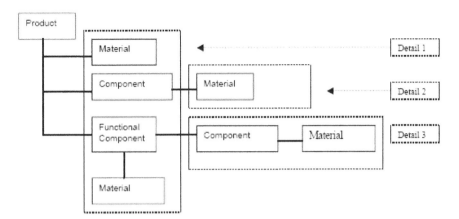

Figure 2.   Product structure.

- Volume: the item's volume;
- Purchase cost: the average purchase price of the item after original manufacture;
- Second value: the average value in the second-hand used market;
- Material cost: the average purchase price at its End-of-Life;
- Suitable for re-manufacturing: can the item be re-manufactured?
- Suitable for reclamation: can any components, functional components be reclaimed from the item?
- Suitable for recycling: can any of the materials be recycled from the item?
- Suitable for incineration: can the item be routed to incineration?
- Suitable for landfill: can the item be routed to landfill?
- Hazardous: does the item contain any hazardous materials or fluids?

The cost and value data of each material will need to be maintained as these prices vary over time and affect the financial calculations. Items must first be defined before they can be used to define any supply or demand streams, or included in an EOL scenario.

### 2.2  *Product builder*

Up to this point the user has defined each item in the database, but has yet to define the structure of any product, functional component or component. The user has three options when defining the product, functional component, or component structures:

1. Define from scratch;
2. Define using predefined item definitions;
3. Use a copy of an existing product, functional component, or component and define a new product etc.

### 2.3  *Step 1: Scenario definition*

Select a product and EOL option(s), which together make up an EOL scenario. The definition of the scenarios is unlimited by any constraints, such as local availability of facilities – the refinement of scenarios occurs later in the *loop*. The EOL options included in the AEOLOS methodology and software are re-manufacturing, component reclamation, material recycling, incineration with energy recovery, incineration without energy recovery, and landfill. The minimum amount of information required for a product definition is the weight of that product.

- Product selection: The user is presented with a list of product types, based on the WEEE hierarchy, and is able to select products that will be used in the scenario and add them to the *product bag*.

- EOL option selection: Each EOL product is linked to an EOL option. The amount of product treated through each option is defined either by quantity or by weight.

## 2.4  Step 2: Scenario assessment

Select which environmental, social and economic indicators are to be calculated. The user has the option of selecting all economic, environmental and social aspects (i.e. issues or impacts) and indicators (i.e. measures of performance) or to make a sub-selection.

- Selection of indicators: The user selects their preferred indicators (social, environmental and economic). A description of each indicator is provided.
- Input of indicator data: Once the indicators have been selected, data are entered for the selected indicators. In some cases, however, data are supplied by the software, for instance, the environmental inputs and outputs associated with consumption of energy.

## 2.5  Step 3: Scenario analysis

Analyse results (and compare scenarios). The user can choose the method of analysis from a *toolbox* of methods, which range from a simple comparison of indicator results, through to more complex multi-criteria decision analysis (MCDA) techniques. In the software prototype, the comparison of EOL scenarios is done by means of a multi-criteria decision-making method, which allows the user to manage the conflicts and trade-offs among the different indicators. Here the user is able to enter weightings for each of the selected indicators, and a number of other parameters.

## 2.6  Step 4: Scenario refinement

Checklist of questions to ensure that preferred scenario is feasible. For instance, the user can:

- Check system boundaries and allow adjustment.
- Collect more data to be able to calculate more indicators, or select indicators that are more relevant.
- Seek expert opinion on feasibility of scenarios and relevance of indicators.
- Check local capacity for implementation of a scenario.

The purpose of this step is to urge the user to address the constraints that might reduce the feasibility of a preferred EOL scenario, and encourage users to address stakeholder acceptance issues and validate the results obtained with stakeholders (especially relating to any value judgements made during the analysis).

## 3  SUSTAINABILITY INDICATORS FOR THE WASTE MANAGEMENT INDUSTRY

A fairly comprehensive list of environmental, social and economic (internal and external) aspects and indicators is provided to the user (Table 1). The AEOLOS project has consulted a wide array of sources for these lists (e.g. (Global Reporting Initiative 2000); (OECD 1999); (WBCSD 2000), and sought expert opinion. Selection will be made from this recommended list of aspects and indicators. Templates of aspects that are relevant to each EOL option are also provided to the user.

In addition, the AEOLOS methodology provides guidance to the user on how to select a list of environmental, social and economic aspects that are critical to their assessment. A number of questions allow the user to assess the *criticality* of an aspect in terms of the level of impact associated with that aspect, and in terms of the likelihood of that impact occurring. The results of this simple assessment are mapped on a matrix (see Figure 3) and indicate the criticality of that aspect.

Table 1. Environmental, social and economic indicators.

| Category | Aspect | Indicator | Selected? |
|---|---|---|---|
| ENVIRONMENT | | | |
| Natural resources | Energy consumption | Total fossil fuel consumption | ✓ |
| Air | Airquality | Air acidification | |
| Water | Water quality | Water eutrophication | ✓ |
| SOCIAL | | | |
| Employees | Health and safety | Investment in workers health and injury prevention | ✓ |
| | Diversity and equal opportunities | Workforce profile | |
| Consumers | Product acceptability | Consumer satisfaction | ✓ |
| ECONOMIC | | | |
| EOL treatment | Net revenue | Profit/loss | ✓ |
| Supply chain | Incomes generated | Financial expenditure on out-sourcing and procurement | |
| Communities | Regeneration impact | Land value changes | ✓ |

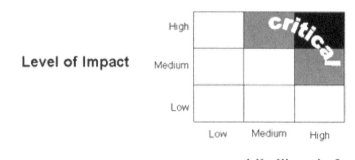

Figure 3.  Criticality matrix that can be used to assess whether an aspect is critical to the assessment undertaken by an AEOLOS user.

## 4  THE AEOLOS *STORY*

The methodology was developed through collaboration among European research and industrial partners. The consortium contains research institutes, universities consulting companies, original equipment manufacturers (OEMs), and also re-manufacturers.

An important initial aspect of the AEOLOS project was the definition of who the potential users of such a methodology and software package could be and what their needs would be. This task was particularly challenging as the new European legislation has yet to be implemented in most member states, and so potential users have yet to address the issue of producer responsibility as it relates to the EOL treatment of waste. The AEOLOS consortium attempted to engage several potential users of the outputs of the project at an early stage, to determine the scope and emphasis of the project (Adda et al. 2001). The two main user groups that were identified were policymakers and the resource recovery/recycling companies and original equipment manufacturers (OEMs).

The outputs of the AEOLOS project should assist decision-makers formulating policy at local authority, regional, national and EU levels by providing the means by which to determine, analyse and evaluate the most appropriate EOL treatment option from an environmental, economic and social perspective for their geographical areas of concern. Similarly, resource recovery/recycling companies and original equipment manufacturers (OEMs) will be provided with decision support on the choice of EOL treatment options, based on a set of environmental, economic and social criteria. The twofold application is necessary to ensure that those determining policy and those who are ultimately responsible for implementing it are acting in concert. In addition, integration among economic issues and the social and environmental aspects of waste treatment is important. Another early exercise was to review current sustainability initiatives around the world, to prevent the AEOLOS project from *re-inventing the wheel*. This exercise involved the review of over 500 sustainability initiatives (Lamvik 2000).

There still is a lack of general agreement among researchers or users of sustainability indicators on which among the existing models is best applicable to measure sustainability. Each of the models has its strengths and weaknesses, which are related to the level of application.

Most assessment models are macro models aiming at establishing options to assess the health of a nation, region or city in terms that are complementary to economic figures. They aim to identify and measure elements other than economic performance that contribute to the wellbeing of a nation, region or city. This means that the models presented here may be difficult to implement as the links among indicators at company or product level and indicators for national statistics are not so obvious. This is one of the main challenges initiated by the scope of the project where focus was put strictly on the product EOL and recycling situation. The search for social indicators was particularly difficult, as a recycling system for industrial products is only one of many systems that affect the human social system.

The introduction of environmental and social issues in addition to economic issues to form the basis for sustainable decision-making regarding EOL treatment will clearly divide responsible companies from the traditionally less sustainable ones.


5   ANALYSING THE RESULTS

As described above, the user has access to a variety of result analyses. The user can simply compare all indicators in a table of results, or choose to represent the results graphically, or make use of a more complex, methodology for ranking the EOL scenarios being assessed.

Some aggregation of indicators is possible, for instance the user can select aggregated environmental indicators such as eco-indicator '99, or eco-points.

The methodology ranks the EOL scenarios, based on mathematical calculations that make use of the weighting of indicators. There are four user inputs to this mathematical process: weighting of the indicators and the specification of an indifference threshold, veto threshold and preference threshold. The thresholds serve to allow refinement of the final ranking of scenarios.

Users can choose to weight the indicators used in the analysis. The manner in which the weighting of indicators is performed will depend on the goal and scope of the project being undertaken by the user. For instance, the user might be content to undertake the weighting process by him. Alternatively, the user might want to consult a wider range of stakeholders who may have an interest in the outcome of the analysis. The process followed to make such judgements should be transparent. In addition, it is recommended that sensitivity analyses be used to test the effects of a change in value judgement on the result. Weighting of indicators can be based on using multiple criteria, the cost to society, distance to a target, or expert opinion, i.e. weighting factors reported in the literature. Where scenarios are being compared, the user will generally end with a ranked list of scenarios. The list can be ranked across three or two dimensions of sustainability, or within one dimension. For example, the user is able to assess the most environmentally sustainable option against the most sustainable economic option.

REFERENCES

Adda, S., Bruce, W., Goggin, K., Ray, R. Lamvik, T. (2001): Goal and Scope. AEOLOS internal deliverable 1.1.
Adda, S., Snaddon, K., Goggin, K., Thurley, A., Lamvik, T., Miljeteig, G., Schnatmeyer, M., Bufardi, A. (2002): The AEOLOS Methodology. AEOLOS internal deliverable 5.2.

Beccali, M., Cellura, M. and Ardente, D. (1998), Decision making in energy planning: the ELECTRE multicriteria analysis approach compared to a fuzzy sets methodology, Energy Conversion and Management, vol. 39, pages: 1869–1881.

Cote, G., Waaub, J.-P. (2000): Evaluation of road project impacts: Using the multicriteria decision aid, Cahiers de Geographie du Quebec, vol. 44, pages: 43–64.

Global Reporting Initiative: Sustainability Reporting Guidelines on Economic, Environmental and Social Performance, June 2000 and Draft Guidelines 2002.

Lamvik, T. (ed.), Hagen, Ø., Aune, M., Goggin, K., Madden, L., Mulligan, D., Adda, S., Bruce, W., Snaddon, K., Kiritsis, D. (2000): State of the Art on Indicators of Sustainability. AEOLOS internal deliverable 1.2.

OECD: Towards sustainable development indicators to measure progress, December 1999.

WBCSD, World Business Council for Sustainable Development. 2000. Measuring eco-efficiency: a guide to reporting company performance.

*Advanced Manufacturing – An ICT and Systems Perspective – Taisch,*
*Thoben & Montorio (eds)*
© *2007 Taylor & Francis Group, London, ISBN 978-0-415-42912-2*

# ECO-efficient LIFE-Cycle technologies: From products to service systems

Trond Lamvik
*SINTEF Technology and Society, Trondheim, Norway*

ABSTRACT: ECOLIFE II focuses on the product-service lifecycle of electrical and electronic products, and involves key players in the electronics and automotive industries among the various stages of the product-service lifecycle – from component suppliers and product manufacturers, to service and logistic suppliers and the End-of-Life processors. The main activities of the Network focus on the environmental and economic aspects of product design, functional innovation and service-system innovation.

*Keywords*: Electronic products, product-service system, End-of-Life, sustainability.

## 1 INTRODUCTION

The European Commission (through a series of Directives – WEEE, RoHS, EuP), Member States and industry in Europe are now addressing a series of important environmental challenges to minimise waste and resource consumption, restrict the use of hazardous substances, and offer more and better products and services to users of electrical and electronic products. ECOLIFE II (ECO-efficient LIFEcycle Technologies) facilitates the co-ordination of European Union and national RTD in this field. It engages in the exchange of information among researchers in the European Union and beyond, provides easy access to state-of-the art RTD development, and reduces the duplication of studies. ECOLIFE II focuses on the product-service lifecycle of electrical and electronic products, and involves key players in the electronics and automotive industries among the various stages of the product-service lifecycle – from component suppliers and product manufacturers, to service and logistic suppliers and the End-of-Life processors. The main activities of the Network focus on the environmental and economic aspects of product design, functional innovation and service-system innovation.

The project has 4 main focus areas: Product Re-Design, Function Innovation, Service System Innovation and Information/Knowledge Management with its sub-tasks. Their goal is to decrease the environmental load of electrical and electronic products by better design and pave the way to the provision of sustainable service systems. Further challenging work will be done in information/knowledge management within the lifecycle, product environmental communication and socio-economic aspects. Finally the project provides a one stop shop for the state-of-the-art including aspects of design, production, operation/use and re-use until the end of the operative life, at technical and organisational levels.

For further information: http://www.ihrt.tuwien.ac.at/sat/base/EcolifeII/index.htm

## 2 STATE-OF-THE-ART TECHNOLOGY IN THE ELECTRONICS INDUSTRY

### 2.1 *Background*

The State-the-Art technology report is following a comprehensive *Innovation System Approach*. The ECOLIFE II report focuses on the product-service lifecycle of electrical and electronic products,

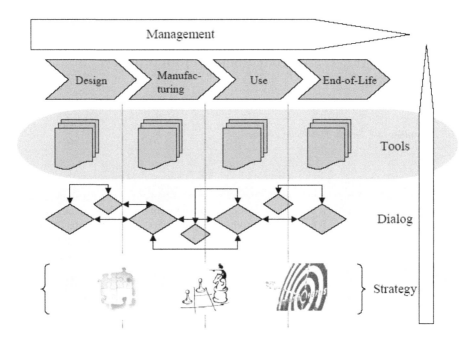

Figure 1. Conceptual Framework of ECOLIFE II State-of-the-Art Technology report.

and reflects the fact that in an innovation system key players in all the various stages of the product-service lifecycle – component suppliers, product manufacturers, service and logistic providers, processing industry etc. – are involved in the innovation process.

ECOLIFE II refers to *Technologies for sustainable Development in the Electronics Industry Innovation System* by defining them as all measures, instruments and (management) tools, both hardware and software, helping to move the Electronics Industry Innovation System towards sustainable growth, i.e. meeting the requirements of the triple bottom line of economic, ecological and social improvements in the Electronics Industry.

## 2.2 *Main areas*

Three main categories – dialogue, strategy and tools – have been embedded into the main stages of the lifecycle of an electronic product. The main stages are design, manufacturing, use, end-of-life, and management (understanding *management* as a horizontal task throughout the lifecycle). Figure 1 shows the lifecycle stages and the underlying categories.

## 2.3 *Technology evaluation methodology*

To evaluate the technologies in view of sustainable development in the Electronics Industry, a Sustainability Scorecard (STS Approach) has been used. Technologies have been evaluated with respect to their contribution to sustainable growth in the Electronics Industry Innovation System. Therefore additional indicators are used beyond technological performance indicators to describe, whether the technology indicates a move towards sustainability. Further technologies have been evaluated regarding their development stage as well as the need for action.

- Stage 1: Design

In the area of dialogue, special attention is paid to the Eco-design relationships to suppliers asked how to secure design requirements within the supply chain of the Electronics Industry

Innovation System. For the Strategy section, the state-of-the-art of DfX is described (X stands for X=Environment, X=chemical content, X=disassembly) as well as strategies to integrate DfX into conventional management systems and into the product development process. On the level of *materials*, some issues of hazardous materials and renewable materials are tackled. Finally in the Tools section, the actual developments in LCA and LCE, databases, teaching curricula, environmental benchmarks, new substrates for PCB, halogen-free and new flame-retardant materials are described.

- Stage 2: Manufacturing

The strategy section describes new strategies for manufacturing gaining competitive advantage (i.e. cleaner production). The new innovative manufacturing technology section describes promising manufacturing technologies that can significantly contribute to a future sustainable manufacturing practice in the electronic industry (i.e. rapid manufacturing). In the tools section, innovative developments in different tools for analysis and improvements of manufacturing systems are described (both individual subsystems and as a whole).

- Stage 3: Use

The main topics in this section are customer information and education on usage, communication of product impacts to the consumer, energy efficiency in use and new business models.

- Stage 4: End-of-Life

ECOLIFE 1 already reported on the state-of-the-art of certain End-of-Life technologies, beginning with identification technologies and separation, as well as touching upon health and safety aspects. The work also summarises financial aspects (covering collection and processing implications) and strategies for re-use and upgrade of discarded electronic products. Finally, new management strategies have been discussed to meet future requirements in the End-of-Life sector. The future needs defined in ECOLIFE I have been taken up in ECOLIFE II. The state-of-the-art of these technologies is presented in the report.

- Management

The role of the management function to green the electronics industry has been analysed.

## 3   EFFECTIVE ENVIRONMENTAL COMMUNICATION

Manufacturers have worked with environmental improvement for a long period of time. They have tried different solutions to lower their environmental impact, e.g. some first looked at the manufacturing processes and improved them, then they went on to improve their products. This work is still going on and many companies are now certified with Environmental Management Systems like ISO 1400 to gain further from good environmental behaviour. Years of research have contributed to an immense volume of information and knowledge, which now can help creating a more sustainable society.

The problem, though, is to transfer this information in an understandable way to the right receivers and help them do their part in a future sustainable society. This is a subject where most companies so far have had little success. There have even been examples of products that have not been launched because of lack of communication of interesting environmental matters, i.e. the positive features were not communicated to a satisfactory level so that customers could understand the extra value of such a product. The main objectives for this work within the ECOLIFE II network is to investigate the information at hand and to try to guide companies in how to use it; thereby defining two appropriate receivers, the right composition of the information and the right way to transfer it.

To be able to understand what actions companies have taken in their environmental communication to date, a questionnaire was made and sent to electronic companies. The answers received have been collected and evaluated. The evaluation has then been used as a basic knowledge of how companies work, think and communicate environmental matters.

Among a broad range of stakeholders interested in environmental information two important areas have been chosen and analysed in more detail: (1) private customer; and (2) supply chain.

ECOLIFE deliverable D9 *Environmental Information Guidelines* gives an comprehensive overview about media which can be used for the various communication purposes (Internet, printing, CDs, manuals, labels, different kind of environmental reports, material declaration tools etc.). Further a chapter is focusing on environmental accounting and emission trading and how these instruments could be used for a more effective communication.


## 4  ANALYSIS OF GAPS AND FORECAST OF FUTURE TECHNOLOGIES, PRODUCTS AND SERVICES

What will the future look like? That is a question that everyone would like to be able to answer. Unfortunately, it is not possible to define a precise scenario of the future, but looking at the past and scientific data, we can imagine what things may look like.

This document offers a general picture of Europe in the near future (2010). The scope of the research includes European Union members and the electronics industry.

The aim of this document is twofold;

1. Provide insight on how Europe will look like in 2010 regarding demography, economy, environment, and legal framework.
2. Define trends in technology, electronic product development, and services.

In the years 2002 – 2004 visible structural changes took place in the EIIS. In most of the Member States (before May 1st 2004), take-back measures according to the WEEE have been implemented, the transposition of the WEEE directive into national legislation is in progress, manufacturers are engaged in DfE programs and are marketing green products, are developing supply chain measures a.s.o.


### 4.1  *Actual legislation context of the Electronics Industry and impacts on innovation*

In Figure 2, the actual regulation framework together with other market and social drivers for the Electronics Industry are depicted. The analysis of the regulation and incentive framework sketches possible impacts on innovation activities in the Electronics Industry, and especially related to new business models like product-service systems.

The **WEEE-Directive** take-back obligation obviously influences different stakeholders of the innovation system by introducing direct and indirect requirements. It does so as a result of different direct legal obligations to be fulfilled, such as collection and recycling *quotas*, the implementation of the '*producer responsibility principle*' and *financial responsibility* for take-back systems, the definition of certain *standards* for the waste management and several requirements concerning *labelling* of products and *monitoring* of data and mass flows. Manufacturers of electrical and electronic equipment are burdened with the costs of collecting their end-of-life equipment leading to pressure on re-structuring the product design (for easy disassembly to decrease disassembly costs), the end-of-life (EOL)-management by establishing new logistical concepts, take-back systems and recycling systems, the innovation management by introducing new environmental oriented requirements like design for environment (DfE) within the supply chain etc. As a driver for new business models, the WEEE serves as a general positive incentive system, since the innovation actors may direct their strategic orientation up the waste hierarchy and may test new business models leading to avoid loops via take-back, recycling, etc. and to increase the volume of re-use of components. For example it can be expected that the WEEE promotes the development of product

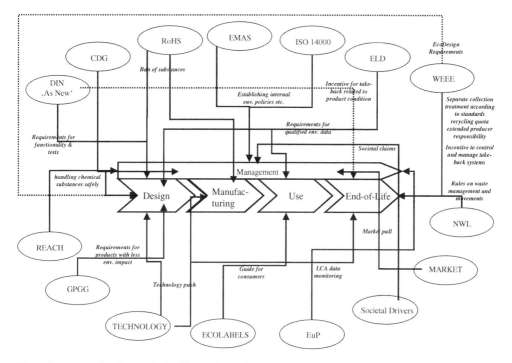

Figure 2.   Innovation Drivers in the Electronics Industry Innovation System.

extension services to avoid take-back costs, since users of electrical and electronic equipment from private households have the possibility of returning WEEE at least free of charge. Also it is expected that the WEEE will promote the development of use-oriented and product substituting services as well as demand side management, because of cost reduction in important WEEE burdens (logistics, EOL) and since the trend to de-materialisation causes less concerns for EOL activities and constraints in general. For the promotion of demand side management the WEEE serves as an incentive for bring-back/take-back on the consumer side.

The **RoHS-Directive** operates with *prohibitions* and restricts the use of certain hazardous substances in electrical and electronic equipment, as for instance lead, mercury and other heavy metals with an impact on the manufacturing process and recycling requirements. In general for the RoHS the same arguments as for the WEEE are applicable, but in the RoHS case with concerns to product inherent properties (materials).

The **EUP-Directive** places a strong burden on companies that produce energy-intensive products to meet *environmental requirements* and targets in the product's design, production, and end of life phase. The EUP requires an assessment of ecological profile of equipment regarding raw material, acquisition, manufacturing, packaging, transport, distribution, installation, maintenance, use, end-of-life. At each phase of this manufacturers are required to assess consumption of materials and energy, emissions to air and water, pollution, expected waste and recycling/re-use. In general the EUP may promote the development of product extension services through a broader dissemination of Eco-Design. The EUP is expected to promote the adding of services to products, because of a systematic analysis of the entire lifecycle. Today the lifecycle is often not known in detail from responsible actors. The consequent structuring of lifecycle brings ideas for new services or lacks of services, which could improve the product. Use-oriented services could be promoted, if the LC is clear and easy to follow up. They could be hindered, if through renting/leasing the still necessary product is too diversified to model its lifecycle. The EUP also may promote the development of demand side management. Since this Directive seeks to achieve a high level of protection for the environment by improving resource efficiency of EUPs, which will ultimately be beneficial to consumers and other end-users.

The upcoming chemical regulation **REACH** (**R**egistration, **E**valuation, **A**uthorisation and Restrictions of **Ch**emicals of the European Union) asks for a registration of all relevant chemical substances in the supply chain. Manufacturers and importers have to demonstrate in a registration dossier that they are managing their chemical substances safely. Companies will be required to register all substances produced or imported in volumes of 1 tonne and more per year per manufacturer or importer in a central database run by an independent agency to be created. This will cause transaction costs of running the database on these substances as well as increase the supply chain pressure on suppliers. The REACH may promote vertical integration, since external service for registration can be offered.

The focus of **ISO 14000** is on establishing internal policies, procedures, objectives, and targets, and on pursuing continual improvement. Compliance with applicable laws is mandatory, but the use of the standard is voluntary, and there is a self-declaration option. It applies to product- and service-related industries. **EMAS** goes beyond ISO 14000: Organisations registering to EMAS must be able to demonstrate that they have identified and know the implications to the organisation of all relevant environmental legislation and that their system is capable of meeting these on a continuing basis. The focus is initial environmental review, involvement of employees and to make available relevant information to the public. EMAS seems to be not directly relevant to the development of new business models. EMAS II states to include all (also indirect) effects not only site specific ones. Current discussion is continuing, if products (and thus offered services) are included in this consideration/certification or not. If products and their effects, e.g. in use, will be mandatory part for certification, EMAS II will have similar effects as e.g. EUP. But tendency now shows the exclusion of products and only to include transport to and from production site.

**Ecolabelling**: There are three types of Ecolabels, Type I-Ecolabelling according to ISO 14024, Self-declared Claims (Type II-Ecolabelling) according to ISO 14021 and Environmental Product Declaration (Type III-Ecolabelling) according to ISO 14040. Type I is a guide for consumers in that it identifies products as being less harmful to the environment compared to other, similar products fulfilling the same function (i.e. German Blue Angel, Nordic White Swan etc.). Type II sets up requirements for self-declared environmental claims including statements, symbols and graphics on products or services which are not certified by an independent third party. Type III requires a set of quantified environmental data consisting of pre-set categories of parameters based on Lifecycle Assessment according to ISO 14040. As all labelling schemes primarily and traditionally refer to products and not to services, up to now the impact of Ecolabelling on new business models remains ambivalent. Accordingly, a promotion towards services could be possible, but the delivery of and proof by a label will also promote the concerned products. Therefore no clear effect for a product–service shift is obvious. However, a positive effect on the development of product extension services may be expected, because external product certification and compliance services may be offered.

Electrical and electronic equipment will be affected by Integrated Product Policy (**IPP**) as well. It represents a new approach for product related environmental policy and advocates lifecycle thinking which means that consideration is given to the whole of a product's lifecycle from cradle to grave. This seeks to minimise environmental degradation by looking at all phases of the product lifecycle and taking action where it is most effective (design, manufacturing, use, disposal). IPP is flexible as for the type of policy measures to be used, working with the market where possible. The commission wants existing instruments to become more market oriented like environmental management systems, labelling and information concerning the product's cycle. Within the IPP the co-ordination among the measures shall be improved to use synergies. The IPP communication of June 18th 2003 is part of the Commissions efforts to achieve the goals set down in the European Union's 6th Environmental Action Programme and to fulfil the commitments made by the European Union at the World summit on sustainable development held in Johannesburg. For further information: http://europa.euint/comm/environment/ipp.

Besides the legislative or institutional framework conditions there are of course market and technology drivers, influencing a company's business strategy. Customer demands, i.e. customer satisfaction, fashions and user requirements rank on top of these drivers placing burden primarily on product and service design but also on corporate information and communication policy. Beyond the product and service price, in the last decade also "green" competitive elements came into place as the result of an increasing awareness of customers on ecological sound products. From the marketing perspective eco-efficiency as a tool to compete on the market was regarded as increasingly important, if competition on prices proved to be without effect.

## 4.2  Trends in innovation for 2010

Technologies influence everyday life in a variety of ways. Some inventions, such as the refrigerator and the internal combustion engine, perform pre-existing functions more cheaply and efficiently than the previous technology. Others, such as radio and the phonograph, perform functions that had not previously been anticipated. Either way, they stimulate increased consumption. Economic analysis tends to focus on these two results of technological change: improving the efficiency of supply of goods and services, and generating new goods and services. But technology and consumption are intertwined in more complex ways. Technology shapes worldviews, which in turn shapes behaviour in all sorts of ways, including feeding back to the direction of technological development.

In considering how recent and continuing technological innovations may change life over the next 30 years, the following *taxonomy of innovations* (derived from Freeman and Perez 1988) could be used as a guide to different types of change.

1. **Incremental innovations** (efficiency gains through scaling-up, learning-by-doing, engineering improvements, end-of-pipe controls, gradual response to market forces). These innovations occur continually.
2. **Radical innovations** (completely new technologies such as photovoltaic power sources, wheat genetically engineered for pest resistance, molecular sieves, etc). These innovations are discrete events. It might be possible to identify several radical innovations in a given year.
3. **Changes of *technology system*** (clusters of radical innovations – e.g. development of a new form of transport, communication or energy supply system). Such changes occur fairly rarely – there might be a few such changes in the economy as a whole in a given decade – but when they do occur they have widespread effects throughout a sector or group of sectors.
4. **Changes in *techno-economic paradigm*** (linked to *long waves* in the economy, or *Kondratieff cycles* – these are revolutions which may include innovations of types 2 and 3 but go much further). Such changes might occur once or twice per century.

With extended life expectations among European citizens, age will become a self-reliant, designable and extended phase of life linked to increasing expectations of the senior clients. A big challenge for service- and technology provider and products for elderly people will be the extremely differentiated ageing process of society and individuals as well as the extreme accentuation of their individuality. This calls for particular product- and service design measures as well as marketing concepts.

### 4.2.1  Forecast of technologies

Each year, or indeed every six months, a new generation of products replaces the previous one. Paradoxically, this field also demands long-term anticipation to a particularly great extent. It can take at least five to ten years to develop the standard and the required technology for an innovative product. For more than one decade, the ICT field has been evolving rapidly based on strong technological breakthroughs: wireless communications, technological integration, low power electronics, high resolution colour displays, and miniaturised digital cameras.

A consequence of Moore's Law was that a strong reduction in costs generated a very large introduction of a wide range of services based on innovative Information and Communication Technologies services while reducing the time-to-market of the latest microelectronics technologies.

Two key trends in the future of electronics;

- The miniaturisation which continues to follow Moore's law thanks to the increased capabilities of the processes and equipment.
- The diversification of the devices (Logic, SRAM, Flash, E-DRAM,CMOS RF, FPGA,FRAM, NEMS, Chemical Sensors, Electro optical, Electro biological) allowing many new applications.

**Silicon** is today largely used as substrate and for the bulk transistor. However some unfavourable properties of silicon indicate that other materials are needed for specific applications. Studies are carried out in the transistor architecture to avoid parasitic effects. Germanium and III/VI family components, new thin films materials combined with heteroepitaxy are experimented with to resolve these issues. However silicon will remain a key element, as it is a *green* material, abundant and non-toxic.

**Device Technology:** An immense progress has been achieved and will continue its sophistication in all following areas:

- **Thin Film deposition:** Layer and pattering on top of the silicon require low k-dielectric obtained with polymers and organosilicates. The transistor gates require high k dielectric which cannot be made anymore with SiO2 and require the usage of Hafnium in the form of HfO2 and Hf silicate. Chemical Vapour Deposition will be increasingly used to deposit these dielectric layers and metalisation even if evaporation and sputtering are still in use.
- **Doping:** Ion implantation will continue to be used to make the transistor structure (N or P doped regions), but this process will be followed by a flash anneal step.
- **Lithography:** Lithography principle will remain the same, but the process will require increasingly complex and expensive tools and X Rays are forecasted around 2010.
- **Dry etching:** Wet etching of dielectrics will almost entirely move to Reactive Ion Etching (RIE).
- **Cleaning (wet and dry):** With the miniaturisation and the sophistication of the devices the problems of particles and surface state are becoming increasingly critical and difficult to resolve. Chemical/ mechanical polishing is used extensively and must be refined to achieve perfect surface states.
- **Packaging:** The chip miniaturisation has progressed at a very fast pace which allows to integrate already several functions but efforts are done to go beyond and aim for complete SOC (System on Chip) or SIP (System in Package). Several choices are considered and will be implemented to fulfil the European Union directive to eliminate LEAD. Other packaging techniques are also under study and will most probably be implemented in the next ten years like the wafer bonding allowing even to assemble several very complex chips containing a complete system. The evolution between the last two generations of mobile phones give a good perspective of the progress which can be forecasted for the nest ten years using the above SOC's and SIP's

### 4.2.2 *Forecasts of products*

Technological developments expressed in the previous section, are translated and integrated in electronic products. The question is: what can be expected from electronics manufacturers in the years to come? Four areas seem to be the cornerstone of upcoming electronics products: IC, Displays, Connectivity, *Infotainment*. Some of these areas are connected to servicing, and some are highly linked to technological developments. A description of each area follows.

- **IC:** Integration of IC (Integrated Circuits) is broad in all families of electronic products. Functionality of IC increases (Moore's law) as also does environmental load during the user phase. On the other hand, production continually requires use of less resource (materials and water).

Although progress is expected to continue for a few more microchip generations, it is certainly understandable that is will not carry on forever. Fundamental physical limits will be reached sooner or later, at which point further evolution will become technically impossible or far too expensive. This fact fuels the motivation to look beyond silicon and current transistor logic, to new areas, such as optical and quantum computing. These may well be the replacements of silicon the same way transistors replaced vacuum tubes, and silicon chips replaced transistors more than two decades ago.

Research in other ways of computation such as the ones mentioned above is well under way and even though non-silicon based processors will not appear for many years to come, it is more than certain that information technology will not be stopped when Moore's law ceases to be valid. Whether any of these next generation technologies will be ready to be put into practice by the time the limits of silicon are finally reached, remains to be seen.

Fortunately the industry does have choices for the future. Even when silicon does become economically and technically unprofitable, the paths opened by new proposals such as the not too distant optical computing or the more radical quantum computing, can carry information technology into the future far faster than any other silicon chip could ever hope to. The application of these technologies may not start immediately after the demise of silicon, but when it does, it will certainly be something to look out for. Moore's law will not hold anymore, since it has nothing to do with these non-silicon based systems, but the notion that computers will continue to become faster and more powerful as time goes by will probably hold true for many years to come.

- **Displays**: The end for CRT (Cathode Ray Tube) is near, LCD (Liquid Crystal Display) is the new kid on the block and comes with good news for the environmental community, but also some room for improvement. While requiring lower energy consumption during user phase, higher environmental load is required during production process. At present, end-of-life treatments are still immature, especially in recycling techniques.
- **Connectivity**: The first decade of 21st century was characterised by the spread of communication devices. An explosion of products like cell phones and computers connected to global networks (Internet) allows millions of people to communicate by voice, video, and sharing documents in real-time. Mobile phones are also significant for social capital; they are accessible to unprecedented numbers of people. And as the number of people connected goes up, the value for social capital goes up much faster.

Key features of mobile phones that make them significant for social capital:

- Portability
- Individual ownership
- Text messaging

From an environmental perspective, large amounts of energy are required to run complex networks of servers, antennas, and electric devices to keep a simple phone conversation. Generally speaking, a lowering of environmental load is not expected.

### 4.2.3 *Forecast of services*

Future service development and offering is driven by consumer needs from changing society and life styles as well as technology opportunities as enablers of service provision. Society is increasingly moving to a 24h society. The curriculum vitae is characterised owing to a mosaic living, means different projects and locations, periods of work alternated with periods of education and regeneration, changing partners and relationships. A growing trend towards cultural homogeneity, fun society and event consumption can be seen. The proportion of elderly and young people in the European countries will be reversed. Life long childlessness, delayed child rearing and greater life expectations are driving forces and result in changes in the structure and function of family and new definition of government, corporate and individual responsibilities.

In terms of information and communication technology there will be richer possibilities of information exchange, education and communication. New forms of networks and framework to support new forms of commerce, economic growth and benefit, to support flexibility and self-organising mechanism, bridging skill gaps, to protect citizens, social care and welfare, personal health and assistance through intelligent environment. Future Service Areas studied are: Information and

Communication; Automation and Control; Personal care and health; Transportation and Mobility; Education and Assistance.

These areas are studied and barriers and opportunities have been identified. These findings have been compared with the current state-of-the-art activity of the ECOLIFE II project and gaps in technology, products and services are identified. These findings have been turned into an extensive list of potential research projects to contribute to close the gaps between current state-of-the-art and future trends within the different areas. Reference is made to deliverable D14 Europe in 2010.

Future deliverables will include a green paper on sustainable service systems comprising a guideline for implementation of sustainable services in the electronics industry in Europe.

## ACKNOWLEDGEMENT

This work has been partly funded by the European Commission through Project ECOLIFE II (ECO-efficient LIFEcycle Technologies; From Products to Service Systems (No. G1RT-CT-2002-05066). The paper has been elaborated with contributions from Bernd Kopacek, Raymond Nyer, Katrin Muller, Markus Stutz, Joachim Hafkesbrink, M. Charter, V. Herman, A. Stevels, O. Pascual and Ecolife II partners.

## REFERENCES

Ecolife II deliverable D7 - State-of-the-Art Technology in the Electronics Industry Innovation System. Dialogue, Strategies and Tools towards Sustainable Development. http://www.ihrt.tuwien.ac.at/sat/base/EcolifeII/index.htm
Ecolife II deliverable D9 Environmental Information Guidelines, http://www.ihrt.tuwien.ac.at/sat/base/EcolifeII/index.htm
Ecolife II deliverable D14 Europe in 2010, http://www.ihrt.tuwien.ac.at/sat/base/EcolifeII/index.htm

*Part IV*
*Production scheduling and control*

Manufacturing has changed radically over the course of the last 20 years and rapid changes are certain to continue. The emergence of new manufacturing technologies, spurred by intense competition, has led to dramatically new products and processes, as well as reconsideration of management and labour practices, organisational structures, decision-making methods, and supply chain relationships. Furthermore, manufacturing system work in a fast changing environment full of uncertainties and where increasing complexity is another feature showing up in production processes and systems, and in enterprise structures as well. For example, advances in Information and Communication Technologies (ICT) have injected *velocity* into the front business activities and enabled companies to shift their manufacturing operations from the traditional factory integration philosophy to a supply chain-based e-factory philosophy: digitally integrating the entire manufacturing chain to support globalisation of production. This consists in forming production networks based on independent companies collaborating by shared information, skills, and resources, driven by the common goal of exploiting market opportunities. Such networks are built on internal (inside companies) and external (among companies) agility, adaptability and flexibility of the various technical and business processes. To support these new features with regards to the manufacturing facilities and resources, highly proactive and fault-tolerant scheduling, planning and control systems are needed, while keeping within cost, reliability, availability, maintainability, and productivity (CRAMP) parameters. Moreover, the natural shift towards collaborative environment encourages the development of (shopfloor) intelligent systems where interoperable and autonomous units with embedded digital intelligence and e-information capabilities transform information flows into product flows by controlling resources flows.

In this way, within the frame of the IMS-NoE, the special interest group 2 (SIG2) was formed with a large group of academic and industrial partners (end-users, vendors) to work on the above issues and challenges. The SIG dealt with investigating new industrial organisation paradigms, new ICT solutions … to potentially achieve *Manufacturing Scheduling and Control in the Extended Enterprise*. Thus, this section of the book addresses, on the one hand, concrete results of the special interest group in the first contribution, and, on the other hand, two selected contributions on future *Production Scheduling and Control* challenges.

These latter contributions cover two complementary sub-areas and require new ways of working to be implemented to face (a part) of future challenges towards global e-manufacturing. The first of these contributions, *Organising and running real-time, co-operative enterprises* by SZTAKI (Hungary) and ITIA (Italy), illustrates some solutions together with open issues for further research on organisational design, planning and execution in mass customisation and customised mass production, based on the *EUROShoe* project funded by the European Commission and the *VITAL* R&D project started in 2004, within the Hungarian National Research and Development Programme.

The second of these contributions is developed jointly by Profactor (Austria), University of Linz (Austria) and University of Applied Sciences Solothurn (Switzerland). It is entitled *Flexible and responsive cross-organisational interoperability*. It addresses needs of SMEs for ICT support for co-ordination, co-operation and collaboration to support production networks. In the paper, interoperability is underlined as a key point within production networks.

The significance of contributions in this section is described relative to existing literature.

*Advanced Manufacturing – An ICT and Systems Perspective – Taisch,*
*Thoben & Montorio (eds)*
*© 2007 Taylor & Francis Group, London, ISBN 978-0-415-42912-2*

# Manufacturing scheduling and control in the extended enterprise

Benoit Iung[1] & László Monostori[2,3]

[1] *Université Henri Poincaré – CRAN – CNRS – UMR, Faculte des Sciences – BP*
*VANDOEUVRE, Franc*
[2] *Computer and Automation Research Institute, Hungarian Academy of Sciences,*
*Budapest, Kende u. Hungary*
[3] *Department of Production Informatics, Management and Control, Budapest University*
*of Technology and Economics, Budapest, Hungary*

ABSTRACT:   In accordance with manufacturing topics emphasised at IMS initiative and *Manu-Future* vision, the IMS-NoE SIG2 investigated industrial organisation paradigms and ICT solutions to potentially achieve intelligent manufacturing scheduling and control in the extended enterprise. This intelligence of scheduling and control encourages the development of architectures based on autonomy and co-operation of the production facilities and processes while keeping within cost, reliability, availability, maintainability, and productivity (CRAMP) parameters. SIG2 was formed with a large group of academic and industrial people (end-users, vendors) who are working on the important processes in the back-end and at border between front-end and back-end systems to deploy, consistently, manufacturing scheduling and control. This paper summarises the main results obtained by the SIG2 over three years.

*Keywords*:   Manufacturing scheduling and control, production networks, extended enterprise.

## 1   TOPIC AND PROBLEM ADDRESSED

Manufacturing has changed radically over the course of the last 20 years and rapid changes are certain to continue. The emergence of new manufacturing technologies, spurred by intense competition, has led to dramatically new products and processes. New management and labour practices, organisational structures, decision-making methods, and supply chain relationships have also emerged as complements to new products and processes (Bollinger et al. 2000). For example, the advances of network computing and Internet technologies injected *velocity* into the front business activities and enabled companies to shift their manufacturing operations from the traditional factory integration philosophy to a supply chain-based e-factory philosophy (Lee 2003): to digitally integrate the entire manufacturing chain. This allows for support of mass customisation of products, to face product variability, and to provide business systems with accurate information concerning product tracking. This transition is dependent upon the advancement of next-generation manufacturing practices in *e-manufacturing* such as initiated by worldwide IMS story (Yoshikawa, 1995), which is focused on the use of Internet and tether-free communications technologies to make things happen collaboratively on a global basis (Koc et al. 2003).

These evolutions (Ollero et al., 2002) underlined the increasing role of information within companies at all levels and more precisely within the manufacturing plant, where interoperable and autonomous units embedded with digital intelligence and e-information capabilities (distributed vs. collaborative architecture) transform information flows into product flow by controlling resource flows (Pétin et al. 1998). Thus the required synchronisation among information and product flows throughout the extended manufacturing plant has justified the development of Manufacturing Execution Systems (MES). They aim at scheduling the product operations according to a given production plan provided by the enterprise front-end systems (i.e. ERP) and to the manufacturing resources controlled by the backend systems (i.e. shopfloor control systems). For SIG2, this materialised concretely the global link between scheduling and control fields implemented through distributed vs. collaborative processes not necessary belonging to the same *production (supply) networks* (i.e. a network of firms is formed by several entities which could share various data regarding their capacity planning and scheduling). Nevertheless MES are often inflexible to operate effectively in a changing environment with uncertainty; ERP systems cannot include the dynamics of the factory floor conditions, and the crucial link among these different processes is hindered by the lack of formalised information (even if the ANSI/ISA-S95[1] provides first normative guidelines). This lack of true system integration capabilities is identified as one challenging obstacle in deploying e-manufacturing systems (Qiu et al. 2003, Bollinger et al. 2000).

In this context, IMS-NoE SIG2 dealt with investigating industrial organisation paradigms, new methodologies, and new ICT solutions, to potentially achieve intelligent manufacturing scheduling and control in the extended enterprise. This intelligence of scheduling and control encourages the development of architectures based on autonomy and co-operation of the production facilities (vs. processes) while keeping within cost, reliability, availability, maintainability, and productivity (CRAMP) parameters. SIG2 was formed by a large group of academic and industrial partners (end-users, vendors) who are working on the important processes at the back-end (horizontal integration), and at the border between front-end and back-end systems (vertical integration) to deploy, consistently, manufacturing scheduling and control.

With this extended participation (about 35 active members; 110 members registered on the IMS-NoE web site) four working groups emerged (WG) focusing on four issues, scientific (for formulating long-term perspective) and pragmatic (for facing the day-to-day challenges of industrial needs), to satisfy the two communities and to encourage discussion and exchange of views. These groups addressed:

- *E-maintenance* topic (leader TEKNIKER, WG1). E-maintenance aims at supplying scheduling and control systems, with current information about component availability, reliability, . . . but also proactive information such as component degradation state, residual life time, expected system performance, . . .
- *Planning and scheduling* topic (leader LAP, WG2) for investigating innovative organisations, methods and software packages related to new production planning and scheduling approaches in Network Supply Chain (NSC).
- *Co-operation and co-ordination in supply network* topic (leader PROFACTOR, WG3) for focusing on technological as well as human aspects of the available information and communication technology that allows companies to join forces for fulfilling customer requests more precisely for satisfying scheduling purposes.
- *E-control* topic (leader UNIKARL, WG4). E-control aims at supporting the foundation of e-manufacturing infrastructure for online business strategies where the integration of control with other back-end processes such as scheduling facilitates the achievement of lean manufacturing systems.

The WGs worked on the complementary processes of the *manufacturing scheduling and control* chain, since WG1 and WG3 focused on the back-end, while WG2 and WG4 addressed the border between front-end and back-end systems. As these WGs had some *interoperability* links with other

---

[1] ANSI/ISA-95 (IEC/ISO 62264), 2000, Enterprise-control System Integration.

processes performed in other SIGs, SIG2 established strong relationships mainly with SIG4 and SIG3 to avoid inconsistent overlapping within IMS-NoE.

The SIG2 can be considered as an *informal* repository with respect to discussions, exchanges, work and research *on Manufacturing Scheduling and Control in the Extended Enterprise* topics, resulting in an updating of the relevance on this topic.

## 1.1 *Industrial needs on SIG2 topics*

To set the basis of the SIG2 scientific effort, a first action was to develop a questionnaire with the aim of defining industrial requirements on production scheduling and control. This questionnaire was based on two complementary areas of expertise and competence identified by relevant people:

- One related to *application fields* (or *user goals*, *user services*) in the domain of Manufacturing Scheduling and Control. What type of needs is addressed by the modern industry world?
- Another one related to the ICT needed to support these services.

The questionnaire was sent to SIG2 members and others outside the SIG, and the main industrial needs extracted from questionnaire are synthesised in Table 1.

In addition to the first needs identified through the questionnaire, other industrial requirements emerged from internal audits made by the other working groups. Specifically those developed in:

- WG1: integration of maintenance strategy into plant strategic business, new maintenance models for cost-effectiveness assessment, efficient maintenance indicators;
- WG2: formalising trust conditions in networks of companies, IT-Support of trust and security adaptable to the specific business-branch the network, normative standards fixing quality requirements for service products, IT-Support of co-operation in multi stakeholder environments);
- WG3: security and thrust related issues in open networks, conflict goal of partners and power differences, costs and time of co-ordination activities, hidden competencies of network partners; and
- WG4: generalised business models for service products covering the complete lifecycle of the products as well as concepts, methods and their technological implementations for establishing and ensuring trust and security among business-partners in the added value chain.

## 2 STATE-OF-THE-ART

The *state-of-the-art* phase in SIG2 activities started by focusing on concept of *extended enterprise* representing the innovative environment requested for new production scheduling and control. In this innovative environment, the *enterprise* acts as a network of companies formed by several entities that should share various data, information and resources to reach the holistic goal.

So, the latest approaches to realise extended enterprises, which involve building virtual enterprise architectures, enterprise integration and supply chain management, are based on the agent theory (Camarinha et al. 1997, Iwata et al. 1997, Macgregor et al. 1996), mobile agents technology (Brugal et al. 1998, Rabelo et al. 1998), Monostori et al. 1997) or neural networks technology (Lau et al. 2000).

## 2.1 *Main functional requirements*

In Camarinha-Matos et al. (1997a) and Camarinha et al. (1997b) the main functional requirements for the design of an architecture for extended enterprises are presented. The work is partly done in the framework of the Prodnet II (Production Planning and Management in an Extended Enterprise – http://www.uninova.pt/~prodnet/) project. The enterprises in the network are viewed as nodes that add some value to the process. The nodes can exchange information at the same time. Every node is

Table 1. Main industrial needs extracted from questionnaire.

| SIG2 Topic | What are the industrial needs? |
| --- | --- |
| *Field of Application* | |
| Capacity planning | Lack of specialists. How to develop capacity planning inside virtual enterprises? Need for algorithms able to schedule capacity and materials at the same time in a flexible way. Availability of precise decision support. How to manage capacity in distributed manufacturing networks? Problems related to finite capacities, time-varying capacities, and building systems for solving capacity planning and master scheduling in the same step. |
| Master scheduling | How to develop master scheduling inside virtual enterprise? Solutions on collaborative forecasting. Detailed temporal relations. Capacity conscious scheduling. |
| Scalable scheduling | How to integrate scalability to industrial applications? Need for incremental non-disruptive scheduling algorithms. Problems on labour and machine scheduling, short response time, and anytime methods. |
| Advanced control | Integrating feedback in adaptive control. How to guarantee desired plant behaviour, regardless of un-modelled terms, including external and unknown disturbances? How to control production systems in uncertain environment? New adaptive control systems. (self) re-configuration of control systems. |
| Logistics | Logistics inside virtual enterprises. Integration of advanced logistics concepts (integrating transport, storage and repair service). Managing distributed virtual enterprises. SCM: efficient order planning and order fulfilment in dynamic circumstances over the supply chain involved. Correct treatment of risk and uncertainty. |
| Production monitoring | Need for collaborative and distributed control tools. Integration of production monitoring in industrial architecture. How to deal with data quality and uncertainty, confidence level? Availability of multi-sensors, sensors with wireless communications, intelligent sensors. Correct analysis of large data amounts (i.e. with statistical methods) – using measured (corrupted) data for control purposes. |
| e-maintenance, e-diagnosis, e-production | From diagnosis to prognosis (predictive capability, trend analysis). New decision making support (learning process, case based reasoning). Embedded diagnosis, intelligent diagnosis systems. New types of architectures to support e-service, Auto-ID technologies. How to merge component and system performances? |
| *ICT* | |
| Techniques | What is the potential of agent-based environments/solutions, HMS, etc. in the industrial world? How to implement co-ordination mechanisms, mechatronic units, etc. in an integrated architecture? How to support stochastic model-reference adaptive control, and Kalman-based estimation? State space design, Active observer design, Hybrid applications, etc. |
| IT | Added value on wireless technologies (interfaces, security), web-services, interoperability. IT-tools and standards for B2B-SCM also suitable for SMEs including web-platforms. |
| | New middleware (i.e. GRID). New Protocols for collaboration (i.e. FIPA ACL). |

extended with a co-operation module. The co-operation module supports the information exchange among the nodes. The internal module is connected to the co-operation module via a *mapping interface* and the co-operation modules of different enterprises are connected to the network via a *network infrastructure interface*. The internal module comprises the complete structure of the company's information and all the internal decision making processes.

Within this architecture two kinds of nodes are defined: network co-ordinator node, as the regulatory component of the enterprise network, and member enterprise node, which will store the information about the enterprise itself, and provide the external connectivity. To support several important information management requirements of the virtual enterprises, federated object-oriented information management system are used.

## 2.2 Agile scheduling

In Monostori et al. (1997) and NIIIP (1999) a prototype of a multi-agent system for agile scheduling and the extension for the operation in a virtual enterprise environment is described. The later system is called MASSYVE (multi-agent agile manufacturing scheduling systems for virtual enterprises) and is part of the MASSYVE INCO-DC KIT Project. MASSYVE uses the HOLOS framework (Rabelo et al. 1999) as a baseline for advanced scheduling and the PEER information management framework for the information integration.

The HOLOS (Rabelo 1997) scheduling system allows a system to be custom-tailored for each enterprise and, at the same time, to be re-configured and adapted whenever new production methods, algorithms, production resources, etc. are introduced or changed. In this framework an instance of a scheduling system is interactively and semi-automatically derived from the HOLOS generic architecture (HOLOS-GA) (Rabelo et al. 1997) reference model, supported by the HOLOS methodology (Rabelo 1997), based on Agent Oriented Programming (Shoham 1993) constructs.

The agent classes used in HOLOS are: the *scheduling supervisor* (*SS*) (agent that performs the global scheduling supervision – it is the unique system's *door* to other systems), the *enterprise activity agents* (*EAA*) (the executors of manufacturing orders), the *local distribution centres* (*LDC*) (represent functional clusters of EAAs to avoid announcement broadcasting, also responsible to select the most suitable agent for a certain order after a negotiation process) and the *Consortium* (*C*) (temporary instances created to supervise the execution of a given order).

HOLOS uses the contract-net protocol co-ordination mechanism to support the task assignments to agents, and the negotiation (Davis et al. 1983), (Rabelo et al. 1996a) method to overcome conflicts taking place during the scheduling phases. The interaction among the agents is only vertical and agents cannot change the set of other agents that they can communicate with.

The HOLOS control hierarchy uses one *global manager* (the agent SS), some *functional managers* (the agents LDC), and assumes that a shopfloor is usually composed of production resources (the agents EAA) with small production capacities. A HOLOS agent can establish communications with four kinds of external entities: other HOLOS agents, the end-user, a CIM Information System and the production resources.

MASSYVE proposes a three-layered federated database architecture to support the sharing and exchange of information within each multi-site enterprise and among different enterprises uniformly, based on PEER: Intra-organisation federated layer; federation of HOLOS systems; federation of virtual enterprises.

## 2.3 Holonic based framework

A framework for virtual enterprises based on holonic principles is presented in Iwata et al. (1997). The framework is presented through two views: the global and the local view. From the global perspective the authors consider two kinds of holons: virtual enterprise (VE) holon and member enterprise (ME) holon.

The VE holon is seen as the global co-ordinator of the virtual enterprise and it is located at the top level of ME holons. The VE holon is not imposing anything to the other holons, but acts as an assistant to them, and provides goals, constraints, warnings, suggestions, knowledge sharing services, etc. The ME holons, from the global view, are located on the same level.

Seen from the local view, holons are distributed on three levels: on the first level the ME holon can be found, on the second level the planning holon and the scheduling holon, and on the third level the task holons and resource holons.

## 2.4  *Partner selection*

In Macgregor et al. (1996) the problem of choosing the enterprises forming the virtual enterprise (VE) in a multi-agent environment is discussed. Besides the enterprises, which are seen as agents, the framework comprises VE co-ordinator agent and information server agents. The goal of enterprises is to form the most favourable group to satisfy a certain need. The co-ordinator agent decomposes the goal in sub-goals and forms a VE goal hierarchy. This hierarchy can be dynamically changed during the process of partner selection. The partner selection is treated as a distributed constraint satisfaction problem.

## 2.5  *Subcontracting*

In production networks different kind of subcontracting may take place (Mezgar et al. 2000) owing to technological reasons, and referred as *technology-driven subcontracting*, or as a result of insufficient capacity, referred as *capacity-driven subcontracting*, or for strategic reasons to keep the co-operation with a certain supplier, or subcontracting among factories of the same company, etc.

Subcontracting between suppliers and producers or between producers and customers is called *classic subcontracting*. To achieve better reliability enterprises in the network should allow and assure some information sharing regarding machine loading, availability, orders progress, planned demands or stocks. To help increase the redundancy in networks, a software tool called FAST/net was developed. The software provides information about the status of orders, resource availability, and partners' stocks. FAST identifies bottleneck systems, marks orders suitable for subcontracting, and makes calculations on security stocks.

## 2.6  *Reference model*

A reference model for virtual enterprises can be found in Bremer (2000). Enterprises searching for a business opportunity form an aggregation named *virtual industry cluster (VIC)*. When a business opportunity can be exploited, a *virtual enterprise broker (VEB)* will search through different *virtual industry clusters* for enterprises with the appropriate competencies. The enterprises in question will form a *virtual enterprise (VE)*. When a VIC has a VEB of its own, the VIC becomes a *virtual organisation (VO)*. The number of enterprises in a VIC is varying, and enterprises may enter or exit the aggregation at any time. The VEB obtains and offers business opportunities to the members of VIC and sets up the VE. The VIC provides information about the competencies of the enterprises involved. The VE will run the business opportunities and take part in its re-configuration. The VO has to manage alliances, and define and manage strategies.

## 2.7  *Agent based architecture*

The collaborative agent system architecture (CASA) and the infrastructure for collaborative agent systems (ICAS) have been developed for implementing Internet enabled virtual enterprises and for managing the Internet enabled complex supply chain for a large manufacturing enterprise, though, initially they were proposed as a general approach for Internet-based collaborative agent systems (Shen et al. 1999). The approach comprises the following elements: co-operation domain servers; yellow page agents; local area co-ordinators; collaborative interface agents; high-level collaboration agents; knowledge management agents; and, ontology server.

The co-operation domain server receives all the messages sent by the agents in the co-operation domain and forwards the message transparently to its destination and may also record all the transactions. Yellow page agents accept messages for registering services and record this information in a local database. A local area co-ordinator acts as a representative of the area to the outside world, and a manager for the local agents within the area. It also provides an interface service to the outside world.

Collaborative interface agents are supposed to be communicative, semiautonomous, collaborative, reactive, pro-active, adaptive, self-aware and mobile. High-level collaboration agents are introduced to increase the basic collaboration services provided by yellow page agents and local area co-ordinators agents. Agents do not communicate with one another directly, but those working together form co-operation domains. Each agent in a particular co-operation domain, routes all its outgoing messages through the co-operation domain server. All co-operation domain servers are connected with the ontology server for translating messages from different agents into a common format.

Five mechanisms developed in previous projects (MetaMorph, MetaMorph I, MetaMorph II) are used: agent-based mediator-centric organisation; task decomposition, virtual clustering, partial agent cloning, adaptation and learning. A hybrid agent-based architecture for manufacturing enterprise integration and supply chain management is proposed in the framework of MetaMorph II project (Rabelo et al. 1998a), (Shen et al. 1998a). The objective of the project is to integrate the manufacturing enterprise's activities such as design, planning, scheduling, simulation, execution, and product distribution, with those of its suppliers, customers and partners into an open, distributed intelligent environment. The architecture was named agent-based manufacturing enterprise infrastructure (ABMEI).

### 2.8 *Generic reusable enterprise model*

The architecture of the integrated supply chain management system presented in Fox et al. (1993) is based on a set of co-operating agents. There are two types of agents: functional agents for planning and controlling activities in the supply chain, and information agents for information and communication services. The functional agents in this architecture are the order acquisition agent, the scheduling agent, the resource management agent, the dispatching agent, the transportation management agent, and the logistics agent. Supply chain agents exist within an enterprise information architecture (EIA), having a generic reusable enterprise model in the centre, to support the integration of supply chain agents.

The EIA is responsible for finding the information that a certain agent asks for, and also for distributing the information an agent wants to share, to the agents that are interested.

### 2.9 *Synthesis*

The involvement in extended enterprises has numerous advantages for the member enterprises, such as:

- increased reaction to changes in market condition and demands, because the participation in the network involves a better flexibility and agility;
- possibility to complement core competencies to be able to share some market opportunities;
- ability to specialise in few areas, representing the core competencies;
- new market opportunities, by increasing the geographic coverage;
- improved quality and responsiveness to market opportunities;
- elimination of capacity bottlenecks and reduction of capital investment through resource sharing;
- making global lifecycle orientation possible;
- increased reuse and recycling in waste management;
- increased redundancy, as more than one partner may produce the same service or product.

To enlarge the *state-of-the-art* on production networks, each WG carried out a specific state-of-the-art analysis related to each WG topic:

- E-maintenance: Using the relevant information coming from IMS centre (http://www.imscenter.net), the workshop 2001 *on tether-free Technologies for e-manufacturing, e-maintenance & e-service* http://www.uwm.edu/CEAS/ims/, the-work of Lee (1998), Iung (2003a), Iung et al. (2003b), and Léger et al. (2001), but also from several projects such as AMOSE, REMAFEX, DIAMOND, PROTEUS, DYNAMITE, initiatives such as MIMOSA, and forums such as CECIMO, CAMATT.

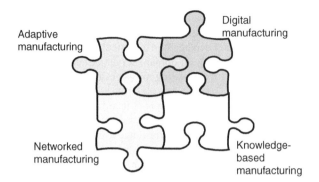

Figure 1.    Four important R&D areas (N.N. 2004).

- Planning and scheduling; Using the review papers of Croom et al. (2000) and Tan (2001), by focusing on sub-topics such as constraint logic programming, constraint based repetitive production planning (Skolud and Zaremba 2000), and transition phase during system emptying (http://zeus.polsl.gliwice.pl/~cim2001/swz.html).
- Co-operation and co-ordination in supply networks: Using the results of the numerous projects on multi-agent systems (HOLOS, MASCOT, DEDEMAS, …), the work related to product identification technologies (Mcfarlane et al. 2003), human factors and human-computer interaction (MacCarthy and Wilson 2001) and computer supported collaborative work (CSCW).
- Advanced control: Using the results of the projects like KRASH, VIDOP, MAGIC, PROSERV, SOFIA, ARIKT, DIAMOND.

## 3    VISION AND TRENDS

Vision related to SIG2 topics are strongly connected to the directions emphasised at the *ManuFuture* Workshop, July 1, 2004, Dortmund, Germany: *adaptive manufacturing, digital manufacturing, knowledge-based manufacturing, and networked manufacturing*. The above areas were also outlined at the *ManuFuture* 2004 Conference, Enschede, The Netherlands, December 6–7, 2004 (http://www.manufuture.utwente.nl/) and in the material *ManuFuture: A Vision for 2020, Assuring the Future of Manufacturing in Europe*.

On this basis, during the IMS-FORUM 2004, SIG2 organised three sessions to formulate more precisely dedicated visions on manufacturing planning and control through:

- Ideal Factory Session on *new models and technologies to select and improve a cost-effective condition-based maintenance policy practic*ally. A white paper on *industrial/scientific challenges to support maintenance in the future* is available on the www.ims-noe.org web site.
- 2 Invited Sessions
  - New methods, models and technologies for manufacturing scheduling and control (5 papers, 40 participants);
  - New models, methods and technologies for maintenance engineering and management (5 papers, 30 participants).

From all these events it became evident that networked manufacturing is of highest importance, however, it must be pointed out that the four important areas shown in Figure 1 represent overlapping domains and can be considered in a holistic way only.

Thus, the newly emerging area of research, representing a challenge for the planning and management of production systems is related to the evolution of the paradigm of *production networks*. The determination of methods for the identification and verification of the manufacturing requirements

Table 2. Challenges and possible answers (Monostori et al. 2005).

| Challenges | Possible answers |
|---|---|
| Complexity | Autonomy and co-operation |
| Changes and disturbances | Intelligence |
| Finding the possible best technical-economic solutions | Integration and optimisation |
| Quick reaction ability | Real-time capability |

of all parties in the network, as well as the specification of the necessary processes and ICT systems, will be central tasks for researchers in this field (N.N. 2004).

### 3.1 Necessity for real-time and co-operative capabilities

At least two important requirements, i.e., the *real-time* and *co-operative capabilities* of the whole system, have to be added as issues of high and increasing importance (Monostori et al. 2005). The first one refers to the ability to recognise and act on internal and external changes and disturbances within the time frameworks required by the given level of the manufacturing-production hierarchy. The second issue highlights that complex production structures – from machine tools, robots, etc. to production networks, including the human beings involved – are increasingly considered and built up as autonomous, but co-operative entities. Table 2 summarises some challenges and possible answers to them.

One of the most important trends in manufacturing is manifested in the paradigm of *customised mass production*, which means accomplishing the difficult task of producing *customised products at a price near to the level in mass production.*

The aim is to research and develop new methods for the real-time management of complex technical and economic systems that work in changing, uncertain environments. Since the methods come from various, novel areas of informatics, operational research and knowledge-based systems, their integration will balance the aspects of optimisation, autonomy, and co-operation.

## 4 FUTURE CHALLENGES

To implement the above vision, SIG2 members underlined some scientific challenges that should promote the IMS activities in the future. The aim is to contribute more to the *manufacturing scheduling and control* area by evolving from local satisfaction of local needs to production patterns able to respond flexibility to global demand in face of system complexity (holistic approach). It is fully in consistence with international IMS objectives, technical themes and priority R&D areas (see Visionary Manufacturing Challenges for 2020 Worldwide IMS initiative report):

- Objective: To respond effectively to the globalisation of manufacturing; to enable greater sophistication in manufacturing operations.
- Technical themes: Process strategy/planning/design tools.
- Priority R&D areas: Sustainable design, products and manufacturing processes; knowledge-based value creation in e-manufacturing; manufacturing on-demand (e-manufacturing); dynamic collaborative value-creating networks.

These challenges are included within the items identified by the IMS challenges:

- Grand Challenge 1. Achieve concurrency in all operations;
- Grand Challenge 2. Integrate human and technical resources to enhance workforce performance and satisfaction;

- Grand Challenge 3. *Instantaneously* transform information gathered from a vast array of diverse sources into useful knowledge for making effective decisions;
- Grand Challenge 4. Reduce production waste and product environmental impact to *near zero*;
- Grand Challenge 5. Re-configure manufacturing enterprises rapidly in response to changing needs and opportunities;
- Grand Challenge 6. Develop innovative manufacturing processes and products with a focus on decreasing dimensional scale;

More precisely, these R&D challenges addressed the following sub-topics (Monostori et al. 2005):

- **Integrated production planning and scheduling:**
  - development of novel mathematical models and related solution methods;
  - combination of the methods of operations research (scheduling theory, linear and integer programming) and artificial intelligence (constraint programming);
  - estimation of solutions' goodness;
  - development of parametric, scalable modules for production optimisation;
  - development of systems to be installed at the enterprises or for use as e-service;
  - formulation and resolution of planning and scheduling problems through algebraic approach, design of decision support systems, multi-criteria optimisation methods.
- **Real-time production control:**
  - modelling of disturbance and change sensitivities;
  - ambient intelligence;
  - intelligent analysis of production data warehouses, data-mining;
  - automatic situation recognition and related problem solving, decision support, machine learning;
  - research on reactive and proactive rescheduling algorithms;
  - research on e-control architecture (dependability of distributed systems, safe systems);
  - new model and technologies to select and improve a cost-effective condition-based maintenance policy practically;
  - how to integrate maintenance strategy with plant strategic business effectively;
  - research on pro-active maintenance, e-maintenance and e-diagnosis;
  - research on decision-making in maintenance for system performances optimising in sustainable environment.
- **Management of distributed, co-operative systems:**
  - research on multi-agent (holonic) systems;
  - development of ontologies for exchanging production-related information;
  - planning of negotiation mechanisms (Bruccoleri et al. 2003) and communication protocols;
  - development of models for describing production networks, behaviour analysis of the networks, development of efficient behaviour patterns;
  - development of an IT framework for supporting co-ordination and co-operation;
  - human factors in cross-organisational collaboration (cognitive and motivational aspects).

The above issues are matters of survival, independent of the size of the firms. New research and development projects of great importance are to be initiated on such important fields as:

- co-operative planning;
- multi-agent based (holonic) approaches for production networks;
- production ontologies;
- e-maintenance, e-diagnosis, e-control;
- integration of advanced maintenance and production processes for supporting sustainability;
- market mechanisms and communication protocols;
- co-ordination mechanisms inspired by social insects, nature and social behaviours in human societies;
- modelling and behaviour analysis of production networks;

- development of effective behavioural patterns;
- IT framework for supporting co-ordination and co-operation;
- portfolios of co-operation mechanisms.

## ACKNOWLEDGEMENT

The authors wish to acknowledge the European Commission for their support. They also wish to acknowledge their gratitude and appreciation to all the members (main, associated or simply interested) that participated to the diverse SIG2 events and more precisely as the leaders of each working group.

## REFERENCES

Bremer, C. F., 2000: From An Opportunity Identification to its Manufacturing: A Reference Model for Virtual Enterprises, Annals of the CIRP, vol. 49/1, 2000, pp. 325–329.

Bollinger J.G. et al.; A vision of Manufacturing for 2020; Proceedings of the IMS vision 2020 Forum, Beckman Center, California, USA, February 24–25; 2000

Bruccoleri, M., Lo Nigro, G., Federico, F., Noto La Diega, S., Perrone, G., 2003, Negotiation Mechanisms for Capacity Allocation in Distributed Enterprises, Annals of the CIRP, 52/1: 397–402.

Brugali D., Menga G., Galarraga G., 1998: Inter-Company Supply Chain Integration via Mobile Agents, In: The Globalization of Manufacturing in the Digital Communications Era of the 21st Century: Innovation, Agility, and the Virtual Enterprise. Kluwer Academic Pub. 1998, http://www.polito.it/~brugali/.

Camarinha-Matos L.M., Afsarmanesh H., Garita C., Lima C., 1997a: Towards an Architecture for Virtual Enterprises, Keynote paper, Proc. 2nd World Congress on Intelligent Manufacturing Processes and Systems, Springer, Budapest, Hungary, June 1997, pp. 531–541.

Camarinha-Matos, L.M., Afsarmanesh H., 1997b: Virtual Enterprises: Lifecycle Supporting Tools and Technologies, Handbook of Lifecycle Engineering: Concepts, Tools and Techniques, A. Molina, J. Sanchez, A. Kusiak (Eds.), Chapman and Hall

Camarinha-Matos, L., Afsarmanesh, H.: Virtual Enterprise Modeling and Support Infrastructures: Applying Multi-agent System Approaches. Lecture Notes in Artificial Intelligence, Vol. 2086. Springer Verlag Berlin Heidelberg (2001) 335–364

Croom, S., Romano, P. and Giannakis, M.; Supply chain management: an analytical framework for critical literature review European Journal of Purchasing & Supply Management, Volume 6, Issue 1, March 2000, Pages 67–83

Davis, R.; Smith, R., 1983: Negotiation as a Metaphor for Distributed Problem Solving, Artificial Intelligence, N 20, pp. 63–109.

Fox M.S., Chionglo J.F., and Barbuceanu M., 1993: The Integrated Supply Chain Management System, Internal Report, Dept. of Industrial Engineering, University of Toronto. The Integrated Supply Chain Management Project, http://www.eil.utoronto.ca/iscm-descr.html.

Iung B.; From remote maintenance to MAS-Based E-maintenance of an industrial process. International Journal of Intelligent Manufacturing. Eds. A. Kusiak. Special Issue on Internet-Based Distributed Intelligent Manufacturing Systems (Ed. Z. Banaszak). Vol.14 n° 1, February 2003a. pp 59–82. ISSN 0956-5515

Iung B., Morel G., Léger J.B., Proactive maintenance strategy for harbor crane operation improvement, Robotica, Special issue on Cost Effective Automation, Eds H. Erbe, vol. 21, issue 3, pp 313–324, June 2003b.

Iwata, K., Onosato, M., Teramoto, K., Osaki, S.: Virtual Manufacturing Systems as Advanced Information Infrastructure for Integrating manufacturing Resources and Activities, Annals of the CIRP Vol. 46/1/1997, pp. 335–338.

Koc, M, Ni, J., Lee, J., Bandyopadhyay, P., Introduction of E-Manufacturing, International Journal of Agile Manufacturing, Special Issue on Distributed E-Manufacturing, Vol. 6, Dec. 2003

Lau H.C.W., Chin K.S., Pun K.F., Ning A., 2000: Decision supporting functionality in a virtual enterprise network, Expert Systems with Applications, 19/ 2000, pp. 261–270.

Lee J, Teleservice engineering in manufacturing : challenges and opportunities. Int. Journal of Machine Tools & Manufacture. 38, 1998, pp 901–910

Lee J, E-manufacturing – fundamental, tools, and transformation, Robotics and Computer Integrated Manufacturing, vol. 19, 501–507, 2003

Léger, J.-B, Morel, G., "Integration of maintenance in the enterprise: towards an enterprise modelling-based framework compliant with proactive maintenance strategy", Production Planning and Control, Vol. 12 N° 2, pp. 176–187, 2001.

MacCarthy B. & Wilson J., The human contribution to planning, scheduling and control in manufacturing industry – Background and context. In: B. MacCarthy & J. Wilson. Human performance in planning and scheduling (pp. 3–14). London: Taylor & Francis. 2001

Mcfarlane D, Sanjay S, Chirn JL, Wong C Y, Ashton K. Auto ID systems and intelligent manufacturing control. Engineering Applications of Artificial Intelligence, 16. 365–376. 2003.

Macgregor, R. S.; Aresi, A.; Siegert, A., 1996: WWW.Security, How to Build a Secure World Wide Web Connection, IBM, Prentice Hall PTR.

Mezgár I., Kovács G.L., Paganelli P. 2000: Co-operative Production Planning for Small and Medium-Sized Enterprises, International Journal of Production Economics, No. 64, pp. 37–48.

Monostori, L., Márkus, A., Van Brussel, H., Westkämper, E., 1996, Machine learning approaches to manufacturing, Annals of the CIRP, 45/2: 675–712.

Monostori, L.; Szelke, E.; Kádár, B.: Intelligent techniques for management of changes and disturbances in manufacturing, Proceedings of the CIRP International Symposium: Advanced Design and Manufacture in the Global Manufacturing Era, August 21-22, 1997, Hong Kong, Vol. 1, pp. 67–75.

Monostori, L.; Váncza, J.; Márkus, A.; Kádár, B.; Viharos Zs.J., 2003, Towards the Realization of Digital Enterprises, 36th CIRP Int. Seminar on Manufacturing Systems, June 3-5, Saarbrücken, Germany: 99–106.

Monostori, L., Váncza, J., Márkus, A., Kis, T., Kovács, A., Erdős, G., Kádár, B., Viharos, Zs.J., 2005, Real-time, co-operative enterprises: management of changes and disturbances in different levels of production, 38th CIRP International Seminar on Manufacturing Systems, May 16–18, 2005, Florianopolis, Brasil, (CD version is available)

NIIIP 1999: Vision of the National Industrial Information Infrastructure Protocols (NIIIP), http://www.niiip.org/vision.html.

N.N., 2004, ManuFuture: A Vision for 2020, Assuring the Future of Manufacturing in Europe, Report of the High-Level Group, EC, November: 1–20.

Qiu R. et al., Extended structured adaptive supervisory control model of shopfloor controls for an e-Manufacturing system, Int. Journal of Production Research, vol. 41, n°8, pp 1605–1620, 2003

Ollero A., Morel G., Bernus P., Nof S. Y., Sasiadek J., Boverie S..; Erbe H. and R. Goodall; From MEMS to Enterprise systems, IFAC Annual reviews in Control, vol. 26, n° 2, pp 151–162, 2002.

Pétin J.F., Iung B., Morel G., Distributed intelligent actuation and measurement system within an integrated shop-floor organisation, Computer in Industry, Special Issue on IMS, vol. 37, issue 3, pp. 197–211; 1998

Rabelo, R.J.; Camarinha-Matos, L. M., 1995: A Holistic Control Architecture Infrastructure for Dynamic Scheduling, in Artificial Intelligence in Reactive Scheduling, Eds. Roger Kerr e Elizabeth Szelke, Chapman & Hall, pp.78–94.

Rabelo, R.J.; Camarinha-Matos, L. M., 1996a: Deriving Particular Agile Scheduling Systems using the HOLOS Methodology, International Journal in Informatics and Control.

Rabelo, R.J., 1997: A Framework for the Development of Manufacturing Agile Scheduling Systems – A Multi-agent Approach, Ph.D. Thesis, New University of Lisbon, Portugal.

Rabelo R., Spinosa M., 1997a: Mobile-agent based supervision in supply chain management in the food industry, Proceedings of Agrosoft'97 - Workshop on supply chain management in agribusiness, Belo Horizonte, Brazil, Sept. 97, pp. 451–459.

Rabelo R., Camarinha-Matos L.M., Afsarmanesh H., 1998a: Multiagent perspectives to agile scheduling, Proc. Of BASYS'98 - 3rd IEEE/IFIP Int. Conf. On Balanced Automation Systems, Intelligent Systems for Manufacturing (Kluwer Academic), pp. 51–66, ISBN 0-412-84670-5, Prague, Czech Republic, Aug 98.

Rabelo, R.J.; Camarinha-Matos, L. M., 1998b: Generic framework for conflict resolution in negotiation-based agile scheduling systems, Proceedings IMS'98 – 5 th IFAC Workshop on Intelligent Manufacturing Systems, Gramado – Brazil, pp.187-192.

Rabelo R., Camarinha-Matos L.M., Afsarmanesh H., 1999: Multi-agent-based agile scheduling, Journal of Robotics and Autonomous Systems (Elsevier), Vol. 27, N. 1-2, April 1999, ISSN 0921-8890, pp. 15–28.

Scholz-Reiter, B., Höhns, H., Hamann, T., 2004, Adaptive Control of Supply Chains: Building Blocks and Tools of an Agent-Based Simulation Framework, Annals of the CIRP, 53/1: 353–356.

Skołud, B., Zaremba, M., Scheduling the Repetitive Production: Constraints Propagation Approach, 2nd Conference on Management and Control of Production and Logistics, 5-8 July 2000, France.

Shen, W., Norrie, D.H., 1999: Implementing Internet Enabled Virtual Enterprises using Collaborative Agents, In Camarinha-Matos, L.M. (ed.), Infrastructures for Virtual Enterprises, Kluwer Academic Publisher, pp. 343–352, (http://imsg.enme.ucalgary.ca/ publicate.htm#ISG publications in 1999).

Shen, W., Xue, D., Norrie, D.H., 1998a: An Agent-Based Manufacturing Enterprise Infrastructure for Distributed Integrated Intelligent Manufacturing Systems, in Proceedings of PAAM'98, London, UK. (A hybrid agent-based approach for integrating manufacturing enterprise activities with its suppliers, partners and customers within an open and dynamic environment. Description of its functional architecture, main features and a prototype implementation.) (http://imsg.enme.ucalgary.ca/).

Shoham, Y., 1993: Agent-Oriented Programming, Artificial Intelligence, N 60, Elsevier, pp.51–92.

Keah Choon Tan, A framework of supply chain management literature European Journal of Purchasing & Supply Management, Volume 7, Issue 1, March 2001, Pages 39–48

Van Brussel, H., Wyns, J., Valckenaers, P., Bongaerts, L., Peeters, P., 1998, Reference architecture for holonic manufacturing systems, Computers in Industry, 37/3: 255–276.

Wiendahl, H.-P., Scholtissek, P., 1994, Management and control of complexity in manufacturing, Annals of the CIRP, 43/2: 533–540.

Wiendahl, H.-P., Lutz, S., 2002, Production in Networks, Annals of the CIRP, 51/2: 1–14.

Yoshikawa H. (1995). Manufacturing and the 21st Century – Intelligent Manufacturing Systems and the Renaissance of the Manufacturing Industry. Technological Forecasting and Social Change. Vol. 49 (2), pp195–213

*Advanced Manufacturing – An ICT and Systems Perspective – Taisch,*
*Thoben & Montorio (eds)*
*© 2007 Taylor & Francis Group, London, ISBN 978-0-415-42912-2*

# Organising and running real-time, co-operative enterprises

László Monostori[1,2], Rosanna Fornasiero[3] & József Váncza[1,2]

[1]*Computer and Automation Research Institute, Hungarian Academy of Sciences, Budapest, Hungary*
[2]*Department of Production Informatics, Management and Control, Budapest University of Technology and Economics, Budapest, Hungary*
[3]*ITIA-CNR, Via Bassini, Milano, Italy*

ABSTRACT:   Mass customisation and customised mass production are important, highly inter-related issues which are to be taken into account when organising and running real-time co-operative enterprises. The main aim of the paper is to illustrate some approaches and solutions for organisational design, planning and execution in this field, based on the *EUROShoe* European research project and the *VITAL* R&D project started in 2004, within the Hungarian National Research and Development Programme. On the basis of the experiences gained in the two projects, important research issues are enumerated, which are to be addressed in the coming future by the European research policy and solved within the frameworks of new research projects.

*Keywords*:   Supply chain planning, mass customisation, customised mass production, production planning, scheduling.

## 1   INTRODUCTION

Manufacturing systems work in a fast *changing environment* full of *uncertainties*. Increasing *complexity* is another feature showing up in production processes and systems, as well as in enterprise structures (Monostori 2003). One of the recent areas of research is related to the *globalisation* of production. *Production networks (PNs)* are formed from independent companies collaborating by shared information, skills, resources, driven by the common goal of exploiting market opportunities (Wiendahl and Lutz 2002).

According to the *ManuFuture* initiative, the ultimate goal of manufacturing is the *general transformation of all resources to meet human needs* (N.N. 2004). In this transformation process, the question all manufacturers have to answer time and again can be put simply as, how to produce what is needed, when required, not more, nor less, not earlier, nor later, just in the required quality. Finding appropriate answers is hard because market demand is uncertain and distributed, while production processes are complex involving geographically dispersed producers of raw materials, components and end products. Furthermore, customers have a tendency to expect their needs to be met almost immediately: typically, acceptable order lead times are much shorter than actual production lead times. This is the case when retailers require shipment of consumer goods within one day, or customers buying vehicles configured to order, expect delivery within a couple of days. All in all, decisions are to be made under the pressure of time, relying also on uncertain and incomplete information.

Taking high service level as their main priority, manufacturers can hedge against demand uncertainty only by maintaining time, capacity or material buffers, or all three. This however, incurs extra equipment, labour, inventory and organisational costs, as well as, especially under dynamic market conditions, the risk of producing obsolete components or products or both. Partners within a production network are legally independent entities, with their own resources, performance objectives and internal decision mechanisms. They have to find their own *trade-offs* between service level and cost that are acceptable for their markets and other partners. Such a solution can only emerge from the interaction of local and asynchronous decisions.

- Hence, there is an inevitable need to *design organisations* that can perceive actual market demand and respond to it by facilitating and sustaining co-ordination and co-operation among network members.
- Essential *planning problems* must be solved, since manufacturers would like to exercise control over some future events based on information that they either know for certain (about products, production technologies, resource capabilities, fixed orders, sales histories) or only anticipate (market demand, resource and material availability).
- *Execution* of production plans and schedules should be supported by real-time production control that is able to adapt plans to changing conditions in a proactive way, with minimal global effects.

The concept of the *digital enterprise*, i.e., the mapping of the key processes of an enterprise to digital structures by means of information and communication technologies (ICT) gives a unique way for managing the above problems (Monostori et al. 2003). By using recent advances in ICT, theoretically, all the important production-related information is available and manageable in a controlled, user-dependent way.

In what follows, examples of two industry-related projects are presented, showing the potential of ICT in realising real-time, co-operative enterprises. One of the cases focuses on the aspects of organisational design, and the other study concerns planning and execution. Both cases have in common that their ultimate goal is to satisfy customised demand on a market of *mass products*.

## 2  ISSUES OF ORGANISATIONAL DESIGN AND PLANNING IN MASS CUSTOMISATION

*Mass customisation* (MC) implies that the ultimate goal for a company is to detect the customer's needs and then to fulfil these needs with efficiency that almost equals mass production, using appropriate enabling technologies (Tseng and Piller 2003). There could be an individual demand for particular products (like footwear, garment, furniture, cars, etc.). *Customised mass production* on the other hand is a case where production is aimed at satisfying demand in a market of mass produced products (like cosmetics, food, mobile phones, light bulbs, low-technology electronics, etc.), but where demand appears for a complex and ever changing variety of products, both for small and large quantities, in hardly predictable temporal patterns (Simchi-Levi 2000). In both cases, a market is typically served by a manufacturer that works in a focal point of a production network where other nodes provide necessary raw materials, components and packaging materials. Market demand is transmitted to the manufacturer by distribution centres (DC). Some of the suppliers (e.g. packaging material providers) serve several manufactures acting in different markets (see Figure 1).

Guaranteeing extremely high service levels and, at the same time, keeping operational costs low, requires the integration of the traditionally extreme principles of mass production and customisation (Selladurai 2004). This requires that manufacturers change their vision on *organisation* and *production*. Some changes are already underway in large companies offering the possibility for the customer to personalise their own product, but for small and medium size producers this transition is much more difficult.

*Organisational re-engineering* was the main issue in the project aiming to develop ICT support for a mass customising company in the *footwear industry*. Here, different network members (shoe manufacturer, external designers, suppliers, component manufacturers, subcontractors and customers) involved in the shoe lifecycle needed to have their relationships re-engineered. The evaluation of the impact of such a strategy is based on the results of a European Union funded project, *EUROShoE* (G1RD-CT-2000-00343) where an integrated production plant for shoe design and mass customisation (DMC Lab) was conceptualised and implemented in an important Italian

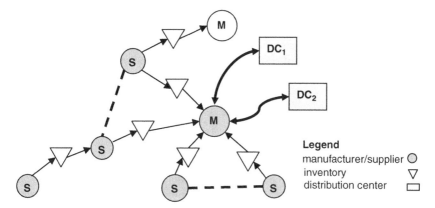

Figure 1. Structure of a production network.

footwear district. Based on the contributions of different European actors, innovative ICT plat-forms were developed and tested to integrate the production process with all the other processes of the shoe lifecycle like sales, design and distribution (Dulio and Boer 2004). In this paper, the different business models are presented for traditional shoe producers willing to implement mass customisation, along with strategic and economic evaluation of other approaches.

The second case study comes from a Hungarian research and development project called *VITAL: Real-time Co-operative Enterprises* aimed at developing ITC solutions for enterprises producing mass customised products in a common network. The main endeavours are as follows: the research and development of solutions from the level of production networks through single enterprises to production lines, which can ensure the optimal or near to optimal behaviour of the whole system, in a *real-time* fashion required by the given level of production.

## 2.1 *Common critical issues*

Mass customisation and customised mass production imply some generic organisational and tech-nological changes at company and production network level. The key common drivers of theses changes are as follows (see also Anderson 2004; Tu et al. 2001; Su et al. 2005):

- Manufacturers' main concern is the design, assembly and distribution of end products. They work in the centre of a focal supply network where suppliers deliver components and parts.
- Members of a production network are, however, autonomous, partly competing, and partly co-operating entities.
- There are difficulties in reaching economies of scale as product variety increases and batch sizes are reduced.
- Production is carried out on production lines that are able to process several products, but the changeover costs between them are significant.
- Changeover costs increase owing to the diversification of products. This should be counterbal-anced by innovative, new production systems.
- Effects of customisation should be postponed far along the production lifecycle as possible to reduce diversification costs.
- Distribution and transportation costs increase together with the increase of the batch numbers.
- Need of new competencies for the usage of new automated machines in addition to the consolidated skills at production level.
- Investments should be made in automating production equipment. At the same time, the workforce has to acquire new competencies and skills.
- Customer's information management requires novel software solutions.

135

These additional costs need to be counterbalanced by appropriate design for variety, product line planning, using a modular product family architecture, increasing the capability of production planning and control with the support of appropriate tools, and postponement strategies. Most important issue is the integration of customers into the value creation process.

Evaluating the impact of MC strategies on these expenditures can be counterbalanced with efficient ICT tools to provide the required autonomy in decision making and flexibility in job sequencing.

## 2.2 *Differences*

Although mass customisation raises some questions that call for common solution approaches, we have to emphasise also the differences. The most substantial differences are in the strategies to respond to market demand and in their consequences to *supply planning* and *inventory management*. The case studies presented here represent two extremes:

- In the footwear sector, customised demand is considered individual that is to be satisfied by a *make-to-order* (assembly-to-order) strategy.
- In the market of customised mass products – like in the second case study, of electric bulbs – demand for large quantities may appear instantly, often in a hardly predictable manner. Such demand can be satisfied only by the combination of make-to-order and *make-to-stock* strategies.

The costs of stock management (raw material, work-in-process and final product) for customised shoes *decrease* compared to standard shoes owing to the elimination of unsold finished product and to a better management of components. It is possible to consider that the amount of stock of raw material and work-in-process (uppers and components) decreases only when a strong investment is undertaken also in the ICT systems supporting the capability of the company to manage production in a more efficient way, reducing the amount of stocks at the end of the period compared to the amount which needs to be stored for standard shoes.

In the second case, a large number of product variants are produced by the manufacturer at the same time. Owing to the highly volatile market conditions, production is based mostly on *forecasts*. Part of the forecast information must be shared with the suppliers to decrease the well-known bullwhip effect (Lee et al. 2004), hence long-term relations and trust are prerequisites for managing the network. As a consequence, it can be considered stable: there are tight relations among the nodes (e.g. key supplier and customer partnering, dedicated warehouses, etc.), and there are few and rare newcomers. *Inventories* are inevitable to provide service at the required level and to enable local resource optimisation, but the partners in the network cannot store too much products or components because the *inventory holding costs* are high, some components are *perishable*, and some products or components may quickly became *obsolete*.

In what follows, the two case studies are presented that stress the organisational design and planning related issues of mass customisation, respectively.

## 3  MASS CUSTOMISATION: A CASE STUDY IN FOOTWEAR INDUSTRY

### 3.1  *Steps towards mass customisation in the footwear sector*

Typically the production process of shoe producers is concentrated on the final step – *making/assembling* – where the assembling of uppers, soles and other components is realised using a traditional production line. The other two main production steps that are preliminary to this – *cutting* and *stitching* – and the manufacturing of main components such as heels and soles are usually outsourced to external partners.

The most important investments which should be undertaken by a shoe producer to implement a MC strategy have been grouped in four scenarios according to the studies previously undertaken within the *EUROShoE* project (Fornasiero et al. 2005). The scenarios represent different degrees of innovation and capability to answer to MC requirements. They are incremental: starting from the minimal investment required for MC to the most complete level of investment. For each of the scenarios, additional costs and benefits for a traditional company are evaluated and estimated.

- **Scenario A**: This scenario represents the very first step that a company can make towards MC. It can be considered the minimum investment to produce customised shoes. To apply this strategy, and given the limited planning capability of an SME, the company can outsource a feasibility study to a service provider not only for the process of customising shoes but also for the product itself. This service supports the company in defining, for each phase of the shoe production, times and methods most convenient for customisation. In this context the company does not have to necessarily change its manufacturing processes but can simply modify the production organisation and the industrialisation of the customised shoes. In the realisation, the service consists of a simulation analysis based on the current production configuration of the shoe producer, itself considering standard and customised shoes (Fornasiero et al. 2004). The second step which is relevant to assure application of MC principles is changing the sales process: each shop, licensed to sell the brand, must be equipped with the scanning tool and the software for the customer's information management, developed within EUROShoE project to gather information about the shape of the customer's foot and their requirements.
- **Scenario B**: In this scenario, the company adds to the previous investments the purchase of machinery for the production of lasts (the solid form around which a shoe is molded), for the realisation of customised shoes. For the standard shoes, last production is outsourced as in the traditional shoe process and they can be reused for many different customers. For customised shoes it is important to have a personalised last taking into consideration the shape and fit of the foot. The fact that the company makes lasts internally, implies that it purchases also the CAD and CAM systems to integrate the shoe design process and the part program definition with the production of the last. The machine has to be developed within the project and integrated with CAD-CAM system.
- **Scenario C**: In this scenario the company is more innovative and decides to further improve the ICT at the company level adding the ERP, MES, PDM, SCM systems to manage and optimise the information flow concerning products, customers and suppliers. These ICT tools have been customised specifically for the footwear sector integrating them in a unique platform. Moreover in this scenario the company installs the new automated production line, dedicated to the customised shoes and needs to hire new workers for the cutting, stitching (which become internal activities) and for the assembling process. The assumption is that the company buys a new plant and the production of standard and customised shoes is run in parallel in two different lines. The new machines developed within the EUROShoE project permit the realisation, internally, of some phases which are usually outsourced and which acquire a new importance owing to the customisation of the shoe.
- **Scenario D**: In this scenario the company adds an innovative logistic system to manage the distribution of materials and components along the different machines and phases of the new production line. The logistic system developed within the project can be devoted not only for the storage of shoe components, processed both internally at the plant and externally by suppliers, but also to dispatch such components to different locations both in the stitching and in the making departments according to the planned schedule (Dulio and Boer 2004). This investment can increase the flexibility of the production system and the capability to answer to urgent orders.

In all the scenarios it is necessary to consider also the costs for external logistics for delivering customised shoes. Note that this cost is higher compared to the standard production, because the company uses courier services to ship everyday shoes to shops or direct to customers.

### 3.2 *Economic evaluation of mass customisation implementation*

All the technologies included in the above scenarios have been evaluated by forecasting cash flows of the suggested novel business models. Investments are assessed using the Net Present Value (NPV),

137

taking risk and uncertainty of the market into consideration when evaluating the probability of success and consistency of each scenario. This enables an understanding and assessment of the best conditions to apply MC in a footwear company according to its investment capabilities. It is assumed that the investment is analysed over a period of five years. In the model discounted cash flows are considered using rates (investment and inflation rate) according to the cost of the entire sector.

Costs have been supposed to be incremental and the model was created regarding data and information collected from:

- a traditional shoe producer (concerning standard costs),
- the production already undergoing at the DMC Lab (concerning production time, amounts, etc. for the innovative production line), and
- simulations carried on at the DMC Lab to forecast the behaviour of traditional and innovative production systems.

### 3.3 *Results*

According to the collected data and their relationships (variable costs are linked to demand level and other direct and indirect costs driven by quantity or price or both), each investment scenario based on the technologies of the project outcome gives the possibility to increase the NPV. The NPV of such investments for a traditional shoe producer is always above zero and comparing the four scenarios it is possible to see that the value of NPV is always growing according to the increase in the demand level which can be satisfied thanks to the introduction of new technologies which can compensate the increase in the cost level.

In the scenario A (Figure 2) the introduction of the scanning system for collecting information on the customer foot and the support received from a service provider like the DMC Lab for the feasibility study permit the company to sell a larger amount of shoes even though some part of the standard shoes is overtaken by the customised shoes. In scenario B, the cost of the machine for last production is compensated by savings on the outsourcing of last production and by the possibility to produce also custom made shoes, increasing a bit more the level of satisfied demand. In scenario C the big investment in ICT and production technologies permit the company to increase the number of customised shoes sold, but the increase of NPV is limited owing to large investments. Cost savings can be registered also in the quantity of initial stocks and in the amount of raw materials and components that the company manages to reduce thanks to a better organisational system.

Even though scenario D requires larger investments at the same time the total demand level that a shoe producer can satisfy is higher thanks to the higher flexibility of the system (gained with the logistics system) and the capability to answer also to urgent orders. The NPV benefits also from

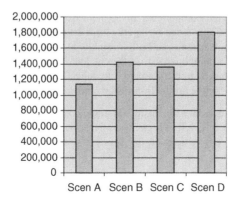

Figure 2. NPV for each scenario.

138

indirect impact on some other variables (like raw material management or shoe delivering time) which allows realising higher revenues.

In any scenario the investment has a payback period which is very dependent on the type of investment: the first two scenarios (A and B) requiring little investment can be quickly recovered while scenario C and D require longer payback period (between 2 and 3 years). The investments in scenarios C and D are more risky in the first year, but in the following years they recover completely and overtake the results of the most risk adverse scenarios, A and B (see Figure 3.).

The risk evaluation was carried on according to Monte Carlo simulation. The most important variables like quantity, price and discounted rate have been given a probability distribution according to some hypotheses on their statistical behaviour based on historical data evaluation.

Results of the four scenarios are shown in Figure 4, where it is possible to see the distribution of the NPV.

All the scenarios have a low risk level since the probability to have a negative NPV is zero. The scenario A and B have lower but more stable mean NPV value since their standard deviation is much lower than in the other two scenarios C and D. Higher variability in the C and D scenarios is linked to the higher level of initial investment required and they guarantee to reach also higher level values. The risk analysis permits evaluation of which are the most important input variables

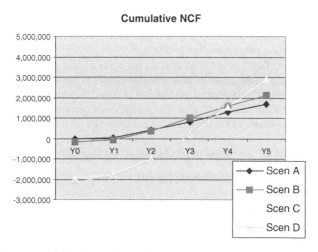

Figure 3. Cumulative net cash flow for each scenario.

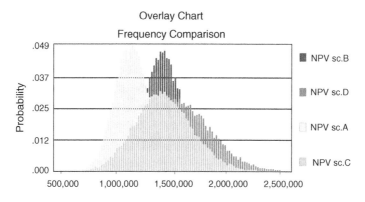

Figure 4. Comparison of the NPV distribution of the four Scenarios.

139

influencing the NPV in each scenario according to their behaviour and their correlation with the NPV distribution.

In scenario A, unit cost of raw materials for customised shoes largely influences the behaviour of the NPV as well as the discounted rate. The variability of quantity level of customised shoes has a positive influence and it is more or less the same for all the years. From scenario B also the unit cost for raw material and components for CM and standard shoes acquire more relevance on the variability of the NPV while in scenario C and D the variability of discounted rate influences the NPV owing to larger variability of the cash flows (see Figure 4 for scenario D). The model here applied is based on specific data that have been collected from the peculiar case of the *EUROShoE* project, but the risk analysis permits to draw general conclusions.

## 4 PLANNING IN A PRODUCTION NETWORK DEDICATED TO CUSTOMISED MASS PRODUCTION

One of the most *vital* features of factories is their ability for *co-operation* and *quick responses* to changes and disturbances. These are matters of survival, independent of the size of the firms.

The goal of the *VITAL* project is to research and develop new methods for the *real-time management* of complex technical and economic systems that work in changing, uncertain environments. Since the envisaged methods come from various, novel areas of informatics, operations research and knowledge-based systems, their integration is expected to balance the aspects of optimisation, autonomy, and co-operation (Monostori et al. 2005).

The orders are to be fulfilled in good quality, on the agreed price and on time. The customers do not necessarily realise that they usually face a conglomerate of firms, i.e., production networks. The importance of the time is illustrated by the watches in Figure 5, which incorporates the different levels (network, enterprise, production line) of the production expected to react on the external and internal changes and disturbances (indicated by thunderbolts) with a reaction time characterising the level in question.

The problems to be solved are as follows (referring to the notations of Figure 5):

- integrated production planning and scheduling (B),
- real-time production control handling changes and disturbances (C),
- management of distributed, co-operative systems (A).

The reason for the above sequence is that the high-level resource-management and scheduling of enterprises can give the basis, on the one hand, for the reliable, optimal or near to optimal

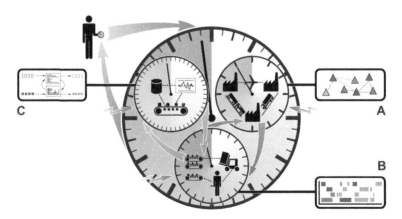

Figure 5.   General concept of the VITAL project (Monostori et al. 2005).

management of supply chains and production networks, and, on the other, for handling changes and disturbances on the shopfloor or in production lines.

*Integrated production planning and scheduling* requires the development of novel mathematical models and related solution methods. There must be found a bridge over the gap between theoretical end empirical scheduling methods: representation power and flexibility, as well as solution efficiency, are of key importance. This calls for the combination of operations research methods (scheduling theory, linear and integer programming) and artificial intelligence techniques (constraint programming). Since generating optimal solutions is almost hopeless, one has to be able to find solutions close to the optimum.

The function of *real-time production control* is to adapt the production system to the changing environment while preserving efficiency with respect to cost, time and quality requirements. Most of the change and disturbance handling subsystems of a shopfloor control system are based on the traditional hierarchical infrastructure of the controllers, and usually, their reporting schemes do not provide support for the disturbance handling process. Hence, this work focuses on proactive disturbance handling approaches for real-time manufacturing control, which not only report the deviations and problems of the manufacturing system, but also suggest possible choices to handle them. On the base of the current state of the system different possible actions can be taken (e.g. rescheduling, stopping the machine, re-buffering, fine-tuning, initiating a new set-up, initiating a correction or a maintenance process, restarting, etc.) Actions like these are going to be identified in the concept of *behaviour based control* (BBC) which applies knowledge-base techniques when making control decisions. A related problem is to find unknown relations in an existing production monitoring database that can be used for predicting faults and their consequences. The challenge is to develop and apply advanced *data mining* methods to find these unknown dependencies among the data in production monitoring data warehouses. Finding these relations, a knowledge-base can be built which allows the recognition of critical circumstances.

The aim of research concerning the *management of distributed, co-operative systems* is to develop planning methods that improve the logistic and production performance of the production network. The basic assumption is that although there exists a host of systems for managing the supply chain (SCM), these systems rather provide means for communication, information sharing and administrative functions and do not support decision-making with system-wide, far-reaching consequences (Fleischmann and Meyr 2003; Stadler 2005). Since production networks are unique and complex, and since local planning at the nodes is done in many different ways, there is no *one-size-fits-all* solution. Instead, there is a need for a portfolio of co-ordination mechanisms, where relations among the partners can be represented on a range of colours: from cold (competitive auctions, single business relations), through warm (co-operative planning), to hot (full integration). The main interest is in *co-operative planning*. Hence, it is assumed that partners have incentives and commitment to co-operate, to share their risks and benefits. The main driver for co-operation is *uncertainty*, which has its roots in market demand, manufacturing, and supply. Uncertainty can be managed only, if:

- Powerful planning systems fill in the various planner roles locally (Márkus et al. 2003; Váncza et al. 2004). Plans which are executable and cost efficient make the future – even market demand – more predictable, and the actual production more profitable.
- Novel information channels are established among the partners to share the results of local planning, from detailed production schedules up to demand plans, on all the horizons and levels.
- Partners possess sufficient inventory or capacity buffers or both. However, inventories are not only costly, but incur also the risk of obsolete production. Co-operation can be seen as a method of managing these buffers.

As a first step towards co-operative planning, a *multi-agent organisational model* has been developed for capturing the planning processes in a production network (Egri and Váncza 2005). The evolution of planning functions in production management, and recently, in supply network management resulted in a planning hierarchy (Fleischmann and Meyr 2003) that was adapted to

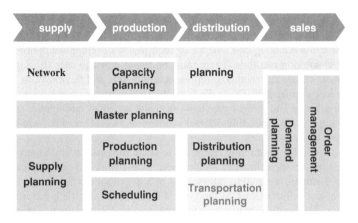

Figure 6.   The planning matrix.

the particular modelling purpose considered. This so-called *planning matrix* contains long-term, medium-term and short-term planning functions organised along the main flow of materials (see Figure 6). The functions are common at each node of a supply network, though, of course, they manifest themselves in different forms and complexity.

In the framework, planning functions (like demand planning, master planning, production planning and scheduling, supply planning, etc.,) and their interactions and information resources can be described on different horizons and aggregation levels. The analysis was focused first to the *master/production planning* and *scheduling* functions of the focal manufacturer in the supply chain. Some of the main findings are as follows: production planning and scheduling (PPS) match future production loads and capacities by determining the flow of materials and the use of resources, over various horizons and on different levels of detail. Although planning and scheduling problems have their own, specific time scale, resource and activity model granularity as well as optimisation criteria, the two levels of PPS are strongly interdependent. On the one hand, planning guarantees on the long term the observance of high level temporal and resource capacity constraints and thus sets the goals as well as the resource and temporal constraints for scheduling. On the other hand, scheduling is responsible for unfolding a production plan into executable schedules; i.e., to detailed resource assignments and operation sequences. No scheduling strategy can improve much on an inadequate plan, whereas a bad scheduling strategy that wastes resources may inhibit the fulfilment of a good plan. All this makes PPS extremely complex and hard to solve. At the same time, PPS calls for efficient decision support methods and intuitive, flexible models with fast, reliable solution techniques that scale-up well even to large problem instances. Hence, even if production planning and scheduling problems are solved in a superior-inferior hierarchy, they have to be treated in an integrated manner.

However, production plans and schedules of the focal manufacturer determine the behaviour of the whole supply network. The production plan of the manufacturer is, from another angle, demand forecast for the suppliers. If the plans are executable, i.e., they can be can be refined into executable schedules, then one main element of uncertainty can be removed from the forecasts towards the suppliers. Further on, the manufacturer needs the materials and components just in time when they are assembled into the final products. Reliable production schedules even with a short horizon can pull the supply, thus relieving the manufacturer from the burden of keeping and controlling inventories.

Finally, controlling the future by extended planning is possible only if accurate status information about past and present affairs are available. Hence, there is a potential for the application of *embedded* and *ambient* ICT to enhance the visibility of logistic and production processes by implementing active, intelligent identifiers and integrating information and material flows.

# 5 CONCLUSIONS

Among other things, *adaptive manufacturing, digital manufacturing, knowledge-based manufacturing*, and *networked manufacturing* were emphasised in the European initiative *ManuFuture*, which are to be handled in a holistic way. Two additional requirements, i.e., the *real-time* and *co-operative* capabilities in every level (plant, enterprise, production network) of the whole system were underlined in the paper. The first one refers to the ability of recognising and act on internal and external changes and disturbances within the time frameworks required by the given level of the manufacturing-production hierarchy. The second issue highlights that the complex production structures – from machine tools, robots, etc. to production networks, including the human beings involved – are increasingly considered and built up as autonomous, but co-operative entities.

*Mass customisation* and *customised mass production* are important, highly interrelated issues which are to be taken into account when organising and running real-time co-operative enterprises. The paper illustrates some solutions together with open issues for further research on organisational design, planning and execution in mass customisation and customised mass production, based on the *EUROShoe* European research project and the *VITAL* R&D project started in 2004, within the Hungarian National Research and Development Programme.

The model applied in the *EUROShoe* project considers not only product related costs but also all the costs related to the implementation of the new *services* for the customer (like sales support) and for the producer (like planning support). In such a context, more than the mere economic data, it is important for a shoe producer to understand which steps to undertake so as not to make mistakes and not to take risky actions. As shown in this work, this can only be realised through the detailed definition of constraints and assumptions for each new scenario and studying and evaluating all the variables concerned. In the model applied, investments are evaluated as convenient according to the degree of change a company is willing to apply.

On the other hand, the *VITAL* project stressed the importance of co-operative planning and real-time execution methods in production networks producing customised mass products. The new developments are not to challenge existing SCM systems. Instead, the goal is to identify and fill in some niches where co-operative logistic and production planning can be performed by building upon existing communication and administrative services.

On the basis of the experiences gained in the two projects, important research issues were enumerated, which are to be addressed in the future by European research policy and solved within the frameworks of new research projects.

## ACKNOWLEDGEMENTS

The authors would like to acknowledge the support of the European Commission for the *EUROShoE* project (G1RD-CT-2000-00343), the support for the *VITAL* project (NKFP grant No. 2/010/2004) as well as for the OTKA grants No. T046509 and No. T049481. This work has been partly carried on in the frameworks of these projects.

## REFERENCES

Dulio, S., Boër, C.R., 2004, Integrated Production Plant (IPP): an innovative laboratory for research projects in the footwear field, International Journal of Computer Integrated Manufacturing, 17/7 :601–611.

David M. Anderson, 2004, Build-to-Order & Mass Customization, the Ultimate Supply Chain and Lean Manufacturing Strategy for Low-Cost On-Demand Production without Forecasts or Inventory, CIM Press, 1-805-924-0200, www.build-to-order-consulting.com/books.htm.

Egri P., Váncza, J., 2005, Co-operative Planning in the Supply Network – A Multiagent Organization Model. In: Multi-Agent Systems and Applications IV (eds. Pechoucek, M., Petta, P., Varga, L. Zs.), Springer LNAI 3690, 346–356.

Fleischmann, B., Meyr, H., 2003, Planning Hierarchy, Modeling and Advanced Planning Systems, In: de Kok, A.G., Graves, S.C.(eds): Handbooks in Operations Research and Management Science, Vol. 11—Supply Chain Management: Design, Co-ordination and Operation. Elsevier, 457–523.

Fornasiero, R., Zangiacomi, A. Avai, A., 2004, Web cost simulation service for footwear sector, International Journal of Computer Integrated Manufacturing, Vol.17, No.7, 661–667.

Fornasiero R., Zangiacomi A., Tamborino C., Dulio S., Boër C.R., 2005, Evaluation of innovative business models for mass customisation in the Shoe sector, Proceedings of the World Congress on Mass Customisation and Personalization, 18–21 Sept 2005, Hong Kong

Lee, H.L., Padmanabhan, V., Whang, S., 1997, Information Distorsion in a Supply Chain: The Bullwhip Effect, Management Science, 43(4): 546–558.

Márkus A., Váncza J., Kis T., Kovács A., 2003, Project Scheduling Approach to Production Planning. CIRP Annals – Manufacturing Technology, 52(1): 359–362.

Monostori, L., 2003, AI and machine learning techniques for managing complexity, changes and uncertainties in manufacturing, Engineering Applications of Artificial Intelligence, Elsevier, The Netherlands, 16 (4): 277–291.

Monostori, L., Váncza, J., Márkus, A., Kádár, B., Viharos, Zs.J., 2003, Towards the realisation of Digital Enterprises, Proceedings of the 36th CIRP International Seminar on Manufacturing Systems, Progress in Virtual Manufacturing Systems, June 3–5, Saarbrücken, Germany: 99–106.

Monostori, L., Váncza, J., Márkus, A., Kis, T., Kovács, A., Erdös, G., Kádár, B., Viharos, Zs.J., 2005, Real-time, co-operative enterprises: management of changes and disturbances in different levels of production, 38th CIRP International Seminar on Manufacturing Systems, May 16–18, Florianopolis, Brasil, (in print, CD version is available)

N.N., 2004, ManuFuture: A Vision for 2020, Assuring the Future of Manufacturing in Europe, Report of the High-Level Group, EC, November: 1–20.

Selladurai, R.S., 2004, Mass Customization in Operations Management: Oxymoron or Reality? Omega, 32: 295–300.

Simchi-Levi, D., Kaminsky, Ph., Simchi-Levi, E., 2000, Designing and Managing the Supply Chain: Concepts, Strategies, and Cases. McGraw-Hill.

Stadtler, H., 2005, Supply chain management and advanced planning – basics, overview and challenges, European Journal of Operational Research 163: 575–588.

Su, J.C.P., Chang, Y-L., Ferguson, M., 2005, Evaluation of postponement structures to accommodate mass customization, Journal of Operations Management, 23: 305–318.

Tseng, M., Piller, F.T., 2003, The customer Centric Enterprise, The Customer Centric Enterprise, Advances in Mass Customization and Personalization, Springer ed., 3–16.

Tu, Q., Vonderembse, M.A., Ragu-Nathan T.S., 2001, The impact of time-based manufacturing practices on mass customisation and value to customer, Journal of Operations Management, 19: 201–217.

Váncza, J., Kis, T., Kovács, A., 2004, Aggregation – The Key to Integrating Production Planning and Scheduling. CIRP Annals – Manufacturing Technology, 53(1): 377–380.

Wiendahl, H.-P., Lutz, S., 2002, Production in Networks. CIRP Annals – Manufacturing Technology, 51(2): 1–14.

*Advanced Manufacturing – An ICT and Systems Perspective – Taisch,*
*Thoben & Montorio (eds)*
© 2007 Taylor & Francis Group, London, ISBN 978-0-415-42912-2

# Flexible and responsive cross-organisational interoperability

Georg Weichhart[1], Stefan Oppl[2] & Toni Wäfler[3]

[1]*Profactor Research, Am Stadtgut, Steyr, Austria*
[2]*University of Linz – Institute for Business Informatics – Communications Engineering,*
*Linz, Austria*
[3]*University of Applied Sciences Solothurn – Northwestern Switzerland, Olten,*
*Switzerland*

ABSTRACT: Inter-organisational co-operation holds a huge potential for exploiting the opportunities of dynamic, global markets – especially for SMEs. Furthermore, today's ICT potentially allows for quick exchange of information and for co-ordinating activities. However, state-of-the-art software architectures and technologies also hamper cross-organisational interoperability. A technologically determined need for extremely detailed modelling of business processes is a significant obstacle for establishing co-ordinated processes, which is a pre-requirement for scheduling and control in organisational networks. An approach that achieves the exploitation of technological potentials for efficient and effective cross-organisational processes without hampering flexibility is still to be developed. This requires simpler modelling and with it an easier adaptability of ICT on the one hand, and a better integration of human factors on the other hand. This paper reflects on existing ICT for co-operation, collaboration and co-ordination and identifies gaps and further research opportunities.

*Keywords*: Human factors, co-ordination, collaboration, cross-organisational processes, modelling.

## 1 INTRODUCTION

Within the frame of the Intelligent Manufacturing Systems Network of Excellence (IMS-NoE 2005) several Special Interest Groups (SIGs) were formed. The work presented here has to be seen within the context of SIG2 (Scheduling and Control in the Extended Enterprise) and Workgroup 3 (Co-ordination and Collaboration in Supply Networks) established within SIG2. The work in this workgroup focused on issues related to ICT support for people from different organisations, which are working together. Co-ordinating and scheduling people requires approaches that take the human factor into account right from the start. Human centred approaches are necessary for small and medium-size enterprises (SMEs).

SMEs account for most employment within Europe. While larger companies can leverage economies of scale for mass-produced goods, SMEs work in niche markets with lower volume and have to be more flexible to cope with changes in demands and technology (ENSR 2004). Around half of the SMEs collaborate with others to counter the pressure of the market. Modern information and communication technology (ICT) can provide support for co-ordination, co-operation and collaboration in organisational networks. Many large-scale enterprises already have strong ICT support for these issues. In contrast SMEs lag behind in the use of ICT in organisational networks (Nicola 2003, e-Gap 2004). One of the reasons is the high set-up cost of the software tools. These costs consist of the price of the product, the costs for customisation and adaptation of the software, the costs for adapting the organisation to the computer system, and training costs. Larger organisations use these systems for a longer period of time – therefore the set-up costs are acceptable. However the tools available today are not suitable to match the flexibility required by SMEs, to support dynamic establishment of joint activities, and least of all to support flexible cross-organisational scheduling and control mechanisms. Typically, inexperienced users also cannot set-up these tools and enter their knowledge without the help of specialised and trained personnel. Additionally, more human-centred approaches are required since in many SMEs co-operation and collaboration is dependent on interpersonal relationships and the capacity of these human resources.

This paper addresses requirements and needs of SMEs for ICT support for co-ordination, co-operation and collaboration in organisational networks. Special attention is given to the needs of the stakeholders and to the intelligibility of the involved business processes in all participating organisations.

First an overview of state-of-the-art ICT systems supporting co-operation, collaboration and co-ordination is given. In section 3 a vision is established for ICT systems that do not hamper the users in their need for flexible collaboration. Section 4 then compares and identifies the gaps between the existing technologies and the established vision. The following section introduces a language for ICT systems that provide better support for flexible collaboration. This work is then compared to results from work of the Information Society Technologies Advisory Group (ISTAG). The final section summarises the vision and future challenges.

## 2 STATE-OF-THE-ART

Several types of ICT systems have been developed, or are under development, aimed at supporting collaboration, co-operation and co-ordination in organisational networks. Although not uniquely distinguished in the literature, these terms are considered to have a different meaning and thus need to be differentiated. The purpose of *co-ordination* is to avoid gaps and overlaps in work that is distributed among single organisations within a network. Co-ordination includes resource planning and scheduling. *Co-operation* goes one step further and aims at obtaining common benefit by sharing or partitioning work. The most advanced form of interaction is *collaboration*, where the goal is to achieve collective results that the participating organisations would be incapable of accomplishing working alone (Pollard 2005).

Within the IMS-NoE SIG 2 on Scheduling and Control in the extended enterprise, Workgroup 2 focused on algorithms to optimise machine utilisation. In contrast to this, the work presented in this paper focuses on co-ordination of people. In addition to this, a distribution of control is assumed to exist within organisational networks (Weichhart and Fessl 2005).

Below, the main types of available ICT systems for co-ordination, co-operation, and collaboration are described.

### 2.1 *IT for virtual organisations*

*Virtual enterprises* (VE) or *virtual organisations* (VO) are temporal collaborative networks of organisations that share skills and resources to allow for fast reaction to business opportunities (Camarinha-Matos and Afsarmanesh 2001, Weichhart and Fessl 2005). The goal of such VOs is to enable the participating organisations to immediately react to customer demands and provide personalised services along the lifecycle of the product. Research on ICT for VOs focuses on the technical infrastructure to provide electronic services, and *ad hoc* and optimised access to these across organisations located at different places (Karagorgios et al. 2003).

### 2.2 *Computer supported collaborative work*

Research on Computer Supported Collaborative Work (CSCW) systems focuses on an ICT infra-structure that enables end-users to jointly work on a particular topic (Grundy and Hosking 1998). Sub-types of these systems can be identified by distinguishing whether the system supports co-operation and collaboration of people being either co-present, like Group Decision Support Systems, or distributed in place or time or both (Sprague and Watson 1995). Email for example is an ICT that supports asynchronous communication of people distant in time and space. Common to these tools is the characteristic that they support joint activities by providing a communication infrastructure, for example, server-based tools like shared virtual desktops.

### 2.3 *Workflow management systems*

A third type of ICT aims at providing infrastructure for a more structured and controlled cross-organisational collaboration. These tools focus on supporting processes within as well as processes among organisations. Workflow Management Systems (WFMS) and Process-Centred Environments (PCEs) support the design and enactment of work processes across a number of organisations (Grundy and Hosking 1998, van der Aalst and van Hee 2002). Currently much research in this domain is in the area of e-business and web-services (cf. Leymann and Roller 2005, OASIS WS BPEL TC2005, Web Services Choreography Working Group 2005). This work is related to the VO concept, as the vision is to establish an infrastructure for dynamic service discovery and enactment across organisations (e.g. Buhler 2004, Vidal et al. 2004). In contrast to VO related research, the ICT infrastructure is not the focus of the work, but instead the representation of the co-ordination itself in form of workflows, including a representation that is human readable.

Tools like Serendipity (Grundy and Hosking 1998) try to bridge the gap between CSCW and WFMS types of systems. Serendipity lets multiple users concurrently work on the design of a common work process with support of a virtual shared work environment.

There are also initiatives to standardise business processes to facilitate cross-organisational collaboration (cf. e.g. the Collaborative Planning, Forecasting and Replenishment model (CPFR) as suggested by the Voluntary Inter-industry Commerce Standards Association (VICS) (http://www.vics.org/committees/cpfr/)). However, such initiatives although being implemented successfully in pilot projects, have been adopted by industries much more slowly than expected (Smaros 2004).

### 2.4 *Enterprise modelling*

Currently software frameworks for enterprise modelling (EM) are in the focus of research on inter-operability of IT systems (Vernadat 2002, Piddington 2005, Berio et al. 2005). Using a common language, users model their organisations, products, technical resources and especially the properties of their internal processes and IT infrastructure. Enterprise models are used as a semantic unification mechanism (Vernadat 2002). Figure 1 indicates this below.

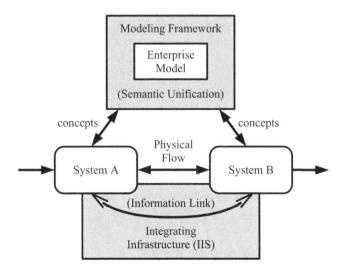

Figure 1. Model-based enterprise integration principles (Vernadat 2002).

EM is a model driven approach where graphical languages allow end-users to build models without the need for a specialised programming language. Again these models also help in communicating requirements of internal processes and IT infrastructure to (potential) partners and establish links among existing IT systems. In contrast to WFMS and PCE approaches as described above, the models are designed independently and are subsequently used to exchange data.

### 2.5 *Electronic performance support system*

More value on integrating end-user and on knowledge transfer is set in the domain of process oriented knowledge management and with electronic performance support systems (EPSS) (Heftberger and Stary 2004). Electronic performance support aims to empower stakeholders to accomplish their work tasks accurately, namely adapted to their individual perception of work.

(Stary and Stoiber 2003) have integrated model-driven and task-based EPSS approaches in a framework for process-oriented knowledge management, which resulted in the software system *KnowIt* (renamed to *ProcessLens* later) (Heftberger and Stary 2004, Oppl and Stary 2005). This work already tries to bring support for co-ordination of activities to the end-users.

### 2.6 *End-user programming*

A different approach, having a similar aim as the one described above, is end-user programming. Mehandjiev (1997) has applied this paradigm to the workflow domain. Members of organisations, who are experts in their domain, are empowered with the ECHOES system to directly control and modify their software systems by the use of a graphical language.

## 3 VISION

Enabling cross-organisational collaboration for SMEs is the vision. The aim of such collaboration is to increase the capacity to act of each participating SME thereby adding to its competitiveness. The vision is clearly not to support the design of highly sophisticated and complex cross-organisational processes. Rather it is the creation of boundary spanning understanding as a source for a co-ordinated individual acting, thereby creating synergies. Furthermore – since uniqueness is considered a source of competitiveness – it allows for and even furthers individual mental models and personal approaches as well within and across organisations. Nevertheless, it creates opportunities for a better tuning of situated individual acting.

Additionally the following issues and features characterise the enabler:

Hard features:

- It is independent of IT solutions used by participating SMEs;
- It only requires minor changes, if any at all, in the specific business processes of participating SMEs;
- It can be easily set-up, implemented, and handled by participating SMEs themselves; no specialised personnel are required;
- It is easily adaptable to changes in processes as well as in the organisation or in the technical infrastructure of participating SMEs;
- It is open; i.e. new partners can easily be integrated.

Soft features:

- It supports trust building among participating SMEs;
- It does not interfere with the given (im)balance of power among the participating SMEs;
- It considers the heterogeneity in the needs of the participating SMEs.

## 4 GAPS

The approaches described in section2 show some success in supporting cross-organisational col-laboration. However, they strongly focus on technical aspects of co-ordination, co-operation and collaboration, aiming at a direct data exchange among IT systems (in VO and EM approaches), at the provision of a shared electronic working environment (in the approaches of the WFMS family), or at the provision of a technical communication infrastructure (in CSCW approaches). This technical bias may also hamper the exploitation of the collaboration opportunities. In the United Kingdom, 80–90% of IT investments do not meet their performance objectives (Clegg et al. 1997). The reasons for this are rarely purely technical. They rather originate in the ways IT is developed and implemented as well as in the negation of a range of human and organisational factors (Clegg 2000). This is supported by (Smaros 2004). In case studies on collaborative fore-casting she found that required investments in technology may slow down collaborative practices, but are not considered to be critical obstacles. Much more critical are different needs of the partner organisations as well as limited resources and a lack of adequate processes even within the single organisations.

In contrast to VO and EM the approaches of the WFMS family may help in integrating such differences, as they are not (only) aiming at a technical integration, but at a common work process. However, there are some significant non-technical obstacles for collaborative design (Mohtashami et al. 2003). Co-operating organisations very often differ regarding communication mode and culture. Also power and trust play important roles. Obstacles like these are hardly ever considered in respective concepts.

Different from the other approaches CSCW supports human-to-human interaction. Nevertheless respective ICT solutions are normally not task specific. As a consequence, CSCW may sup-port cross-organisational co-operation and collaboration as it provides additional and probably more efficient communication possibilities. However it does not consider specific needs of cross-organisational co-ordination such as a communication of anticipated events; a common notification of options, and a communication of opportunistic planning and the like (cf. Windischer and Grote 2003).

EPSS aims at enabling users to model their individual perception of the task to be accomplished. However, existing approaches lack an intuitive intelligibility of the used visualisation methods. As a consequence the modelling of processes which fully represent (subjective) reality is hampered. This causes a limited usefulness of the models.

End-user modelling allows users to use a graphical interface to accomplish programming tasks. However, this type of systems can cope with limited complexity only. Furthermore, users build applications for themselves. Consequently gaining understanding across company borders is not achieved.

Common to all these ICT systems is that they provide either no direct support for collaboration or that they support the establishment of a common bird's eye view of the (cross-organisational) processes. The latter may support users in reaching an overview about the involved process and hence facilitates the flow of knowledge across partners (Heftberger and Stary 2004, Weichhart and Fessl 2005). Such a bird's eye view of a supply chain is depicted in Figure 2 on the left-hand side. Respective models assume that all enterprises have a common conceptualisation and shared mental models (e.g. the SCOR model (Supply Chain Council 2004)). Tools supporting these centralised views allow for a better communication of the underlying structure of the joint process. Consequently they support the identification of structural problems, like critical paths. As a common picture for all participants is drawn to reach an understanding of the structure, all participants use the same model. Therefore these models are also used for resource co-ordination and schedule optimisation within the supply chain.

What is missing with such models is to provide the end-users with support for their individual and actual circumstances and personal styles. Abstract centralised models of a single supply chain ignore for example the fact that most suppliers are taking part in a number of supply chains (cf. Figure 2 right hand side).

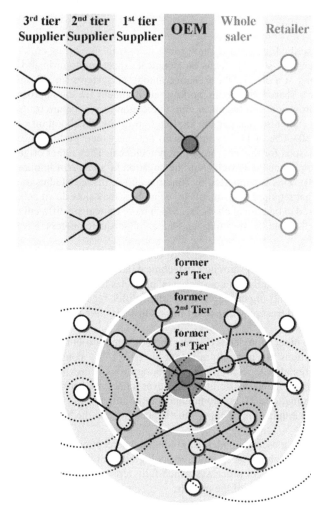

Figure 2. Bird's eye view of a single supply chain versus the individual view of all supply chains in which a particular organisation is involved (Weichhart and Fessl 2005).

The centralised approaches also force users to map their individual mental models and their organisational knowledge to representations, which often lack the capability to explicitly express this information. Hence, users on the one hand do not find their view on the process captured in the model and on the other hand are hardly able to identify issues crucial for cross-organisational process tuning that are not directly educible with traditional process notation. This leads to a setting, where it is hard to develop a common understanding as a basis for an effective and efficient cross-organisational business process (cf. Figure 3).

Furthermore, if such systems are used, each new project partner causes the current cross-organisational processes to be outdated and enforces the communication, design and set-up of these processes from scratch. This contradicts the competitive advantage of SMEs: their flexibility and reactivity.

In general what is missing to support SMEs are ICT tools and accompanying methodologies, which are closer to the end-users needs. They should be easy to learn, easy to set-up, and

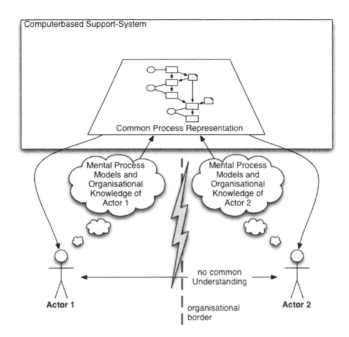

Figure 3.   Current ICT support for exchange of knowledge.

easy to manage. Software developers need to give the end-user more focus, since collaboration, co-ordination, and co-operation is among humans, where ICT only plays a supportive role.

## 5   METHODOLOGY APPLIED

In this section an abstract architecture is described, which provides support for the properties described in the vision section and tries to bridge the identified gaps.

### 5.1   *Introduction*

In contrast to the setting depicted in Figure 3 an ICT-supported system as shown in Figure 4 is envisioned.

Every involved user holds his individual set of processes representing his view on the processes and knowledge of his organisational sphere. A (graphical) notation is needed to allow the introduction of contextual symbols to represent knowledge unique to the respective organisation's philosophy and concepts. The information has to be captured in an abstract form (e.g. using formalised rules and constraints) by an ICT back-end system, so the end-users do not have to build large detailed models.

### 5.2   *Language*

The language and the according notation presented in detail in Oppl and Stary (2005) have these desired properties. The figure below shows an exemplary abstract process description using this notation.

There are four fundamental modelling elements, which are common to all diagrams:

- activities (red);
- roles (green);

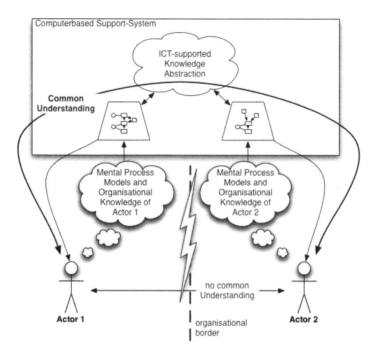

Figure 4.  Proposed approach for human-centred Knowledge exchange.

- data (yellow);
- goals (orange).

The first three elements enable modelling of the relevant aspects of the user- and data-perspective of a process. The goal enables users to remain at a level of abstraction, with no need to detail information about sub processes immediately. The explicit modelling of goals also supports evolutionary modelling and supports users to keep the objective in mind.

In addition to the fundamental modelling elements this language supports the creation of user dependent modelling elements. These elements allow the inclusion of additional information to represent context. In the example shown in Figure 5, the *shamrock* element shadowing the *product manager* role represents such a context information type.

Using the fundamental elements the shamrock could be modelled as a set of five roles. But this (syntactical) simplification does not convey the meaning (semantics) of the shamrock symbol. The stakeholders working in the process sketched above easily realise that activities where the shamrock is placed involve all five roles as a whole. The context elements allow more complex issues to be represented as a single symbol.

The possibility to let users customise symbols enhances the expressiveness of the notation and the resulting representations, since tasks and work activities can be captured in their *organisational context*. Contextual information symbols consist of two building blocks: *structural information* and *normative information*, which complement each other in representing a *contextual information symbol*.

That part of the symbol that can be represented using fundamental modelling elements forms its *structural information*. In the example considered, these are the five roles representing the involved departments.

*Normative information* represents those parts of the symbol that can be defined using rules and constraints. In the example above this is the information that all departments have always to be consulted in a concerted way. The formal representation of these rules and constraints at both layers is still subject of continuing research.

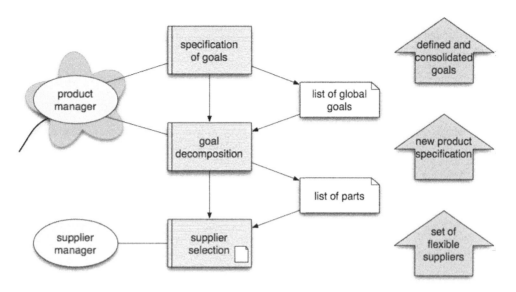

Figure 5. Example application of the language for abstract process description.

### 5.3 Co-ordination support

In case of collaboration between two organisations, the respective internal processes are revealed to each other including a description of the contextual symbols, which enables the development of a common understanding of the partner's processes. However, the extent of the revealed internal process is controlled by the owning organisations horizontally (number of revealed processes) as well as vertically (level of abstraction of revealed processes). The task of tuning the internal processes to form a cross-organisational process can be supported by the ICT system using the knowledge (rules and constraints) derived from formalised contextual symbols. A method for formalisation of the contextual elements is subject to continuing research, to allow ICT support for co-ordination of processes modelled using this graphical language.

The use of separated individual models and a back-end ICT system allows the implementation of all hard features requested in section 3:

- *It is independent of IT solutions used by participating SMEs.*
  This can be ensured by providing legacy interfaces to the framework. An additional ICT-driven abstraction layer simplifies the implementation of such interfaces. The approach to enable modelling close to the mental models of end-users can also be applied to legacy systems.
- *It only requires minor changes, if any at all, in the specific business processes of participating SMEs.*
  By explicitly starting with the internal processes of an organisation and tuning them to form a cross-organisational process as well as keeping the users in control of their processes and the respective public visibility (and thus alterability), this requirement can be fulfilled.
- *It can be easily, set-up, implemented, and handled by participating SMEs themselves; no specialised personnel are required.*
  This has to be ensured by the implementation and not by the concept. However, an intuitively intel-ligible visualisation raises the acceptance level within organisations and lowers the probability of faulty operation.
- *It is easily adaptable to changes in processes as well as in the organisation or in the technical infrastructure of participating SMEs.*

With the different levels of abstraction available in this approach, this requirement can be easily met through hiding the concrete, detailed process implementation under an abstracted view on the process for all other participants.

- *It is open; i.e. new partners can easily be integrated.*
Using the individual models as a starting point to form a cross-organisational process easily allows the integration of new partners in contrast to approaches, where cross-organisational processes are designed from scratch.

However the soft features, which are rarely considered in traditional approaches, are taken into account, by providing each single organisation with control over its knowledge and the amount of information to be revealed dynamically.

The language presented here is a response to the demand for a more *intelligent*, more sophisticated back-end ICT system, which goes further than just providing the technical infrastructure for cross-organisational process modelling. Part of this is subject to continuing work.


## 6 RESULTS

Visions and projects in the realm of Ambient Intelligence (AmI) have been already established years ago. Already there it was recognised, that personalised support for people working with ICT is required (ISTAG Scenarios for AmI 2001, Shadbolt 2003).

The ISTAG report on IST research content (ISTAG 2003) has defined nine ambient intelligence application areas (ISTAG, ftp://ftp.cordis.lu/pub/ist/docs/istag-wg1-final_en.pdf). The one being of interest for this work is on *improvement of business processes*:

> *"...opportunities for AmI in this area are considerable particularly with regard to the set-up, operation and dissolution of virtual enterprises and to the fluid configuration of business processes and the seamless inter-operation of underlying information systems. Enhanced interoperability of ICT platforms for commercial collaboration and mechanisms to allow these environments to be easily re-configured on the fly will allow companies to participate easily in several networks simultaneously without the need to radically alter their company cultures and preferred methods of working."*

The importance of these statements needs to be highlighted, and the thoughts and concepts presented in this paper are in line with the general request in the field of AmI, that the user is at the centre of the research work. However the recommendations of the ISTAG focus on single users and their interaction with their individual ICT platform ("...allow these environments to be easily re-configured on the fly"). Regarding collaboration and co-ordination it seems that the issue is only seen as a technical one, namely the *seamless inter-operation of underlying information systems*.

Research results show that more research is needed regarding the factors involved in reaching a common understanding and seamless collaboration *of people* distributed across different organisations. While the ISTAG also has put forward recommendations regarding trust and related issues, in general, the collaboration aspects have also to be seen in their context of use, for example a cross-organisational process in a supplier network.


## 7 FUTURE CHALLENGES

The vision is to enable cross-organisational collaboration. The suggested abstract architecture is suggested as a suitable basis for achieving this vision. However, there are still many open questions. As described above, current research work aims to provide answers to some of these questions. Other questions still remain to be tackled. The largest gap concerns the human factor in cross-organisational collaboration. Although the suggested approach incorporates human factor aspects there is still too little known about it. What are required are a deeper understanding of cognitive and motivational aspects of cross-organisational collaboration as well as much more knowledge about its consequences on the design of processes and the technical infrastructure. This includes the need for further research in the use of ICT for manufacturing in networks. There is, especially in this field, a high potential for ICT support for SMEs.

ACKNOWLEDGEMENT

The authors wish to acknowledge the European Commission for their support. The work has been taken place within the frame of the IMS-NoE (http://www.ims-noe.org).

REFERENCES

Berio, G.; Mertins, K.; Jaekel, F. W.: Common Enterprise Modelling Framework for Distributed Organisations. In: Horáček, P.; Šimandl, M.; Zítek, P. (Eds.): Preprints of 16th IFAC World Congress. Praha, 2005

Buhler, P.: A Software Architecture For Distributed Workflow Enactment With Agents And Web Services. PHD Thesis, Department of Computer Science and Engineering, College of Engineering and Information Technology, University of South Carolina, 2004

Camarinha-Matos, L. M.; Afsarmanesh, H.: Virtual Enterprise Modeling and Support Infrastructures: Applying Multi-agent System Approaches. M. Luck et al. (Eds.): ACAI 2001, LNAI 2086, pp. 335–364, Springer-Verlag Berlin Heidelberg, 2001

Clegg, C,: Sociotechnical principles for system design, in: Applied Ergonomics, 31, pp. 463–477, 2000.

Clegg, C.; Axtell, C.; Damodaran, Leela; Farbey, B.; Hull, Richard; Lloyd-Jones, Raymond; Nicholls, John; Sell, Reg; Tomlinson, Christine: Information technology: a study of performance and the role of human and organizational factors. Ergonomics, 40 (9), pp.851–871, 1997.

European Network for SME Research (ENSR): 2003 Observatory of European SMEs. http://europa.eu.int/comm/enterprise, European Communities, 2004

eGap: European Union-funded research project, http://www.egap-eu.com/, 2004

Grundy, John, C.; Hosking, John: Serendipity: Integrated Environment Support for Process Modelling, Enactment and Work Co-ordination. Automated Software Engineering: An International Journal Vol. 5, No. 1. pp. 27–60. Kluwer Academic Publishers, 1998

Heftberger, S., Stary, C.: Partizipatives organisationales Lernen - Ein prozessbasierter Ansatz. Wiesbaden : Deutscher Universitäts-Verlag, 2004.

IST AG: Information Society Technologies Advisory Group Reports 1999–2004. http://www.cordis.lu/ist/istag-reports.htm, 1999–2004

Karageorgos, A.; Mehandjiev, N.; Weichhart, G.; Hämmerle, A.: Agent-based optimisation of logistics and production planning. In: IFAC journal Engineering Applications of Artificial Intelligence Volume 16, Issue 4, Pages 271–393, June 2003

Leymann, F.; Roller, D.: Modelling Business Processes with BPEL4WS. Information Systems and e-Business Management (ISeB), Springer, 2005

Mehandjiev, Nikolay D.: User Enhanceability for Information Systems through Visual Programming. PHD Thesis, University of Hull, 1997

Mohtashami, M.; Fadi, P.; Il, Im: Critical factors of collaborative software development in supply chain management. In: Seuring, Stefan; Müller, Martin; Goldbach, Maria; Schneidewind, Uwe (Eds.): Strategy and organization in supply chains, pp. 257–272, Physica-Verlag Heidelberg 2003

Di Nicoal, P.: ICT and e-Work in European SMEs. In Cunningham, Paul; Cunningham Miriam; Fatelnig, Peter (Editors): Building the Knowledge Economy, IOS Press, 2003

Nunes, N. J.; Falcão e Cunha, J. (Eds): Interactive Systems. Design, Specification, and Verification, Springer LNCS 2844, Pages 258–272, 2003

OASIS Web Services Business Process Execution Language (WSBPEL) Technical Committee. http://www.oasis-open.org/committees/tc_home.php?wg_abbrev=wsbpel, 2005

Oppl, S.; Stary. C. : Towards Human-Centered Design of Diagrammatic Representation Schemes. In Proceedings of the 4th International Workshop on Task Models and Diagrams for User Interface Design (TAMODIA 2005), Gdansk, Poland

Piddington, C.: SME Interoperability in the Global Economy: A Discussion Paper. In: Horáček, P.; Šimandl, M.; Zítek, P. (Eds.): Preprints of 16th IFAC World Congress. Praha, 2005

Pollard, D.: Will that be co-ordination, co-operation, or collaboration?, In: How to save the world. http://blogs.salon.com/0002007/2005/03/25.html, 2005

Shadbolt, N.: Ambient Intelligence. IEEE Intelligent Systems, July/August, 2003

Smaros, J.: Forecasting collaboration in the European grocery sector: observations and hypothesis. Internet publication: http://www.tuta.hut.fi/library/working_paper/pdf/working_paper_smaros.pdf, 2004.

Sprague, R. H., Jr.; Watson, H. J.: Decision Support for Management. Prentice-Hall Inc., Upper Saddle River, 1995

Stary, C.; Stoiber, S.: Model-based Electronic Performance Support. In Jorge, Joaquim A.; Supply Chain Council: SCOR Version 6.0 Reference Guide. http://www.supply-chain.org, 2004

van der Aalst, W.M.P.; van Hee, K. M.: Workflow Management: Models, Methods, and Systems. MIT press, Cambridge, MA, 2002

Vernadat, F.B. Vernadat: Enterprise Modeling And Integration (Emi): Current Status And Research Perspectives. Annual Reviews in Control Volume 26, Pages 15–25, 2002

Vidal, J. M.; Buhler, P.; Stahl, C.: Multiagent Systems with Workflows. IEEE Internet Computing, January–February 2004

Web Services Choreography Working Group: http://www.w3.org/2002/ws/chor/, World Wide Web Consortium, 2005

Weichhart, G.; Fessl, K.: Organisational Network Models and The Implications For Decision Support Systems. In: Horáček, P.; Šimandl, M.; Zítek, P. (Eds.): Preprints of 16th IFAC World Congress. Praha, 2005

Windischer, A.; Grote, G.: Success factors for collaborative planning. In: Seuring, S.; Müller, M.; Goldbach, M.; Schneidewind, U. (Eds.): Strategy and organization in supply chains, pp. 131–146, Physica-Verlag Heidelberg 2003

*Part V*
*Benchmarking and performance measures*

In the 21st century, industry will continue to be about creation of value through innovation and improvement of products and processes. Manufacturing will have to deal with greater degrees of complexity, uncertainty and change as product lifecycles become shorter; multi-science products will start to emerge (products with knowledge, IT, biotech, chemical, mechanical, electrical, etc., content); the value content of the manufactured artefacts will be relatively small compared to the value of the service or knowledge content or both, associated with the artefact; sustainability of the production system (including product design, recycling, returns management) and agility will become important challenges.

All these developments will lead to the emergence of knowledge-based extended enterprises: their competitive strategy will not be based on the ability to make a specific artefact, but rather on the proper exploitation of shared competencies and capabilities in developing a product, making the product, packaging it within a service proposition and customising it to specific customer needs, i.e. to support mass customisation.

The collaborative nature of an enterprise together with the need for mass customisation will lead to the development of more distributed industrial systems facilitated through scalable technologies and knowledge-based products and services which will move manufacturing departments close to the customers, or even within the customer. Collaboration provides a mechanism by which risks can be shared and opportunities maximised by bringing together the right mix of skills and competencies and creating a critical mass.

Researches to date have identified several reasons why collaborations fail and one of the main reasons is the absence of an integrated performance management system that takes a holistic approach to extended enterprises. It is believed that there is a lack of mature knowledge and practical guidelines on *how to measure and manage global manufacturing networks*.

Within the IMS-NoE, the Special Interest Group on *Benchmarking and Performance Measurement* centred its activities and discussion among members along two main research streams linked to the area of performance measurement systems:

(a)  Performance Measurement Systems in Extended Enterprises
(b)  Performance Measurement Systems and Benchmarking of Production Scheduling Systems

In particular, with regards to the latter area, the discussion on PMS has been instrumental for the design and development, within the activities of IMS-NoE, of a Benchmarking Service. This service would provide a virtual environment where production scheduling solutions, created by scheduling researchers, but also by commercial scheduling software vendors, could be tested and evaluated on an *emulated* production system, provided by members of the industrial community, to identify the best scheduling solution for a specific manufacturing scenario the company has to deal with.

Two contributions are included in this section of the book. The first contribution, *Performance measurement systems in manufacturing: state-of-the-art and future trends*, by Sergio Cavalieri, Umit Bititci, Paul Valckenaers, Romeo Bandinelli and Kepa Mendibil provides a comprehensive state-of-the-art on performance measurement systems along the two main research streams investigated within the SIG. The contribution also envisions future trends and highlights the main issues and challenges for the topic. The paper emerged from the discussion and activities carried out by researchers, experts, and practitioners attending the seven meeting organised by the SIG and papers presented at several international conferences.

The second contribution *PRODCHAIN – Supporting SMEs Participating Successfully in Production Networks*, by Robert Roesgen, reports the main findings and results of the IMS project PRODCHAIN (IST-2000-61205). The paper is in particular focused on the description of the PRODCHAIN toolbox, a tool that intends to support SMEs in reorganising their customer supplier business processes to the needs of modern production networks.

*Advanced Manufacturing – An ICT and Systems Perspective – Taisch,*
*Thoben & Montorio (eds)*
*© 2007 Taylor & Francis Group, London, ISBN 978-0-415-42912-2*

# Performance measurement systems in manufacturing: State-of-the-art and future trends

Sergio Cavalieri[1], Paul Valckenaers[2], Umit Bititci[3],
Romeo Bandinelli[4] & Kepa Mendibil[3]

[1] *University of Bergamo, Dalmine (BG), Italy*
[2] *K.U.Leuven, Leuven, Belgium*
[3] *University of Strathclyde, Glasgow, UK*
[4] *Università di Firenze, Italy*

ABSTRACT: Several research activities and industrial projects have been undertaken in the area of performance measurement systems. The main reason for this is acknowledgement by practitioners and researchers that to improve it is first necessary to understand the present position and circumstances. The rising strategies, policies and methodologies of the last two decades, even from distant fields, such as Business Process Re-engineering, Strategic Benchmarking, Total Quality Management, Balanced Scorecard, Six Sigma, Total Productive Maintenance, have strongly highlighted the need, before acting, to assess the performance of the processes to be controlled, whichever the level they belong to, whether at supply chain or extended enterprise stage or, at the other extreme, at single production or simple machine level.

Indeed, what is often missing in the industrial practice is a sound alignment among the performance metrics, designed and adopted at the different levels of the decision making processes of a manufacturing company. This gap is mainly owing to the fact that a consistent and effective Performance Measurement System (PMS) requires, in its development, the consistent contribution of different expertise, encompassing the complex, multifaceted and interdisciplinary nature of an industrial company.

This is also what the Special Interest Group on Benchmarking and Performance Measurement, established within the IMS-NoE, has experienced during its activities and discussed in its meetings and workshops. Having started mainly with the purpose to focus on a particularly important, yet specific issue, *how to evaluate the level of quality and effectiveness of production schedules* which is quite compelling in the scheduling industrial and research community, the Group progressively extended its scope, gathering also contributions from experts working in the area of strategic management and supply chain management.

This paper provides a synthesis of the various contributions gathered from SIG4 members, assessing in particular the state-of-the-art of PMS in the industrial environment, understanding the current needs from industry and the main gaps, and providing a vision on future challenges and research directions.

*Keywords*: Manufacturing, scheduling, benchmarking, supply chain management, extended enterprise, performance measurement system, performance management, state-of-the-art, future challenges.

## 1 ROLE OF PMS IN MANUFACTURING COMPANIES

Manufacturing companies have to deal with greater degrees of complexity, uncertainty and change as product lifecycles become shorter, multi-science products start to emerge (products with knowledge, IT, biotech, chemical, mechanical, electrical, etc., content), and mass customisation becomes more widespread, especially enabled through scalable technologies. Industry and retailer have to co-operate in the management of end-to-end fulfilment processes, which include product development, supply chain management, reuse, recycling and end-of-life disposal.

All these developments lead to the emergence of knowledge-based extended enterprises, where collaboration among organisations plays a key role. Several terms have been used to refer to collaboration among organisations, including collaborative supply chains, virtual enterprises, strategic alliances, extended enterprises and clusters. A review of current literature on collaboration suggests that the differences between these types of collaboration remain unclear. For consistency purposes, the term extended enterprise (EE) will be used in this paper. This includes all types of manufacturing networks.

In the EE a number of enterprises collaborate leveraging their collective skills, resources and competencies to create competitive advantages Collaboration provides a mechanism by which risks can be shared and opportunities maximised by bringing together the right mix of skills and competencies and creating a critical mass. It allows groups of SMEs to play a central role in global supply chains and projects.

Some pioneering examples of various types of collaborative enterprises can already be seen. However, success rates are not encouraging with a high number of failures (up to 70%) being reported (Zineldin and Bredenlow, 2003). Researches to date have identified several reasons for the failure of collaborations (Lambert and Knemeyer 2004; Bruner and Spekman 1998; Drago 1997) and one of the main reasons is the absence of an integrated performance management system that takes a holistic approach to the EE. It is believed that there is a lack of a mature knowledge and practical guidelines on *how to manage global manufacturing networks*.

Since the recognition of the limitations of financially focused performance measures in the late 1980s, there has been an abundance of research and development looking at integrated, balanced and dynamic performance measurements systems. This has led to development of non-financial performance measures and their integration with financial measures in frameworks or models such as the Balanced Score Card (Kaplan and Norton, 1996), Integrated Performance Measurement Reference Model (Bititci and Carrie 1998) and the Performance Prism (Neely et al. 2002).

However, many of these models and frameworks for performance measurement have focused on a single enterprise considering *how to measure and manage the performance of an enterprise*.

In the 1990s the popularity of supply chain management (SCM) concepts developed. SCM became a strategic focus of many industries worldwide. This emphasis on SCM also led to development of measurement frameworks with a supply chain focus; the SCOR (Supply Chain Management Reference) Model includes a number of hierarchical measures that try to extend the scope of an enterprise to its customers and suppliers. However, all these models and frameworks still take one single enterprise perspective and fail to measure the performance of a supply chain as a single system.

The traditional approaches to performance measurement such as Balanced Score Card (BSC), Performance Prism etc. are no longer valid in the EE. Although, over the past ten years many academics, consultants and practitioners have contributed their expertise and knowledge in to this important area, the focus has been on single, stand-alone organisations. Even the SCOR Model is inadequate as it tries to maximise the performance of a single enterprise rather then maximising performance of the collaborative enterprise. The issue of *how to measure and manage performance of extended enterprises* remains unresolved.

The cited SCOR is also an evident example of the efforts currently carried out in providing a reference model, which could guarantee the proper alignment among processes at the different decision making levels. Collaborative planning practices and the corresponding performances need to be properly considered with aligned execution practices and metrics at every level.

Desired behaviours of the whole industrial enterprise towards the market, such as flexibility, robustness, and effectiveness, and internal economic targets, such as efficiency, can be implemented only if also the company's operational levels, such as the shopfloor, are monitored and, accordingly scheduled and controlled, with a proper set of metrics.

From the industrial side, the adoption of highly reactive and efficient scheduling and control systems strongly determines the level of productivity of a manufacturing enterprise, particularly under the pressure of shortened product cycles, reduced batch sizes and a broad variety of items to be produced. Of primary importance is the definition of a measurement system for evaluating the capability of a scheduling system in handling the current dynamic and changeable manufacturing scenarios.

From a questionnaire sent to all the members of the Special Interest Group, industrial users showed their specific need to adopt a comprehensive framework for evaluating the performances of their systems, able to provide a clear and consistent connection between strategic and operational decisions.

As a result, given also the broad scope of the topic, the Special Interest Group centred its activities and discussion among members along two main research streams linked to the area of performance measurement systems:

(c) Performance Measurement Systems in Extended Enterprises
(d) Performance Measurement Systems and Benchmarking of Production Scheduling Systems

In particular, with regards to the latter area, the discussion on PMS was instrumental for the design and development, within the activities of IMS-NoE, of a Benchmarking Service. This service would provide a virtual environment where production scheduling solutions, created by scheduling researchers, but also by commercial scheduling software vendors, could be tested and evaluated on an *emulated* production system, provided by members of the industrial community, to identify the best scheduling solution for a specific manufacturing scenario the company has to deal with.

The next sections will provide a further insight on the industrial requirements, state-of-the-art, trends and visions related to the two research domains considered..

## 2   STATE-OF-THE-ART AND GAP ANALYSIS IN THE PMS RESEARCH DOMAIN

To help companies to ensure the achievement of their goals and objectives, performance measures are used to evaluate, control and improve their processes. Performance measures are also used to compare the performance of different organisations, plants, departments, teams and individuals. The following words of Lord Kelvin (1824–1907) are commonly used in performance measurement literature:

> *"When you can measure what you are speaking about and express it in numbers, you know something about it ...(otherwise) your knowledge is a meagre and unsatisfactory kind; it may be the beginning of knowledge, but you have scarcely in thought advanced to the stage of science."*

With the adoption of appropriate metrics, manufacturers can quantitatively measure the effectiveness of the manufacturing system. Metrics can help to define goals and performance expectations.

Performance Measurement is often discussed but rarely defined. So, taking cue from Neely et al. (1995) some definitions are provided:

- Performance measurement can be defined as the process of quantifying the efficiency and effectiveness of action.
- Performance measure can be defined as a metric used to quantify the efficiency or effectiveness, or both, of an action.
- Performance measurement system (PMS) can be defined as the set of metrics used to quantify the efficiency and effectiveness of actions.

The background to this research extends back to the mid-1980s when the need for better-integrated performance measurement systems was identified (Johnson and Kaplan 1987, Kaplan 1990, Russell 1992). Since then, there have been numerous publications emphasising the need for more relevant, integrated, balanced, strategic and improvement-oriented performance measurement systems.

In terms of frameworks and models, the SMART model (Cross and Lynch 1988/1989) and the Performance Measurement Questionnaire (Dixon et al. 1990) were developed in the late 1980s. In the 1990s the Balanced Scorecard (Kaplan and Norton 1996) made a significant impact by creating a simple, but effective, framework for performance measurement. During the 1990s, the European Business Excellence Model (EFQM 1998) also made a significant impact on what measures companies used and what they did with these measures. The EPSRC funded research on Integrated Performance Measurement Systems tested the feasibility of developing an auditable reference model. This work built upon the Balanced Scorecard and EFQM Models, using the Viable Systems Structure (Beer 1985) and resulted in the development of the Integrated Performance Measurement Systems Reference Model (Bititci and Carrie 1998).

Other research programmes, and to a certain extent consultancy organisations, also developed approaches, procedures and guidelines for developing and designing effective performance measurement systems (Doumeingts et al. 1995, Krause 1999). The Performance Measurement Workbook (Neely et al. 1996) and more recently the Performance Prism (Neely et al. 2002) encapsulates the contents of the previous models. Both these are now widely published and used.

There have been several other initiatives for developing and defining performance measures for various business areas and processes, including performance measures for production planning and control (Kochhar et al. 1996, Oliver Wight 1993), performance measures for the product development process (Oliver 1996, O'Donnel and Duffy 2002), performance measurement for Human Resources (Kelly and Gennard 2001, Gibb 2002) performance measurement for service management (Wilson 2000, Fitzgerald et al. 1991), and so on.

Bititci et al. (2005) summarise the common themes emerging from the literature on performance measurement as follows:

- be balanced – i.e. the requirements of various stakeholders (shareholders, customers, employees, society, environment) need to be included;
- be integrated – i.e. relationships among various measures need to be understood;
- inform strategy – i.e. not be driven by strategy but provide an input to strategy;
- deploy strategy – i.e. propagate and translate strategic objectives throughout the organisation and to the critical parts of the organisation;
- focus on business processes that deliver value;
- be specific to business units;
- include competencies – i.e. capabilities and competencies that determine how value is created and sustained;
- include stakeholder contribution – i.e. the role of the stakeholders and the contribution they can make to the success and failure of a business.

The following sections will address an in-depth analysis of the current findings in the PMS research applied to the two research streams aforementioned.


### 2.1  *Performance measurement systems in the extended enterprise*

Bititci et al. (2003) define an EE as a knowledge-based organisation which uses the distributed capabilities, competencies and intellectual strengths of its members to gain competitive advantage to maximise the performance of the extended enterprise.

While a supply chain is a customer-supplier chain of individual enterprises, each operating as an individual enterprise trying to maximise its own corporate goals, thus sub-optimising the performance, an EE is a chain of enterprises, which essentially behave as a single enterprise trying to maximise the corporate goals of the extended enterprise, thus optimising the performance of each individual enterprise. This difference is illustrated in Figure 1.

To identify the most appropriate performance measurement system for extended enterprises, a review of performance measurement systems in Supply Chains, Extended Enterprises and Virtual Enterprises has been conducted.

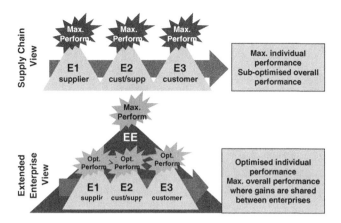

Figure 1.   Extended enterprise v. supply chain.

Gunasekaran *et al.* (2001) propose a series of performance metrics for performance evaluation of supply chains. The measures and metrics are arranged in three levels (strategic, tactical and operational) and along the five elements of an integrated supply chain: plan performance, source performance, production performance, delivery performance and customer satisfaction. This is rather similar to the hierarchical performance measurement structure used within version 5 of SCOR model (Supply Chain Operations Reference model – www.supply-chain.org) which views a supply chain as six key processes (plan, source, make, deliver, return and enablers) and presents a set of performance measures which can be broken down from the entire supply chain (Level 1) down in to individual processes (Level 2) and into specific activities within each process (Level 3).

On the other hand Beamon (1999) proposes an alternative framework, comprising of three types of performance measures: resource measure, output measures and flexibility measures. She argues that supply chain performance measurement system must contain at least one individual measure from each of the identified types.

Kochhar and Zhang (2002) in studying the performance measurement systems of virtual enterprises identified that each individual enterprise has its own performance measurement system, part of which relates to its activities related to the virtual enterprise, which tends to be co-ordinating type measures to ensure that the necessary level of co-ordination and synchronisation is achieved among individual enterprises.

All four works (Gunasekaran et al. 2001, SCOR v5, Beamon 1999, and Kochhar and Zhang 2002) propose a range a performance measures that may be appropriate in supply chains, extended enterprises and virtual enterprises. Closer study of these works reveal that most of the measures proposed are not any different from measures traditionally used in a single enterprise, but they are organised in a fashion to correspond to the supply chain (i.e. plan, source, make, deliver, etc). The exceptions to this are the works by Gunasekaran et al. (2001) and Kochhar and Zhang (2002) who identified the need for *Supply Chain Partnership Measures* and *Co-ordinating Measures* respectively.

In a review of the literature on performance measurement in extended enterprises Bititci et al. (2005) concluded that:

- None of the current strategic models and frameworks for performance measurement, such as Balanced Scorecard, Performance Prism, IPMS, Smart Pyramid etc. considers performance measurement and management from an extended enterprise perspective.
- Other works on performance measurement in supply chains, extended enterprises and virtual enterprises specify a range of performance measures, which should be used in managing supply chains and virtual organisations but fail to integrate these within a strategic performance measurement framework.

163

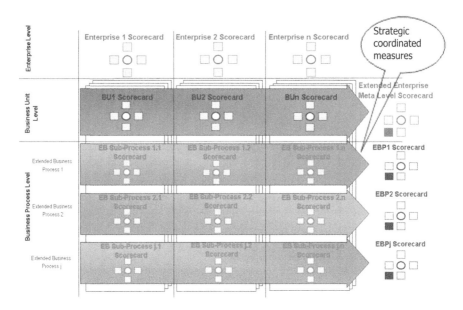

Figure 2.    The extended enterprise performance measurement model.

- Inter-enterprise co-ordinating (or partnership) measures are essential to ensure that various partners within an extended enterprise co-ordinate effectively and efficiently to ensure that the performance of the extended enterprise is maximised.
- None of the current strategic models and frameworks for performance measurement (such as balanced scorecard, performance prism, IPMS and so on) explicitly considers the need for inter-enterprise co-ordinating measures.

Based on these conclusions they suggested a possible vision of an EE Performance Measurement Model for supporting collaborative strategies among enterprises (Figure 2).

In this model, the EE performance measurement system comprises of a series of scorecards. These are:

- Enterprise Scorecards, which are specific to each enterprise collaborating in the extended enterprise. Essentially, these are conventional strategic scorecards;
- Business Unit Scorecard, corresponding to the collaborating business unit of an enterprise;
- Business Process Scorecards (EB Sub-Process Scorecard), these are operational scorecards internal to each enterprise;
- Extended Enterprise or Meta Level Scorecard, which includes strategic inter-enterprise co-ordinating measures;
- Extended Business Process Scorecards (EBP Scorecard), which includes operational inter-enterprise co-ordinating measures.

In spite of the studies described above, the key conclusion from this review of the literature is that the body of knowledge on EE performance management is still in its early days and therefore, further research is required. Section 3 will discuss the key gaps and areas for further research on the subject of EE performance management.

## 2.2   *Performance measurement systems in production scheduling*

In the last years, Production Planning and Control (PP&C) has increasingly become a critical activity, since competition in the markets is leveraging a multitude of factors ranging from product quality, to delivery times and pre-sales and after-sales services. Among all PP&C activities, scheduling decisions are the final temporal decision-making phase where plant managers have to act for fixing any short notice variations and maintaining production system performances, *assigning scarce resources to competing activities over a given time horizon to obtain the best possible system performance* (Kempf et al. 2000).

In the literature, several approaches exist for scheduling problems, from the traditional off-line scheduling systems, which simply elaborate a production plan, to the most sophisticated *intelligent* production scheduling systems, which try to continuously elaborate new production plans accounting for incoming events (e.g. Multi-agent systems, Genetic Algorithms…). With this multitude of possible approaches (*intelligent* or otherwise), industrial practitioners often experience difficulties finding an effective problem-solving scheduling solution for their plant management. Furthermore a comprehensive identification of which performances will be favoured by a given control architecture is a matter that has been neglected in literature during the last decades.

In the literature, basic performance scheduling indicators are well accepted and used (e.g. make-span, tardiness, lateness, flow time, set-up time, working time…), and usually, even if most of them are defined for a single job, they are used in an aggregate way, to calculate mean and total value among all jobs. Nevertheless, although the understanding of what constitutes a *good* production schedule is central to the development and evaluation of scheduling systems, few works have given contributions on this aspect in a more comprehensive way. In particular, the work of Kempf et al. (2000) provides an exhaustive and theoretical approach to scheduling evaluation.

The conducted analysis could be defined along three main categorisation classes, which address diverse scheduling evaluation dimensions:

(a) Stochastic measurement and facing disruption – As is well known, the production floor is not a static environment, but a large variety of dynamic events occurs (e.g. machine breakdowns, deliveries delayed, workers absent), affecting the feasibility of the proposed schedules. The inability to accurately respect proposed schedules is referred to as scheduling nervousness or disruption. To take into account this dynamic dimension, the simulation approach has been applied in the literature. Facing disruptions, a scheduling/control technique can be defined as good if it is capable to guarantee the maintenance of certain desired system characteristics despite fluctuations in the behaviour of its component parts or its environment. In the literature, diverse measures have been proposed. In Bongaerts et al. (1999), the concept of predictability is defined as the degree to which something is known in advance. It may refer to the certainty with which a variable will have a certain value or belong to a certain range. Hence, the stochastic distribution of a value is one way to characterise the predictability. Like Bongaerts et al. (1999), Mignon et al. (1995) proposed measures for investigating schedule robustness under uncertainty. A measure of schedule robustness has been computed as a function of the variability of the objective function value. The lower this variability, the more robust the schedule is. Jensen (2001) proposed the concepts of Robustness and Flexibility. A schedule expected to perform well relatively to other schedules, when facing some set of scenarios and when right-shifting is used for rescheduling is said to be robust. A schedule expected to perform well relative to other schedules, when facing some set of scenarios and when some rescheduling method using search is used is said to be flexible with regard to that rescheduling method.

(b) Rescheduling effort – Schedules generated in practice cannot be used for a long time period because of unexpected disruptions and random events. Thus, it is necessary to revise the existing schedule at some points in time. A question arises: when-to-schedule? This when-to-schedule decision determines the system responsiveness to various kinds of disruptions. There are several ways to decide on timing of schedule decisions: (i) the periodic scheduling approach (the period length can be constant or variable), (ii) the continuous scheduling approach (the schedule is revised after a number of events occur that change system state), or (iii) adaptive scheduling or controlled response approach (a scheduling decision is triggered after a predetermined amount of deviation from the original schedule is observed), or, finally, (iv) hybrid approaches. Regardless what method is employed, the scheduler has many options to react to the presence of real-time events. It can generate a new schedule from scratch, almost a complete reschedule, or make alterations to the previous schedule (a schedule repair). There is a need for measures capable to evaluate the changes caused by rescheduling, to have an idea of rescheduling shopfloor efforts. A plant manager, during scheduling solutions comparison, could prefer a scheduling policy that has little worse general performance, against very low rescheduling efforts. In Jensen (2001) two types of distance measures for schedules are presented. Firstly, the Hamming distance, originally used in the computer world to evaluate the difference between two strings of characters; the absolute Hamming distance $D(S1,S2)$ between schedules $S1$ and $S2$ is equal to the number of precedence relations that differ from $S1$ to $S2$. Secondly, the schedule overlap, as an attempt to measure the likeness between two schedules as seen on the processing floor. The measure is defined such that two identical schedules will have an overlap measure of one, while two completely different schedules will have an overlap measure of zero.

(c) Flexibility – Even if flexibility is one of the most quoted competitive objectives, a literature search for flexibility in scheduling problems did not provide relevant results. Generally, flexibility could be defined as: (i) the ability of a manufacturing system to cope with changing circumstances or instability caused by the environment, (ii) the quickness and ease with which plants can respond to changes in market conditions, (iii) the ability of the system to quickly adjust to any change in relevant factors like product, process, loads and machine failure, (iv) the ability to change or react with few penalties in time, effort, cost, or performance, (v) the capacity of a manufacturing system to adapt successfully to changing environmental conditions and process requirements.

In addition to the above challenges of selecting a proper (reactive/adaptive) scheduling system and relevant performance measures, shopfloor managers face a further problem. When an IT manager has to select computer hardware to run the company database, tests on candidate solutions (i.e. benchmark) can be executed. This IT manager can run standard and customised tests. A shopfloor manager can follow the same procedure if there is a need to buy some machine tools. However, this procedure is too expensive and time-consuming for the selection of a scheduling system. Indeed, it is economically infeasible to use an entire factory during several months to execute the required tests, and it would take a lot of manpower to guarantee identical conditions (including generating disturbances) when choices need comparing. If these choices include a number of new plant layouts, it becomes completely impossible to perform the required tests and measurements. Consequently, there is an urgent need for a virtual equivalent in cyber space, on which these tests can be performed cheaply and with adequate accuracy.

Summing up, two main gaps in the research on PMS in scheduling solutions can be highlighted.

(a) First, there is a need for a benchmarking platform on which testing and evaluation can take place. Such a platform incorporates support for comprehensive testing campaigns and detailed reporting of all relevant indicators.

(b) Second, there is a need for systematic reporting of the benchmarking results. This comprises the proper processing of the raw and detailed results in usable and more concise formats. This second aspect also includes the creation of a consensus on how comprehensive testing and evaluation must take place while using a suitable common platform.

The need for suitable benchmarking platforms is not unique to production scheduling; other research communities also recognise similar needs. However, it is heavily needed in production scheduling because research and developments limited conventional communication media – journal papers and conference presentations – are failing and make static scheduling problems, *ad hoc* toy test cases and token evaluation campaigns the norm. Researchers and developers, who are unable to share developments adequately, need to restart too much from scratch and generally lack resources and time to thoroughly address the benchmarking subtask within their project. Likewise, industrial developments are unable to benefit from a large community to further their technology. What is missing today is:

- A set of emulations of underlying production systems that is representative for industry. This set cannot be restricted to the typical OR models but addresses issues such as the handling of empty containers, batching, matching, uncertain processing outcomes etc.
- A set of scenarios for those underlying systems that adequately reflects the dynamics of industrial systems. This includes breakdowns, maintenance, processing time variations, inaccurate data, missing data, late data, and rush orders, cancellations etc.
- Standardised interfaces to connect control/scheduling systems to such emulations of underlying systems. Importantly, this interface avoids constraining the range of possible underlying production systems in potentially harmful manners.
- A benchmark management system that allows the user to define and execute benchmarks. This includes a user-friendly GUI-based subsystem that significantly lowers the threshold for novel users and includes more advanced facilities in which expressiveness is the main concern.

Systematic reporting of the benchmarking results complements the above platform. These reports should not only provide quantitative evidence of the performance of a scheduling solution (in terms of measure of the most widespread and traditional metrics, such as lead times, throughput time, WIP, ....) but also provide users with qualitative factors related to efficiency, predictability, robustness under uncertainty, stability, flexibility... . These factors turn out to be quite important if we consider that scheduling orders in a production system is an activity which implements at an operational level, and often in real-time, strategic and medium-term planning decisions. Thus, it is quite important to preserve a consistency between company's vision and strategic attitude (whether more devoted to pursuing low-cost production or striving for high flexibility and reactivity to the market) and the resulting operational actions. Experience shows that if there is no clear understanding of the objectives to achieve, an all-in-one solution in the scheduling field can turn out to be dramatically counterproductive. Summing up, the following issues have to be adequately considered in providing a systematic report:

- the report has to be consistent with user's needs and expectations; this means that, during the problem description phase, the user must have the faculty to define which category of quantitative or qualitative, or both, measures that are of most interest or to focused on;
- the report should provide an exhaustive but limited number of measures, since otherwise the user would be overwhelmed by an abundance of figures which would make difficult a prompt and effective evaluation of results;
- a comprehensive reporting facility should enable to analyse not only the performance of the production system (scheduler and underlying system) but also the behaviour of the underlying system (e.g. event log browser) as well as the behaviour of the scheduler (e.g. forecasting accuracy).

## 3 VISION AND MAIN CHALLENGES FOR FUTURE RESEARCH IN PMS FOR EXTENDED ENTERPRISES

It is now an accepted that in the 21st century competition will be among networks of collaborating enterprises (Bititci and Carrie 1998; Browne et al. 1999; Harland et al. 1999). Industry will continue to be about creation of value through innovation and improvement of products and processes, which will be achieved through closer collaboration among organisations. In today's highly competitive environment collaboration can provide a framework to maximise business opportunities and minimise risks at a global scale.

The collaborative nature of an enterprise together with the need for mass customisation will lead to the development of more distributed industrial systems. That is:

- On the one hand collaboration among enterprises will lead to more distributed manufacturing systems;

- On the other hand the need for mass customisation facilitated through scalable technologies and knowledge-based products and services will lead to certain manufacturing (such as final assembly, configuration, etc.,) moving closer towards the customers, or even to the customer. The customer will buy the product from the OEM, but the product will be manufactured by a local small manufacturing facility or configured by the local retail outlet to the specification and quality levels of the OEM.

The collaborative enterprise will consist of distributed enterprises where each individual enterprise has autonomy in two ways:

- They will be able to develop new products and services locally, try to test these locally, and then incorporate successful (viable) ones across the collaborative enterprise where appropriate.
- They will be acting as partners in more than one collaborative enterprise. Therefore, each part of the single enterprise will be autonomous of one another where each part will be operating as an integral part of a larger collaborative enterprise.

The 20 year vision and the research agenda of the IFIP's working group 5.7 strongly supports this view and lays down a research agenda which includes the theme, *how to manage global manufacturing networks.*

As discussed in previous sections, one of the key challenges to create successful EE is the development of approaches to measure and manage the performance on the whole EE. To identify the specific challenges on Collaborative Enterprise Performance Management a similar process to the one used by the CE-NET consortium [CE-NET, 2004] was followed. The key areas that were studied were as follows:

- Actual circumstances: what is the state-of-the-art in EE Performance Management (already described in the previous section)?
- Vision: What is the ideal scenario in EE Performance Management?
- Gaps: What areas of research and practice on EE Performance Management are still to be understood and further developed?
- Research needs: How can research facilitate the transition from the actual circumstances to the vision?

Based on these dimensions, the subject of EE performance management was analysed from 4 perspectives using a similar characterisation to the one suggested by Bal (1998). The following are the perspectives:

- EE Performance Measurement Systems;
- EE Infrastructure;
- EE People and Organisation;
- EE Culture and Behaviour.

### 3.1 *EE performance measurement systems*

Table 1 highlights the key research needs in terms of EE performance measurement systems. These research needs are based on the fact that the extensive knowledge on performance measurement now available focuses on the needs of single organisations and as a result it does not entirely suit to the needs of EE. Although there have been a number of studies looking at performance measurement from a collaborative perspective (Gunasekaran et al. 2001, SCOR v5, Beamon 1999, and Kochhar and Zhang 2002, Bititci et al. 2005) further research is needed to develop the body of knowledge in this area.

### 3.2 *EE infrastructure*

The importance of ICT on performance management has been highlighted by several authors (Kueng et al. 2000; Bourne et al. 2002; Nudurapati and Bititci 2003). As a result, EE

Table 1. Vision, gaps and needs of knowledge on EE performance measurement system.

| Actual circumstances | Vision | Gap | Research need |
|---|---|---|---|
| Extensive knowledge on performance measurement systems on single organisation | Extensive knowledge about EE performance measurement systems | Lack of body of knowledge on performance measurement systems in EE | Develop a better understanding on how to design, implement, use and review performance measurement system in a EE |
| Most organisations are unable/unwilling to measure/manage performance collaboratively | Performance measurement systems are used to translate the vision and objectives of the EE at all levels of each individual organisations | Lack of a generally applicable approach to collaborative performance measurement and management | A dynamic process for managing the performance of EE |
| Most metrics measure local performance | Collaborative performance measurement systems use local measures and inter-enterprise measures to maintain the relevance and effectiveness of the collaborative enterprise business model | Lack of empirical studies on the application of collaborative performance measurement systems | |
| | | Little understanding of relevant measures for EE | Define performance measures which are relevant and useful |
| The limited visibility of demand is a significant problem in Collaborative Enterprises | Collaborative performance management helps accelerate order-to-cash cycles, free up resources and expedite the execution of routine processes across the EE | Current performance management systems do not allow a more complete and accurate analysis | Design performance management systems that enable real-time planning and control of the operations of the EE |

will need to develop ICT infrastructures that facilitate critical aspects for the success of performance management including interoperability, process integration, knowledge sharing, and strategic conversation. Table 2 addresses the key research needs identified during this analysis.

### 3.3 EE people and organisation

For an EE performance management system to be successful it is critical to carefully design organisational and people practices. The appropriate definition of the structure of the processes that cut across enterprise boundaries and the teams responsible for managing them will support the implementation of the EE performance management system.

In the last few years there has been a growing interest on the study of inter-enterprise teamwork, more commonly known as virtual or distributed teamwork (e.g. Lipnack and Stamps 1997). Still, there are a several aspects of EE organisational design that need to be further studied owing to their

Table 2.   Vision, gaps and needs of knowledge on EE infrastructure.

| Actual circumstances | Vision | Gap | Research need |
|---|---|---|---|
| Current information and communication technologies do not meet the needs of EE | EE have shared applications with high level of meaning and interoperability | Lack of understanding on how to develop, implement and use enterprise systems that support the performance management process of a EE | Develop enterprise systems that facilitate interoperability, process integration, knowledge sharing and strategic conversation |
| | | Lack of integrated management systems | An open platform allowing implementation of business processes, performance measures and collaboration strategies |

Table 3.   Vision, gaps and needs of knowledge on EE people and organisation.

| Actual circumstances | Vision | Gap | Research need |
|---|---|---|---|
| Extensive knowledge on organisational design in single organizations | Extensive knowledge on organisational design in EE | Lack of body of knowledge on organisational design on EE | Develop a better understanding on how to design the organisational structure of EE |
| Teamwork practices mainly focus on single organizations | Inter-enterprise teams are responsible for managing the strategy and operations of the EE | Effective inter-enterprise teamwork not fully understood | Inter-enterprise teamwork dynamics fully understood |
| Performance appraisals and rewards used at a single organisational levels | Performance appraisal and reward systems used across the EE | Lack of understanding of the impact of appraisal and reward systems on the performance management process of EE

Lack of understanding on how to develop effective appraisal and reward system for EE | Understand the impact of appraisal and reward system on the performance management process of EE

Develop effective appraisal and reward systems for EE |
| Team performance measured at a single organisational level | Inter-enterprise team performance is continuously measured and managed | Lack of understanding on how to measures and manage inter-enterprise team performance | Improve the understanding on how to measures and manage inter-enterprise team performance |

impact on the success of performance management systems. These include aspects such as inter-enterprise team performance management, performance appraisal and reward systems. Table 3 addresses these research needs.

### 3.4   *EE organisational culture and behaviour*

Recent studies on performance management have highlighted the importance of organisational culture and behaviours on performance management (Bourne et al. 2002, Franco and Bourne 2003, Mendibil 2003, Scott et al. 2003, Bititci et al. 2004). These studies also suggest that further research work is needed to understand the relationship among organisational culture, behaviours and performance management.

The multicultural aspect of a CE is another area that requires further study. Currently seen as a barrier to performance, the vision is that multicultural environments will play a key role in fostering innovation (CE-NET, 2004).

Table 4 highlights the key research areas in relation to culture, behaviours and performance management in EE.

### 3.5   *Key research questions on EE performance management*

From the research needs identified in the previous sections we can identify a number of research questions that will drive the EE performance management research agenda during the coming years.

At a generic level, the key research need is to develop a better understanding of effective and efficient performance management processes in EE. Understanding the characteristics of these processes will be a key step towards successful collaboration.

At a more specific level, there are a number of areas and issues that need to be studied. Table 5 highlights some of the key research questions.

## 4   VISION AND MAIN CHALLENGES FOR FUTURE RESEARCH IN PMS FOR PRODUCTION SCHEDULING SYSTEMS

Manufacturing industries are aware that designing the right kind of flexibility into their forthcoming production systems is vital to their future competitiveness. However, industry is lacking the tools and the methodology to evaluate the performance of other flexible designs, which are being considered for the next production facility, regardless whether this is a green-field or a brown-field development.

Indeed, it no longer suffices to consider operations under nominal conditions to assess performance. The *raison d'être* for the flexibility in a factory design is to have robustness in the face of abnormal circumstances. As a consequence, the analysis of models that make significant simplifications of the envisaged manufacturing plants is no longer capable of predicting actual plant performance.

The alternative is to build simulation models of the choices for a forthcoming flexible production system. These simulation models have to be accurate. In particular, they must include the advanced manufacturing control, which exploits the flexibility in the underlying system and which is instrumental to achieving the desired robustness in the face of change and disturbances.

What can be observed today is that the above performance measurement and benchmarking problem is slowly but steadily being recognised. The recognition that simulation is perhaps the only option to assess performance is not yet widespread in industry and probably requires technological advancements that make this less time-consuming and cheaper.

Benchmarking will evolve toward the construction of a virtual version of the manufacturing system(s) on which the benchmarking procedures are to be executed. Indeed, systems have become too complex and sophisticated for approaches based on simplified models to work properly. Likewise, performance under a nominal conditions is becoming increasingly important, which further invalidates benchmarks that fail to model the manufacturing system in full.

Table 4. Vision, gaps and needs of knowledge on EE organisation culture and behaviour.

| Actual circumstances | Vision | Gap | Research need |
| --- | --- | --- | --- |
| People do not generally behave proactively in collaborative environment<br><br>EE is not implemented as an enterprise wide culture | People behave proactively in collaborative environments<br><br>Collaborative culture embedded in the organization | Lack of collaborative behaviour<br>Collaborative culture unevenly implemented in the EE | Improve the understanding on collaborative behaviour in EE<br><br>Develop EE wide educational systems |
| The impact of organisational culture and people's behaviour on the performance management process is not fully understood | The impact of culture and behaviours is understood, which facilitates the development of effective performance management processes for the EE | Lack of understanding of the impact of organisational culture and people's behaviour on the performance management process | Develop a better understanding of the impact of organisational culture and people's behaviour on the success of a performance management process |
| Multicultural environment perceived as a negative influence to the performance of the EE | Multicultural interaction perceived as a competitive advantage | Lack of understanding of the impact of multicultural interaction on the performance management process | Understand the influence of multicultural interaction on the performance management process of a EE<br><br>Understand how to increase the potential benefits of multicultural interaction from using the performance management process |
| Decision-making based on performance information from single organisations | Decision-making based on EE performance management process | Lack of understanding on how the EE performance management process affects decision-making | Understand the influence of the performance management process on the decision-making of a EE |

The future lies with building models in cyberspace that capture all relevant aspects of the manufacturing system, its control system and its environment.

This comprehensive approach needs cost-effective implementations. Therefore, the technology to address this challenge needs to have the following properties:

- Duplicate development efforts need to be avoided, and especially, verification that two implementations of the same functionality are indeed the same, needs to be avoided;
- Model validation must be reliable and easy. Model correctness must not be tricky and must not require extremely skilled and experienced people;

Table 5. Key research questions on EE performance management.

| | Research need | Research question |
|---|---|---|
| | **Develop a better understanding of effective and efficient Performance Management Processes in Extended Enterprises** | **What are the characteristics of an effective and efficient Performance Management Process of an Extended Enterprise?** |
| | **Specific research needs** | **Research questions** |
| **Performance Measurement Management** | Develop a better understanding on how to design, implement, use and review performance measurement system in a EE | **How can to effectively and efficiently design, implement, use and review an EE performance measurement process?** |
| | A dynamic process for managing the performance of EE | Can to develop a performance measurement framework to accommodate the architecture of EE? |
| | Define performance measures which are relevant and useful | What are the relevant measures for EE? What are their characteristics? |
| | Design performance management systems that enable real-time planning and control of the operations of the EE | What are the relevant inter-enterprise co-ordinating measures? |
| **Infrastructure / Information and Communication Technologies** | Develop enterprise systems that facilitate interoperability, process integration, knowledge sharing and strategic conversation | **How to effectively and efficiently design, implement, use and review the appropriate infrastructure to support the PMP of an EE?** |
| | An open platform allowing implementation of business processes, performance measures and collaboration strategies | What are the specifications of enterprise systems to support the PMP of a EE? How to develop EE systems that facilitate interoperability, process integration, knowledge sharing and strategic conversation? |
| **People and Organisation** | Develop a better understanding on how to design the organisational structure of EE | **How to effectively and efficiently design, implement, use and review the appropriate people and organisational practices to support the PMP of an EE?** |
| | Inter-enterprise teamwork dynamics fully understood | |
| | Improve the understanding on collaborative behaviour in EE | What organisational structures are required to support the PMP of an EE? |
| | Develop EE wide educational systems | How to develop effective teamwork practices in an EE? |
| | Understand the impact of appraisal and reward system on the performance management process of EE | What roles and responsibilities are required to support the PMP of an EE? |
| | Develop effective appraisal and reward systems for EE | How to identify and develop people's competencies in an EE? |
| | | How do performance appraisal systems influence the PMP of an EE? How to develop effective appraisal systems for an EE? |
| | Improve the understanding on how to measures and manage inter-enterprise team performance | How do reward systems influence the PMP of an EE? How can we develop effective rewards systems for an EE? How to successfully measure and manage inter-enterprise teamwork performance? |
| | Research need | Research question |
| | **Develop a better understanding of effective and efficient Performance Management Processes in Extended Enterprises** | **What are the characteristics of an effective and efficient Performance Management Process of an Extended Enterprise?** |
| | **Specific research needs** | **Research questions** |
| **Organisational Culture and Behaviours** | Develop a better understanding of the impact of organisational culture and people's behaviour on the success of a performance management process | **How to effectively and efficiently develop the appropriate organisational culture and people behaviours to support the PMP of an EE?** |
| | Understand the influence of multicultural interaction on the performance management process of a EE | How does organisational culture and people behaviour's impact on the success of the PMP in EE? |
| | Understand how to increase the potential benefits of multicultural interaction from using the performance management process | How can the development of the behaviours required to support PMP in EE be facilitated? |
| | Understand the influence of the performance management process on the decision-making of a EE | How does multi-culture influence the success of the PMP in EE? Can approaches be developed to maximise the benefits of multi-cultural interaction from using the PMP in an EE? How does the PMP influence the decision-making processes of an EE? |

173

- Model reuse must be supported. It must not be necessary to model a piece of manufacturing equipment nor to validate such model every time it is used in a new setting.
- It must be easy to transform the control system in the simulation for the control of the real system as well. And, verification that both control versions are identical must be easy.

These requirements favour an approach in which the simulation consists of a manufacturing control system connected to an emulation of the underlying manufacturing system. There is no aggregation of control functionality with the emulation, which is mimicking the physical production equipment. The emulation of a piece of equipment has a one-to-one relationship with the physical counterpart. This makes the validation task for the model much simpler, and the validation is valid in all possible settings.

Since the control system is the same in simulation and real production, validation is automatic. Further requirements for the technologies are:

- It must be possible to simulate at speeds faster than real-time so that it is possible to simulate months of production in a couple of hours of computer time;
- It must be possible to develop the simulation model in a few weeks. In particular, the simulation of any configuration of production equipment must be generated automatically, given simulation code for the equipment itself. The development of code for production equipment probably will take some (incompressible) time, but should be fast if the equipment to be modelled is just some variant of equipment for which the code already exists.

The latter requirement is a matter of solid software engineering and developing an initial set of components to be able to handle most common manufacturing systems. The former requirement represents a subtler problem. Advanced manufacturing control systems need time to think, interact, negotiate and communicate. Rapid discrete event simulation implementations jump from the current event to next, regardless how long this jump would take in reality. This mechanism needs enhancements if the time-to-think of the control system is to be accounted for (especially if the code is to remain reusable for controlling the real system as well).

Within the IMS-NoE a first step towards the accomplishment of such a vision has been done. The Benchmarking Service would provide a virtual environment where one (or more) online production scheduling approaches, created by scheduling researchers, but also by commercial scheduling software vendors, could be tested and evaluated on an *emulated* production system, provided by one industrial community member, to identify the best scheduling solution for the due test case, but also to evaluate how (and if) a scheduling approach could be applied in different production systems.

The execution of the plant emulation would be elaborated in a distributed manner: the plant emulation code would be on the server of the Benchmarking Service, while the execution of the scheduling and control logic could be resident on a client computer, physically distributed on the web.

This remote emulation approach could be implemented in larger applications exceeding the boundaries of the single enterprise. It could be adopted for improving the management of a multi-site company or collaborative enterprises, providing decision makers to view in advance the effects of scheduling decisions taken in a collaborative distributed environment. The emulation models would be resident in each real plant and maintained by the technical personnel directly working on each plant, so that the model is updated whenever the real plant is subjected to any re-configuration (e.g. new machines installed, lay-out modified).

5 CONCLUSIONS

Performance measurement in manufacturing is a broad topic, ranging from the shopfloor up to the extended enterprise. This paper discusses two extremes of this range. In an extended enterprise setting, suitable performance measurement is the key in achieving global optimisation, resulting in better results for the members of the extended enterprise amongst which the benefits are distributed. In production scheduling, the need for virtualisation is recognised (and its implementation requirements) next to the need for succinct performance reporting fitting a specific production context (e.g. cost-sensitive or customer-satisfaction oriented). Research has produced useful results toward addressing these matters but there is no shortage of work-to-be-done. In particular, the web-based benchmarking service for production scheduling is work-in-progress and its development will continue beyond the IMS-NoE funding period.

Finally, some words of caution are offered to finish the discussion. The developments discussed here produce numbers and figures related to system performance. These are measurements, not axioms, idols, or irrefutable facts, and thus inevitably the measurements have limited accuracy and finite validity ranges. Therefore, proper usage of such numbers requires some basic skills, which too often are neglected in discussions on important matters (often thoroughly changing peoples' life). First, proper usage always requires a thorough understanding what the numbers mean. Secondly, numerically unstable operations – ranking, comparing/differentiating – must never be performed on noisy measurements without making sure that ensuing decisions are not based on noise instead of information. If needed, modified tests are needed that measure the relevant property more directly. Performance measurements and especially using these measurements is not for *accountants*.

## REFERENCES

Ali, S., A. A. Maciejewski, H.J. Siegel, and J. Kim. (2003). Definition of a Robustness Metric for Resource Allocation. In: Proceedings of the 17th International Parallel and Distributed Processing Symposium (IPDPS 2003), Nice, France, 22–26 April.

Bal (1998), Process analysis tools for process improvement, TQM Magazine, VOL 10. No 5, pp. 342–354.

Beamon, M., (1999), Measuring supply chain performance. International Journal of Operations & Production Management, vol. 19, pp 275–292.

Beer, S. (1985) Diagnosing the system for organisations, Wiley, Chichester, England.

Bititci, U. S. and A. S. Carrie, (1998), Integrated Performance Measurement Systems: Structures and Relationships, EPSRC Final Research Report, Grant No. GR/K 48174, Swindon.

Bititci, U. S., K. Mendibil, V. Martinez, and P. Albores (2005), Measuring and Managing Performance in Extended Enterprises, International Journal of Operations and Production Management, Vol. 25, No. 4, pp. 333–353.

Bititci, U. S., (1995). Modelling of Performance Measurement Systems in Manufacturing Enterprises. In: Int. J. Production Economics 42, pp.137–147.

Bititci, U. S., V. Martinez, P. Albores and K. Mendibil (2003), Creating and sustaining competitive advantage in collaborative systems: the what and the how, International Journal of Production Planning and Control, Vol. 14, No. 5, pp. 410–424.

Bititci, U. S., K. Mendibil, S. Nudurapati, T. Turner and P Garengo (2004), The Interplay Between Performance Measurement, Organisational Culture and Management Styles, Measuring Business Excellence, Vol. 8, No. 3, pp.28–41.

Bodner, D. A., Reveliotis, S. A. (1997). Virtual factories: an object-oriented simulation- based framework for real-time FMS control, In: Emerging Technologies and Factory Automation Proceedings, ETFA '97, pp. 208–213.

Bongaerts, L., Indrayadi, Y., Van Brussel, H. and Valckenaers, P., (1999). Predictability of Hierarchical, Heterarchical and Holonic Control. In: Proceedings of the 2nd International Workshop on Intelligent Manufacturing Systems 1999, Leuven, Belgium, pp. 167–176, 22–24 September.

Booth, A. W. (2000). Object Oriented Modeling for Flexible Manufacturing Systems. In: The Journal of Flexible Manufacturing Systems, pp 301–314.

Bourne, M., Neely, A., Platts, K. and Mills, J. (2002). The success and failure of performance measurement initiatives: Perceptions of participating managers, International Journal of Operations and Production Management, Vol. 22 No. 11, pp. 1288–1310.

Brandl, D. (2002). Business-To-Manufacturing (B2M) collaboration between business and manufacturing using ISA95. In: Proceedings of the ISA/SEE conference, Nice, France.

Brennan, R. W. (2000). Performance comparison and analysis of reactive and planning-based control architectures for manufacturing. In: Robotics and Computer Integrated Manufacturing, 16, pp. 191–200.

Browne, J., P. Sackettand and H. Worthman (1995). Industry requirements and associated research issues in extended enterprises, in P. Ladet and F. Bernadat: Integrated Manufacturing Systems Engineering, Chapman and Hall, London.

Bruner, R. and Spekman, R. E. (1998), The dark side of alliances: lessons from Volvo-Renault, European Management Journal, Vol.16 No.2, pp.136–150.

Cavalieri, S., Taisch, M., Garetti M., and Macchi, M., (2000). An experimental benchmarking of two multi-agent systems for production scheduling and control. In: Computers in Industry, 43, pp. 139–152.

CE-NET (2004), A Roadmap towards the Collaborative Enterprise – CE Vision 2010, CE-NET Consortium (ISBN 0 85358 130 4)

Chakravarty, A. K., H. K. Jain, and J. J. Liu (1997). Object-Oriented Domain Analysis for flexible Manufacturing Systems, In: Integrated Computer Aided Engineering, 4, pp. 290–309.

Chen, D., Vernadat, F. (2001). Standardisation on enterprise modelling and integration: Achievements, on-going works and future perspectives. In: Proceedings of 10th IFAC Symposium on Information Control Problems in Manufacturing, INCOM'01, Vienna, Austria.

Choi, K.H., S.C. Kim, S.H. Yook (2000). Multi-agent hybrid shopfloor control systems. In: Int. J. Production Research, 38, 17, pp. 4193–4203.

Cross, K. F. and Lynch, R. L. (1988–1989), The SMART way to define and sustain success, National Productivity Review, vol. 9, no 1, 1988–1989.

Daniels, R., and P. Kouvelis (1995). Robust Scheduling to Hedge Against Processing Time Uncertainty in Single Stage Production. In: Management Science, Vol. 41, No. 2, pp. 363–376.

Dixon, J. R., A. J. Nanni and T. E. Vollmann (1990). The New Performance Challenge – Measuring Operations for World-Class Competition, Dow Jones-Irwin, Homewood, Illinois, 1990.

Doumeingts, G., Vallespir, B. and Chen, D. (1998). Decision modelling GRAI Grid. In: Handbook on architecture for Information Systems, (P. Bernus, K. Mertins, G. Schmidt (Eds.)) Springer-Verlag.

Drago, W. A (1997). When strategic alliances make sense, Industrial Management & Data Systems, Vol.97, No.2, pp.53–57.

Drummond, M. (1995). Scheduling Benchmarks and Related Resources. URL: http://ic-www.arc.nasa.gov/ic/projects/xfr/papers/ benchmark-article.html

EFQM (1998). Self-assessment Guidelines for Companies, European Foundation for Quality Management, Brussels, Belgium.

Fitzgerald, L., Johnston, R., Brignall, S., Silvestro, R. and Voss, C. (1991) "Performance Measurement in Service Businesses", CIMA Publishing.

Fox, B. and M. Ringer (1995). Planning and Scheduling Benchmarks. Benchmarks Secretary of the AAAI Special Interest Group in Manufacturing. URL: http:// www.neosoft.com/

Franco, M. and M. Bourne (2003). Factors that play a role in managing through measures, Management Decision, Vol. 41, No. 8, pp. pp. 698–710.

Gibb, S. (2002). Learning and Development; process, practices and perspectives at work, Palgrave.

Gören, S. (2002). Robustness and Stability for Scheduling Policies in a Single Machine Environment. M.Sc. thesis, Bilkent University, Ankara, Turkey.

Gunasekaran, A., C. Patel, and E. Tirtiroglu (2001). Performance measures and metrics in a supply chain environment. International Journal of Operations & Production Management, vol. 21, pp 71–87.

Hanks, S., Pollack M. E., and Cohen, P. (1993). Benchmarks, testbeds, controlled experimentation, and the design of agent architectures. In: AI Magazine, 14 (4) pp. 17–42.

Harland, C. M., Lamming, R. C., Cousins, P. D. (1999). Developing the concept of supply strategy, International Journal of Operations & Production Management, Vol.19 No.7, pp.650–673.

Jensen, M. T. (2001). Robust and Flexible Scheduling with Evolutionary Computation. Ph.D. thesis, Department of Computer Science, University of Aarhus, Denmark.

Johnson, H. T. and Kaplan, R. S. (1987). Relevance Lost – the rise and fall of Management Accounting, Harvard Business School Press, Boston MA.

Kaplan, R. S. (1990). Measures for Manufacturing Excellence, Harvard Business School Press, Boston MA 1990.

Kaplan, R. S. and Norton, D. P. (1996). The Balanced Scorecard – Translating Strategy into Action, Harvard Business School Press Boston, MA, USA.

Kelly, J. and Gennard, J. (2001). Power and influence in the boardroom, London: Routledge.

Kempf, K., Uzsoy, R., Smith, S., Gary K. (2000). Evaluation and comparison of production schedules. In: Computers in Industry, Vol. 42, pp. 203–220.

King, R. E. and Kym, K.S. (1995). AgvTalk: An object-oriented simulator for Agv system, In: In: Computers & Industrial Engineering, Vol. 28, No. 3, pp. 575–592.

Kochhar, A. and Zhang Y, (2002). A framework for performance measurement in virtual enterprises, Proceedings of the 2nd International Workshop on Performance Measurement, 6–7 June 2002, Hanover, Germany, pp 2–11, ISBN 3-00-009491-1.

Kochhar, A., Kennerley, M. and Davies, A. (1996). Improving Your Business through Effective Manufacturing Planning and Control, Workbook produced by researchers at UMIST as part of an EPSRC Funded research programme.

Krause, O. (1999). Performance Management, Global Production Management edited by Mertins, K., Krause, O. and Schallock, B., Kluwer Academic Publishers, ISBN 0-7923-8605-1

Kueng, P., Meier, A. and Wettstein, T. (2000). Computer-based Performance Measurement in SMEs: Is there any option?, Proceedings of the International Conference on Systems Thinking in Management, Greelong, Australia, 8–10, pp 318–323.

Lambert, D. M. and Knemeyer, A. M. (2004). We're in This Together, Harvard Business Review, December, pp.114–122.

Lipnack, J. and Stamps, J. (1997). Virtual Teams: Reaching Across Space, Time and Organizations with technology. John Wiley & Sons, New York.

Maskell, B. (1989). Performance measures of world class manufacturing, The Journal of Applied Manufacturing System, Winter 1992, pp. 19–26.

Mendibil, K. (2003). Designing Effective Team-based Performance Measurement Systems: An Integrated Approach, PhD thesis, University of Strathclyde, Glasgow.

Mignon, D. J., Honkomp, S. J., Reklaitis, G. V. (1995). Framework for Investigating Schedule Robustness Under Uncertainty. In: Computers in Chemical Engineering, Vol. 19, Supplement, pp. S615–S620.

Neely, A. D., Adams, C. and Kennerley, M. (2002). The Performance Prism, The Scorecard for Measuring and Managing Business Success, FT Prentice-Hall, London.

Neely, A. D., Gregory, M. J., Platts, K.W. (1995). Performance measurement system design: A literature review and research agenda, International Journal of Operations & Production Management, Vol. 15 No. 4, pp. 80–116.

Neely, A. D., Mills, J., Gregory, M. J., Richards, H., Platts, K. and Bourne, M. (1996). Getting the measure of your business, University of Cambridge, Manufacturing Engineering Group, Mill Lane, Cambridge.

Neely, A. D., Mills, J., Platts, K., Richards, H, Gregory, M., Bourne, M. and Kennerley M. (2000). Performance Measurement System Design: Developing and Testing a Process Based Approach, International Journal of Operations and Production Management, Vol. 20, No. 10, pp. 1119–1145.

Nudurupati, S. S. and Bititci, U. S. (2003). Impact of IT enabled Performance Measurement on Business and Management, Proceedings of EurOMA and POMS Joint International Conference, Como Lake, Italy, 16–18 June 2003.

O'Donnel, F. and Duffy, A. H. B. (2002). Modelling Design Development Performance, International Journal of Operations and Production Management, vol. 22, no. 11, 2002.

Oliver, N., (1996), Design and Development Benchmarking, 5th Operations Strategy and Performance Measurement Workshop, Loughborough University, 8 May 1996.

Wight, O., (1993). The Oliver Wight ABCD Check List - 4th Edition, John Wiley & Sons, Inc, New York.

Ouelhadj, D., Cowling, P. and Petrovic, S. (2003). Utility and Stability Measures for Agent-Based Dynamic Scheduling of Steel Continuous Casting. In: Proceedings of the IEEE International Conference on Robotics and Automation, Taipei, Taiwan, 14–19 September.

Parunak, V. D. (1993). MASCOT: A Virtual Factory for Research and Development in Manufacturing Scheduling and Control. In: ITI Tech Memo.

Russell, R. (1992). The Role of Performance Measurement in Manufacturing Excellence, BPICS Conference, Birmingham, UK, 1992.

Scott, T., Mannion, R., Marshall, M. and Davies, H. (2003). Does organisational culture influence health care performance? A review of the evidence, Journal of Health Services Research and Policy, Vol. 8, No. 2, pp. 105–117.

Shen, W., Norrie, D. H. (1999). Agent based systems for intelligent manufacturing: a state of the art survey. In: Knowledge and Information Systems, 1, pp. 129–156.

Shewchuk, J. P., Moodie, C.L. (1997). A Framework for Classifying Flexibility types in Manufacturing. In: Computers in Industry, 33, pp.261–269.

Suwignjo, P., Bititci U. S. and Carrie A. S. (2000). Quantitative Models for Performance Measurement Systems, International Journal of Production Economics, vol. 64, pp 231–241, March 2000.

Venkatesh, K., Zhou M. (1998). Object Oriented Design of FMS Control Software Based on Object Modeling Technique Diagrams and Petri Nets. In: Journal of Manufacturing Systems, Vol. 17, No. 2.

177

White, G.P. (1996), A Survey and Taxonomy of Strategy-Related Performance Measures for Manufacturing, International Journal of Operations & Production Management, Vol. 16, No. 3, pp. 42–61.

Wilson, A. (2000). The Use of Performance Information in the Management of Service Delivery, Marketing Intelligence & Planning, Vol. 18 No.7 pp. 127–134.

Wu, B. (1995). Object-oriented systems analysis and definition of manufacturing operations. In: Int. J. Production Research, Vol. 33, No. 4, pp. 955–974.

Yeung, W. H. R., Moore C. L. (1996). Object-oriented modelling and control of flexible conveyor systems for automated assembly. In: Mechatronics, Vol. 6, No.7, pp. 799–815.

Zhang, D., Zhang, H. C. (1997). An object-oriented integration test-bed for process planning and production scheduling. In: Integrated Computer-Aided Engineering, Vol.4, No.4, Wiley Interscience.

Zineldin, M. and Bredenlow T. (2003). Strategic alliances: synergies and challenges, International Journal of Physical Distribution & Logistic Management, Vol.33, 5, 449–464.

*Advanced Manufacturing – An ICT and Systems Perspective – Taisch,*
*Thoben & Montorio (eds)*
*© 2007 Taylor & Francis Group, London, ISBN 978-0-415-42912-2*

# PRODCHAIN: Supporting SMEs participating successfully in production networks

Robert Roesgen

*Forschungsinstitut für Rationalisierung (FIR) an der RWTH-Aachen, Pontdriesch, Aachen, Germany*

ABSTRACT: Companies are increasingly acting globally, and are integrated in supply chains and production networks. Caused by the very complex structure within production networks, the requirements for the co-operation and co-ordination among the involved partners are increasing continuously. Therefore, easy adoption without great effort to changing markets and customer requirements is only possible with customisable manufacturing resources belonging to different, financially or organisationally independent partners, or both. Although there are various advanced communication and database technologies to support such dynamic re-configuration and customisation, a significant problem remains in the organisational ability of each partner and the network as a whole to link and detach from one another to form efficient networks. The PRODCHAIN toolbox intends to support SMEs in reorganising their customer supplier business processes to the needs of modern production networks.

*Keywords*: Integrated supply chains, benchmarking, dynamic re-configuration, toolbox.

## 1 INTRODUCTION

Today's markets force production networks to deliver the finished products to the final customer within extremely short lead times. The increasing complexity of business processes, owing to individual customer requirements, makes this task extremely difficult. The need for efficient supply chain management is obvious and the support through tools, which help improve network performances while ensuring cost efficiency for all partners involved, is desirable.

The PRTM supply chain trends study has shown that the integration of inter-enterprise business processes is still lacking in most companies, although the process integration is a prerequisite to co-operate in networked environments (PRTM 2003). Within this study PRTM differs between four stages of supply chain maturity. The *functional focus* means, that planning and optimisation takes place within functions or departments or both. *Internal integration* includes the cross-functional integration of planning processes and systems within the company. The implementation of an ERP system is a sign for this stage. Point-to-point planning integration within the extended enterprise is part of the *external integration* stage and *cross-enterprise collaboration* is characterised by cross-enterprise planning, collaboration and optimisation in many-to-many relationships.

The evolutionary step from the internal integration towards the external integration has only been performed by 20% of the companies that took part in the study. Most of these companies are larger corporations based in the United States. This lack of external integration is an opportunity for SMEs to achieve competitive advantages by integration of external business processes, considering latest standardisation developments like RosettaNet or ebXML.

Therefore, one of the main objectives of the IMS project PRODCHAIN (IST-2000-61205) was to develop a decision support technique and methodology to analyse and improve the logistic performance of globally acting production and logistics networks by guiding SMEs to successfully

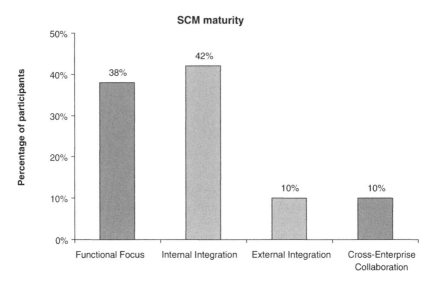

Figure 1. Little external integration has been achieved, (PRTM 2003).

integrate business processes into production networks. This guidance takes place with the help of performance measurements.

Consequently, the key questions answered by PRODCHAIN are:

- What measures can improve the logistic performance of a production network?
- How can the current circumstances of a production network be analysed in terms of logistic performance?
- How can customer-supplier relationships in production networks be modelled with regard to the processes?
- How can customer-supplier relationships be described and standardised?
- How can appropriate measures for the improvement of the logistic performance be derived for a specific customer-supplier relationship?

While pursuing the improvement of the logistic performance of globally acting production and logistics networks, the focus of PRODCHAIN was set on the measurement and improvement of the inter-company logistic performance, i.e. the logistic performance of customer-supplier relationships and of the network itself, and not only the isolated performance of a single company.

## 2 MARKET ANALYSES

One of the first tasks of the PRODCHAIN project was to analyse why supply chain management concepts are hardly implemented in Europe and in doing so identify the obstacles. Therefore, 200 logistics and IT-leaders from companies of various sizes and branches were questioned using a standardised questionnaire. 8000 companies were asked whether they would like to participate in the survey. As a result, 275 experts expressed an interest to participate in the survey.

The survey, which in the end involved 200 experts, revealed interesting results. Nearly two-thirds of the companies stated that no appropriate software suiting their requirements is available, and about half of them said that no structured approach is available to implement supply chain management (SCM) (cf. Figure 2). Although, European companies still see a high potential in supply chain management for their business success, they still do not know how to implement

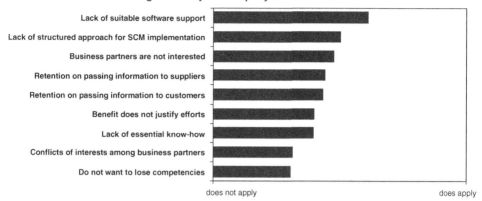

**Are there any difficulties constricting the introduction of Supply Chain Management in your company? Which ones?**

Lack of suitable software support
Lack of structured approach for SCM implementation
Business partners are not interested
Retention on passing information to suppliers
Retention on passing information to customers
Benefit does not justify efforts
Lack of essential know-how
Conflicts of interests among business partners
Do not want to lose competencies

does not apply       does apply

Figure 2.    Obstacles in introducing SCM (Schönsleben et al. 2003).

SCM and what kind of software and functionality is available and appropriate for their businesses (Schönsleben et al. 2003).

## 3   BENCHMARKING STUDY

To develop an efficient support tool, a typology for companies in networks and for customer-supplier relationships needed to be elaborated. Therefore, 46 companies took part in a benchmarking study. The benchmarking study took place in two steps. First, the companies needed to fill out a questionnaire to identify company characteristics and the customer-supplier relationship characteristics. For this purpose determinants have been defined. The values of the determinants allowed a cluster analysis, which identified company clusters similar to the SCOR definitions and five different kinds of customer supplier relationships.

In the next step, the values of selected performance indicators have been collected in a benchmarking database. The performance indicators have been allocated to typical critical customer supplier relationship processes. This way, on the one hand, a identification of best practices for the different types of and, on the other hand, the identification of problematic business processes becomes possible. The detailed procedure is described in (Sennheiser 2003).

## 4   PRODCHAIN TOOLBOX

The determinants, which allow a classification of the company in the network and its customer-supplier relationship, and the performance indicators, are starting points for the use of the PRODCHAIN Toolbox (cf. Figure 2). Once the company characteristics have been identified, the company can be allocated to a company and customer-supplier relationship class. The value of the performance indicators allows, in combination with the benchmarking database, an identification of the problematic business processes of this company.

The best practice repository, which is connected to the typology and the performance indicators in frameworks, will then give the company suggestions on which practice to use. For each best practice, the requirements for introducing them are given as well. (see Schnetzler (2003) for details). The frameworks itself, consider the different standardisation activities like SCOR, RosettaNet and ebXML. This is important to ensure that the suggestions given are in line with these standards, so that the business processes and complying protocols enable collaboration with other companies also following these standards.

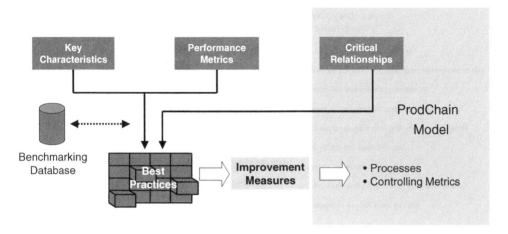

Figure 3.    Infrastructure of the PRODCHAIN Toolbox.

The framework is the PRODCHAIN model. The PRODCHAIN model, based on SCOR, shall ensure that the proposed measures are contributing to the benefit of the network. SCOR has been chosen for this purpose because it has established itself as the most used *de facto* standard at a high level. For the lower-level customer-supplier relationship processes, which are included in the frameworks of the typology, the business processes of RosettaNet and ebXML have been considered, since they are focused on inter-company processes and protocols.

Once the best practice has been implemented, controlling metrics are given to control the effectiveness of the improvement measure. Owing to the integration of the processes and controlling metrics into the PRODCHAIN model, the improvement of the business processes in the company and the network is ensured.

Another way to enter the toolbox is to identify the critical processes directly. For this approach, the supply chains need to be mapped according to the SCOR methodology. The SCOR model itself has been found very useful for the mapping of the supply chains (Stich and Weidemann 2002). Especially, the hierarchy of SCOR starting on level 1 and becoming more precise towards level 4 has made the supply chains very transparent, so that the other partners, who were not involved in the same supply chain can easily understand its characteristics, complexity and current requirements and problems. Thus critical processes can be identified and, in combination with the typology, the best practice repository can give improvement measures as described above.

## 5   DISCUSSION

The survey among logistics and IT-experts clearly revealed the necessity of supporting SMEs in implementing concepts and strategies, which allows them to easily participate in production networks and to adapt to changing requirements within networks. The survey results of PROD-CHAIN are in line with the results of previous studies (META Group 2000, Frost and Sullivan 2001). Unfortunately, the circumstances in Europe have not changed since these studies have been performed.

## 6   CONCLUSION

Most SMEs do not have the time and the financial resources to inform and to reorganise their processes according to the latest developments and standards (e.g. RosettaNet; ebXML, SCOR) to successfully participate in production networks. For most companies, missing information on available software and the lack of a structured approach prevents them from implementing new concepts and standards that have been identified as important for their business success. On the other hand, increasingly, global operating companies implement these standards and expect their suppliers to reorganise their processes accordingly. The PRODCHAIN toolbox guides these companies in dependency to their company characteristics and logistics performances. The usability of the determinants and performance indicators has been validated in a European-wide benchmarking study. The toolbox therefore is a useful support for SMEs in Europe to successfully participate in production networks (PRODCHAIN).

## REFERENCES

PRTM: Supply Chain Trends 2003: What is on the management agenda? Report, PRTM 2003.

Schönsleben, Paul; Nienhaus, Jörg; Schnetzler, Matthias; Sennheiser, Andreas; Weidemann, Martin: SCM – Stand und Entwicklungstendenzen in Europa. In: Supply Chain Management, No 1, 2003, pp.19–27.

Sennheiser, Andreas: Lean Benchmarking with clustered company prototypes. In: Conference proceedings of eChallenges 2003.

Schnetzler, Matthias: Identification of performance improvement strategies in production networks. In: Conference proceedings of eChallenges 2003.

Stich, Volker; Weidemann, Martin: Decision Support for improvement of logistics performance in production networks. In: Challenges and Achievements in E-business and E-work, Brian Stanford-Smith, Enrica Chiozza, Mireille Edin (Eds.), ISBN 1 58603 284 4. IOS Press, Amsterdam 2002, pp. 638–645.

META Group: META Group: Supply Chain Management und Collaboration in Deutschland – Technologien und Trends für das erweiterte Unternehmen. 2000, (URL http://metagroup.de/studien/scm2000/).

Frost & Sullivan: Europamarkt für Supply Chain Management Software – Frost & Sullivan Report 3848, Frost & Sullivan, 2001. (URL http://www.frost.com).

PRODCHAIN: Development of a decision support methodology to improve performance in production networks. (URL: http://www.prodchain.net).

*Part VI*
*Industrial services*

Competing in a global economy forces manufacturing organisations to consider manufacturing as a collaborative activity. While mainly relying on their core competencies to improve their performance, in-sourcing, supporting capabilities through collaboration with external service suppliers, is now crucial. Given the increase in such collaborations, the challenge for manufacturing organisations and industrial service suppliers is how to successfully integrate the organisation's core competencies and related production structures with the in-sourced industrial services. This requires enhancing organisational production performance as well as industrial service performance across organisational boundaries to drive total production performance. This process requires management capabilities in a variety of areas such as collaborating; designing, negotiating and managing new forms of service agreements; and integrating control systems and performance measurement.

It is a disconcerting fact that many if not most service organisations have difficulties taking full advantage of the opportunities offered through such collaborations. Often they do not accomplish desired objectives and fall short of targeted performance improvements owing to the complexity of the problems they face in managing complex industrial service relationships. There is a glaring need to develop management guidelines for industrial service providers.

IMS-NoE SIG intended to establish a multidisciplinary platform of researchers and practitioners that were willing to share existing competencies, to research and develop a better understanding of the provision and integration of industrial services, and to specify approaches that assist in managing total production performance by integrating more effectively and efficiently organisational production systems and industrial services across organisational boundaries. The specific focus of this SIG was on service organisations so that they can enhance their capabilities to deliver service solutions of high value that are tailored to the demands of existing production systems.

Part VI contains two contributions. The first contribution, written by Gudergan and Garg, provides a framework which enhances the existing scope of intelligent manufacturing by specifically addressing industrial services provided by external suppliers as a part of today's networked supply chain structures in the manufacturing area. Existing research shows that sourcing industrial services from specialised service organisations establishes complex and unique interdependencies and links total production performance to the performance of the external service suppliers. Therefore, an integrated framework and understanding of service organisations and production interactions and interdependencies is crucial for today's integrated manufacturing organisations to succeed. An appropriate framework is demonstrated and the most relevant research areas to improve integration of industrial services are proposed.

The contribution from Salminen and Kalliokoski introduces some generic frameworks created and understanding achieved during the BestServ project. Thirty Finnish companies, all operating in the global market, and several universities and research institutes participated in this project and roundtable work. BestServ acted as an interest group and knowledge community for companies, research organisations, and financiers, to activate and guide industrial service development. The purpose of this analysis work was to establish the state of industrial service business globally, in various business sectors and also in individual enterprises. During the roundtable work, several generic frameworks were created to help enterprises face future challenges.

*Advanced Manufacturing – An ICT and Systems Perspective – Taisch,*
*Thoben & Montorio (eds)*
*© 2007 Taylor & Francis Group, London, ISBN 978-0-415-42912-2*

# Industrial services: Challenges for integration and global co-operation with supply chains

Gerhard Gudergan & Amit Garg

*Forschungsinstitut für Rationalisierung (FIR) an der RWTH-Aachen, Germany*

ABSTRACT: Competing in a global economy is forcing manufacturing organisations to view their worldwide production facilities as a collaborative activity. While mainly relying on their core competencies to improve their performance, in-sourcing, supporting capabilities through collaboration with external service suppliers, is now crucial. Given the increase in such collaborations, the challenge for manufacturing organisations and industrial service suppliers is how to successfully integrate the organisation's core competencies and related production structures with the in-sourced industrial services. This requires enhancing organisational production performance as well as industrial service performance across organisational boundaries to drive total production performance. The aim of this paper is to provide a framework which enhances the existing scope of intelligent manufacturing by specifically addressing industrial services provided by external suppliers as a part of today's networked supply chain structures in the manufacturing area. Existing research shows that sourcing industrial services from specialised service organisations establishes complex and unique interdependencies, and links total production performance to the performance of the external service suppliers. Therefore an integrated framework and understanding of service organisations and production interactions, and interdependencies, is crucial for today's integrated manufacturing organisations to succeed. An appropriate framework is demonstrated and the most relevant research areas to improve the integration of industrial services are proposed.

*Keywords*: Industrial Services, Integration, Collaboration, Manufacturing Supply Chains.

## 1 INTRODUCTION

Competing in a global economy is forcing manufacturing organisations to view their worldwide production facilities as a collaborative activity. While mainly relying on their core competencies to improve their performance, in-sourcing, supporting capabilities through collaboration with external service suppliers, is now crucial. Given the increase in such collaborations, the challenge for manufacturing organisations and industrial service suppliers is how to successfully integrate the organisation's core competencies and related production structures with the in-sourced industrial services. This requires enhancing organisational production performance as well as industrial service performance across organisational boundaries to drive total production performance. Management capabilities in a variety of areas such as collaborating; designing, negotiating and managing new forms of service agreements; and integrating systems for co-ordination, planning and control, knowledge management and performance measurement are required to perform these tasks.

It is a disconcerting fact that many if not most service organisations have difficulties taking full advantage of the opportunities offered through such collaborations. Often they do not accomplish desired objectives and fall short of targeted performance improvements owing to the complexity of the problems they face in managing complex industrial service relationships.

Thus, there is a glaring need to develop management guidelines for industrial service providers and to establish a multidisciplinary framework which allows to the development of a better understanding of the provision and integration of industrial services, and to specify approaches that assist in managing total production performance by integrating more effectively and efficiently organisational production systems and industrial services across organisational, legal and regional boundaries. The specific focus of this paper is on service organisations and their capabilities to deliver service solutions of high value that are tailored to the demands of existing manufacturing chains.

The aim of this paper is to provide a framework which enhances the existing scope of intelligent manufacturing by specifically addressing industrial services provided by external suppliers as a part of today's networked supply chain structures in the manufacturing area. Existing research shows that sourcing industrial services from specialised service organisations establishes complex and unique interdependencies, and links total production performance to the performance of the external service suppliers. Therefore an integrated framework and understanding of service organisations and production interactions, and interdependencies, is crucial for today's integrated manufacturing organisations to succeed.

While mainly focusing on service organisations in the manufacturing supply chain, the framework coherently links to service operations with production supply chain activities to establish a foundation for business process scheduling and control; benchmarking and performance measures and information system design that are of relevance for the design of interfaces within service organisations and manufacturing organisations.

## 2 RELEVANCE OF THE INDUSTRIAL SERVICE SECTOR

Business related services constitute the largest sector of the economy employing around 55 million persons in 2001 – or nearly 55% of total employment in the European Union market economy. The importance of the services sector can also be justified by its sheer weight in the economy (around 70% of European Union GDP) and the increasing consumption of services by manufacturing industry, thus affecting the cost, price and quality of manufactured goods. Since business related services are the dominant part of the European market economy, the sector is important in its own right for the European economy. However, the most important feature of business related services is that they are present in and integrated into every stage of the value adding supply chain. They are a fundamental necessity for the existence of all enterprises, whether in manufacturing or services, micro or large enterprise. They are inextricably linked to manufacturing industry.

All goods contain elements of services and their contribution to the value added of any manufactured product often determines its attractiveness to the market. For example, the automobile industry uses pre-production services such as design services and research and development, production related services (such as engineering and IT services), after production services (transport and distribution services) and financial services and finally other business services such as accounting or legal services. Analysis of the service sector shows that it consumes more than half of the output going to intermediate demand from business related services, compared to a share of less than one-third consumed by manufacturing as shown in the Figure 1 below.

This figure illustrates that manufacturing industry is an important user of business related services, as nearly 30% of the intermediate output from the sector is consumed by manufacturing companies. Nevertheless, crucial for understanding the growth of business related services in the last decades is the demand for business related services created by the sector itself as a consequence of the penetration of these services into the value chain of all enterprises.

Growth of business related services is usually explained by the migration of employment from manufacturing industry to services owing to the outsourcing of the services functions previously produced in-house. The process of externalisation of service functions has been an important driver of the growth in the services sector. Outsourcing decisions are not solely driven by labour cost aspects, but frequently by the need to gain access to specialised skills (quality aspects) to increase flexibility. An enterprise has to make strategic and often long-term decisions: whether to produce

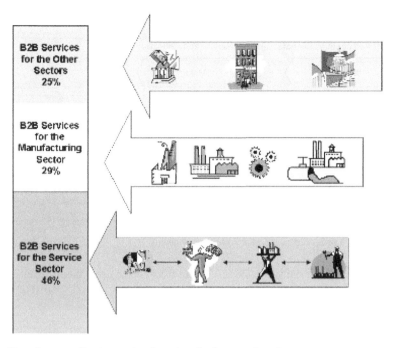

Figure 1. The relevance of business related services for the manufacturing sector.

the necessary services by itself or to contract these services out to specialised companies. Part of the economic performance of manufacturing and service enterprises related to price, quality or market positioning is linked to *make or buy* decisions. In addition to the advantages related to quality and cost, outsourcing allows concentration on core business. To fully benefit from the advantages related to outsourcing, the service purchaser needs to implement the necessary organisational and managerial changes. Especially for SMEs, the lack of appropriate skills amongst employees can hamper reaping the benefits of outsourcing.

But the reasons for the growth are multiple and not just restricted to continuing outsourcing of business processes. Changes in production systems, more flexibility, stronger competition in international markets, the increasing role of ICT and knowledge, and the emergence of new types of services are other important factors. Taking into account the growth envisaged in this sector, the Lisbon meeting of the European Council duly highlighted the role of services in the European economy and their potential for growth and employment creation. But at the same time, the Council realised that without competitive business related services, it would be difficult to fulfil the Lisbon objective of making Europe, by 2010, the most competitive and dynamic knowledge-based economy in the world capable of sustainable economic growth and more and better jobs and greater social cohesion.

Thus it has been well realised that the development of business services is crucial for the competitiveness and catching up of the economies in Europe. There are many challenges to address to facilitate rich interactions and cohesion among the business services and the manufacturers. The most challenging aspect in this domain is to enable and facilitate the interactions among diverse players to ensure flexibility and re-configurability. The process of interactions among services and manufacturers requires an in-depth analysis of the business requirements and constraints. A step further towards the Lisbon goal will be the development of relevant guidelines and framework for the service supply chain domain to enable collaboration and co-ordination in conducting the business. In this direction, standards provide a solution to the development of these interface and co-ordination mechanisms. Standards would benefit the service providers by enabling them to focus on the internal process of service production and also obtain some economies of scale. They

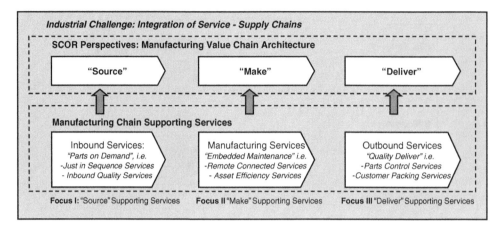

Figure 2.   Integration of services – manufacturing supply chains.

would also be able to efficiently provide services to their clients using the ICT tools supporting them in exchanging information and making decisions.

## 3   CONCEPTUAL FRAMEWORK

Within the European funded initiative *IMS Network of Excellence IMS-NoE*, a group was established focussing on industrial services and their role in the manufacturing value chain. This special interest group in particular focussed on the operation and integration of business related services in manufacturing supply chains. Figure 2 shows the conceptual boundaries chosen and outlines how existing reference architectures from the supply chain management domain serve as the basis for the specification of the work developed. The SCOR reference architecture for supply chain management was applied as a guiding framework for structuring the domain from the manufacturing chain perspective.

The SCOR framework as illustrated above allows classifying services in distinct groups according to the SCOR perspectives to which they are linked:

- *Source* perspective: Services such as Just in Sequence Delivery Services,
- *Make* perspective: Services such as Embedded Systems Enabled Remote Maintenance and Asset Efficiency Control services,
- *Deliver* perspective: Services such as Customised Outbound Packaging and Quality Control.

This general view of manufacturing supply chain activities and supporting services and their inter-relations serves as a general framework for the development of a more complete and comprehensive picture of industrial services and their role in manufacturing supply chains as described in the following. Work presented in the following sections is the result of case analyses, workshops and expert studies conducted within the special interest group in the IMS-NoE.

## 4   INDUSTRIAL CHALLENGES

As illustrated above, nearly 30% of the intermediate outputs from business related services are consumed by manufacturing companies. This dependence offers great opportunities for the services industries to improve the competitiveness in a global environment. Unfortunately, managers of manufacturing and supporting service organisations are facing tremendous difficulties in the efficient integration and synchronisation of their joint activities. The high degree of integration and synchronisation needed is not achieved so far. Moreover, the complexity of the highly interdependent processes and interactions is not understood. Thus, competence in integrated manufacturing – support services system development is not achieved so far. The business / technological challenge for companies can therefore be summarised as follows:

- The challenge for increasingly global acting manufacturing companies is to integrate and synchronise services from external providers into their existing and continuously evolving manufacturing chains and to successfully measure, control and evaluate the performance of these joint activities, at the operational and strategic levels.
- The challenge for supporting service companies is to meet production requirements irrespective of place and time through co-ordinated and synchronised activities and to remain a reliable and innovative partner for global manufacturing organisations in a long term.

The service or manufacturing firms lack mechanisms and tools to measure the performance of individual partners in the emerging business environment. To ensure transparency for enabling the *win-win* circumstances, the partners need to have an explicit and holistic view of the performance of the supply chain as the whole including the service partners.

There are manifold challenges for industry related to services driven manufacturing chains.

- First, the service manufacturing supply chain interaction is poorly equipped with enabling ICT tools that support interactions amongst diverse players. Most of the partners have their own IT solutions which need to be mapped with the individual requirements of the customer. Often such mapping is limited by the lack of standards over the definition of business processes and interfaces across the information systems. Therefore, there is an urgent need to further stimulate the integration of IT into the business processes. Of special importance for services is the improvement of framework conditions for operating and participating in such interactions. To ensure high adaptability, it will be very critical to base these new business processes on existing frameworks well accepted by the supply chain domain.
- Second, focus in industry is on information and knowledge requirements from different partners and to encapsulate the same into the new information systems.
- Third, requirements for performance measurement have also to be dealt while designing the information systems for the respective partners. The possibilities of having centralised or decentralised and distributed information systems need to be evaluated before proposing a solution for this domain. There is a need for empirical analysis; simulation and industrial case study research as the methodology to develop information solutions; and to strengthen seamless integration of the services-manufacturing value chain.
- Moreover, owing to the lack of benchmarking studies in this area, there is a strong need for the development and testing of a comprehensive and prescriptive framework concerning the management of co-ordination and innovation in collaborative support service networks. Critical structures and processes have to be identified from the perspective of entire support service network crucial for the sustainability and competitiveness of the extended enterprise.

Beside these structural challenges, industrial managers are facing very specific problems and challenges when exploring the potential of their service offerings in an increasing international environment:

- These cover professionalisation of industrial service marketing: transformation of concepts, models, measures, strategies from traditional service research into industrial service domain.
- There is a tremendous body of knowledge available in traditional end consumer oriented service research. Little work has been done in the area of business to business service interaction, in particular in the area of manufacturing related services. Thus, questions related to reorganisation of industrial service supply structures, (value chain integration and ownership of services), and emerging business models (performance responsibility, responsibility for safety and health, legal aspects and contracting, risk sharing, profit sharing) are mentioned frequently.
- Networking and collaboration: economies of scale, bundling services, standardisation, innovation, co-ordination and governance structures are mentioned often from practitioners who are

exploring the potential of their industrial services business in the international manufacturing area. Technology support in service- manufacturing interactions, including remote systems, with ERP integration, is considered to be a field of tremendous importance in this context as well.

Other, urgent needs communicated by practitioners and researchers in the field are related to leadership and employee management in dispersed, multicultural firms and working environments. Knowledge management in large scale and virtual service supplier customer organisations is one of the most mentioned challenges.

Most of these areas listed by practitioners are considered to be a problem area in itself. ICT support is of tremendous importance, and is considered to have a key enabler function. ICT is considered not to be the core problem solution in all the areas such as employee leadership, cultural integration, and standardisation for the trade in services, contractual issues and international marketing concepts. However, ICT is considered to be a key element in service-manufacturing chain integration.

## 5   STATE-OF-THE-ART

### 5.1   *Industrial practices and scenarios*

According to SCOR, in a typical organisation there are the three distinct phases of source, make and deliver to fulfil the customer order (as developed in the SCOR Reference Model). These three phases are co-ordinated by the plan phase that seeks co-ordination of all the actions of all phases in a holistic manner. At the same, each of these individual phases may outsource some auxiliary services which are necessary to facilitate the flow of material (typical examples are inbound and outbound logistics) and production of products (including maintenance and spare parts services). In a very traditional scenario, the source phase will opt for inbound logistics and quality control services from the service providers as show an the earlier figure, Figure 2. Furthermore, the manufacturing, or the make phase, will outsource services related to the maintenance of machines and in-process quality control. Finally the Deliver phase outsources the outbound logistics to transport the end product from the warehouse to various distribution centres. These individual service providers interact primarily with the intended users of the services, but there is very little interaction/co-ordination with the other phases or with other service providers. Hence a holistic view of the services required and provided by different service providers is not available in any of the phases within the manufacturing organisation or within the service providers.

The lack of a holistic view over the provision of diverse services to the manufacturer hinders any effort to synchronise the operations of these service providers with internal functions. For example, when inbound logistics fails to deliver the material required for immediate production, then the events related to outbound logistics and machine maintenance needs to be re-worked.

### 5.1.1   *Industrial scenario 1: Services co-ordinated by manufacturer*
In the first scenario, where there are not many interdependencies among individual service providers, it is feasible that the manufacturer develops a single interface (encompassing the requirements of source, make and deliver phases) to interact with the diverse service providers. The focus will be thus on enhancing the manufacturer's control over the activities of service providers to achieve a high level of synchronisation. In this scenario, the individual service provider will still be isolated and will lack a holistic view of the complete supply chain. But at the same, the manufacturer will be able to synchronise its own internal operations thus enhancing the co-ordination between these three phases. A higher level of co-ordination internally within an organisation will facilitate a more precise description of the service requirements. The manufacturer needs to develop suitable compensation and penalty strategies to steer the service provider towards enhanced collaboration in the dynamic event driven environment.

### 5.1.2   *Industrial scenario 2: Single manufacturer interface for all service providers*
The second scenario can be distinguished from the first, since there is now a single interface provided by the manufacturer for all the service providers. The service providers interact primarily

with this manufacturer interface to receive requests/orders for providing services and updating of plans as and when required. Internally, the manufacturer synchronises the requirements of source, make and deliver phases and consolidates their requirements and builds the interdependencies in business processes. These interdependencies are then further integrated into the constraints and capabilities of the service providers.

### 5.1.3 *Industrial scenario 3: Services co-ordinated by a third service provider*

The preceding two business scenarios are based on a couple of assumption regarding the inter-dependencies among various service providers. The first assumption is that there exist very elementary interdependencies among the individual service providers. When the service providers are providing services that require little synchronisation, then these independent service providers need not co-ordinate amongst themselves. Furthermore, the first business scenario also assumes that sufficient inventory buffers exists between the various phases so that they can continue with isolated planning (without integrating all the service providers). These two assumptions face a significant challenge when the service providers are very closely related and interdependent. Some typical examples here could be the services related to plant maintenance and provision of spare parts or interdependencies between inbound logistics and the inbound quality control.

In this third scenario, the relevant service providers interact amongst themselves to further integrate their planning to provide a complete service solution to the manufacturer. These service providers interact with the single interface of the manufacturer and determine the key requirements from the manufacturer's perspective. Furthermore, close integration is achieved amongst the most critical service providers in the service supply chain. The result being that the service supply chain is able to provide an integrated service solution to the manufacturing organisation. Any change at the manufacturer's site will be dealt with more smoothly because of higher co-ordination amongst the key service providers. This third scenario in particular requires a degree of integrated information and knowledge transfer and distributed co-ordination, planning and control capability that has not been achieved so far.

### 5.2 *State-of-the-art in service research*

Over the last years aspects addressed in the last section have been subject to research in several areas. The following outlines the body of knowledge available in different domains:

### 5.2.1 *State-of-the-art: Networked business research*

On a general level, research into networked business relationships deals with two types of organisational relationships and networks: vertical (i.e. the interaction among support companies or suppliers and production companies, e.g., Frazier et al. (1988), Heide and John (1992), Spekman and Strauss (1986)), and horizontal networks (i.e. the interaction amongst the individual companies within a business network, e.g. Bucklin and Sengupta (1993), Lo-range and Roos (1991)). The resultant frameworks and models have contributed significantly to the understanding of organisational relationships and networks (e.g. Anderson and Narus (1990), Anderson et al. (1994)). However, the relationship with supporting services linked to manufacturing chains is not considered so far in the general network research.

### 5.2.2 *State-of-the-art: Service research*

The extensive body of service research provides several approaches for modelling of services (Meffert (1994) Backhaus and Schlüter (1994), Sontow (1997)); much work originates from marketing and industrial engineering (Donabedian (1980), Edvardson and Olson (1996), Jaschinski (1998)) or from quality management (Hentschel (1992), Benkenstein (1993), Meffert and Bruhn (1995), Bullinger, Haischer amd Renner (1994)). An analysis of these approaches shows concentration on single aspects within the respective models (e.g. quality management), a certain scientific discipline (e.g. engineering) or a certain service industry (e.g. financial services), or all four concentrations. No model focuses on the synchronisation of supporting services with the manufacturing chain.

Several studies examined more specifically certain aspects of services or service aspects in supply chain contexts. For example, a special issue of the *International Journal of Service Industry Management* (Vol. 11 (4), 2000), edited by Satish Mehra, dealt with issues in supply chain management in services. The relationship between information technology and service quality in the dual-direction supply chain has been investigated by Zsidisin, Jun and Adams (2000). Van Hoek (1998, 2000) examined the role of third party logistic services in customisation through postponement. The aspect of customer supplier duality, and bi-directional supply chains, in service organisations has been addressed by Sampson (2000). A review of these studies reveals that they are not generally applicable to services in the manufacturing chain. Youngdahl and Loomba (2000) looked at a conceptual framework of service driven supply chains based on the SCOR model. This work however differs in the perspective as compared to the conceptual framework chosen here.

### 5.2.3 *State-of-the-art: Manufacturing chain integration models*
SCOR, as one of the most dominant models applied in supply chain management, is limited in scope to five primary supply chain processes: The distinct processes (source, plan, make, and deliver) of SCOR do not consider external service provision in the supply chain. A benchmarking study based on the SCOR model done by the PRTM and KPMG consulting (1997 PRTM ISC Benchmark Study), limited to measures of logistics and costs, however, clearly shows the applicability and benefit of the SCOR model for this purpose. Conceptualising supporting services functions and their impact on performance, however, is not matter of the SCOR reference architecture so far.

## 6 AREAS FOR ACTION IN INDUSTRIAL SERVICE – MANUFACTURING CHAIN INTEGRATION

Recently, information technology has been driving change in the European business landscape. ICT enables new collaborative ways of doing business and creating value. Value is migrating away from existing business designs to newer business designs that are increasingly characterised by collaborative arrangements rather than single entities. The European challenge is increasingly to continuously innovate and to co-ordinate diverse participants and their resources and competencies into synthesised activities. This is difficult, though, as the resulting business structures frequently cross the boundaries of markets and legal, technical, and cultural areas. Figure 3 illustrates some of these issues.

To better enable competitiveness within a global economy, networks of organisations in the European business system are progressively focusing on production chains and in-sourcing crucial support activities from production service networks. However it seems that many, if not most, production service networks experience difficulties in taking full advantage of technology enabled environments.

The main question to be answered in the future is how can ICT support services- manufacturing chain integration, along with the required capabilities: co-ordination, planning, decision making, and information and knowledge management. The second question is how these aspects can be integrated with cultural and personnel leadership aspects which are considered to be of increasing importance in international service relationships and networks.

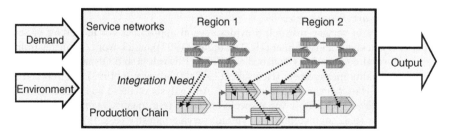

Figure 3.   The service network driven manufacturing supply chain.

194

Analysing the existing research related to service-manufacturing chain integration reveals the following: despite the comprehensive body of literature, in the service research and general network theory domain, no holistic frameworks are available that guide supporting services providers and manufacturing companies to more effectively integrate and synchronise their value collaboration. The review of existing models and studies reveals that:

- they are not applicable to services in the manufacturing chain and to be applied in the context of industrial service-manufacturing chain integration, or
- work is not suitable in the domain of supporting services in manufacturing chains.

Moreover, manufacturing chain integration models and metrics originally developed for specifying the business networks and benchmarking the performance of co-ordination strategies in manufacturing and logistics environments are inadequate and lack critical competencies to support hybrid systems which are focussing on in-service supply chain integration and which are spanning horizontal and vertical networks of integrated service-manufacturing systems.

## 7 ELEMENTS OF A ROADMAP TOWARDS THE INTEGRATED SERVICES MANUFACTURING CHAIN

To respond to these challenges of increasing complexity and interdependencies in the service supply chain sector, the process of interactions among business services and their clients require much more in-depth analysis and the availability of standard business models with much enhanced integration and governance mechanisms compared with what is available today. The complex interdependencies among the service and manufacturing firms need to be captured, analysed and re-engineered to develop new decision support systems and business models.

Within this context, the existing body of scientific knowledge in the production and service domain has to advance in the future by addressing the following research questions:

- What are the mechanisms of co-ordination and integration in the service-supply chain and what are the formal and informal governance structures needed to satisfy the needs of the different parties involved? Key challenges are to design a) new models for transfer, distribution and application of distributed information and knowledge, to design b) new models of collaboration, co-ordination, planning and control supported by enhanced information systems to enable further innovation and integration, to design c) interfaces that allow seamless integration and to include d) innovative methods for personnel management and leadership in particular to well consider interregional aspects in an increasingly global trade with services. Standardisation is an issue of increasing importance within this context.
- Furthermore, there is a need to identify the impact of integration of embedded technology with remote control into the existing decision making process and information system. The goal is to answer the question concerning how the mechanisms of co-ordination can be improved by the integration of embedded systems with remote control and existing information systems within a dynamic context.
- In terms of measuring the performance of the individual partners and the impact on the complete supply chain, new metrics need to be developed. The key research challenge is to consider how to support services activities and performance impact supply chain performance over the lifecycle and what are the relevant performance drivers and measures, even if formal organisational structures are increasingly exchanged with loose affiliations and virtual structures.

To develop a useful roadmap a research framework has been developed based on the key questions and issues addressed above. This framework is illustrated in the Figure 4.

The framework is based on cases, workshop analyses and expert studies and contains the elements considered to be of importance for the integration of support service networks across market borders and legal, technical, and cultural areas in an increasingly global and service driven economy. The

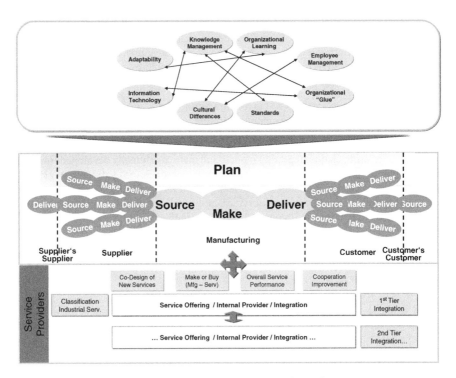

Figure 4.    Service driven manufacturing chain: Framework and roadmap elements.

framework as proposed has to be detailed and extended subsequently. Future achievements in the areas addressed will facilitate radical and evolutionary improvements in the service-manufacturing chain network and more specifically, partners' integration competencies. Addressing the aspects as listed will consequently allow structured improvement in areas that include co-ordination capability, emerging business designs, and the implementation of existing intelligence. Community learning and innovation across regional, legal and cultural boundaries will be increased accordingly.

## REFERENCES

Anderson, J. C. and Narus, J. A.: (1990) "A Model of Distributor and Manufacturer Firm Working Partnerships," Journal of Marketing, 54 (January), 42–58.

Anderson, J. C., Hakansson H. and Johanson, J.: (1994) "Dyadic Business Relationships within a Business Network Context," Journal of Marketing, 58 (October), 1–15.

Backhaus, K. und Schlüter, S.: Wettbewerbsstrategien und Exportorientierung deutscher Investitionsgüterhersteller – Eine empirische Studie. Projektbericht aus dem Betriebswirtschaftlichen Institut für Anlagen und Systemtechnologien. Nr. 94-3, Münster 1994.

Benkenstein, M.: Besonderheiten des Innovationsmanagements in Dienstleistungsunternehmen, in: Bruhn, M./ Meffert, H. (Hrsg.): Handbuch Dienstleistungsmanagement: Von der strategischen Konzeption zur praktischen Umsetzung, Wiesbaden 1998, S. 689–703.

Benkenstein, M.: Dienstleistungsqualität. Ansätze zur Messung und Implikationen für die Steue-rung. In: Zeitschrift für Betriebswirtschaft 63 (1993) 11, S. 1095–1116.

Bleeke J. and Ernst D.: Collaborating to compete: Using Strategic Alliances and Acquisitions in the Global Marketplace, New York, John Wiley and Sons, Inc. 1993; J. Levine and J. Byrne, Corporate Odd Couples, Business Week, July 21,1986.

Bucklin, L. P. and Sengupta S.: (1993) "Organizing Successful Co-Marketing Alliances." Journal of Marketing, 57, April , 32–46.

Bullinger, H.-J., Haischer, M. und Renner, T.: Wer ist der Schlankste im ganzen Land? Dienstleistungsunternehmen auf dem Weg zum Total Quality Management. In: Technische Rundschau 86 (1994) 33, S. 34–38.

Donabedian, A.: Explorations in Quality Assessment and Monitoring. Volume I: The Definition of Quality and Approaches to its Assessment. Health Administration Press, Ann Ar-bor 1980.

Edvardsson, B. und Olsson, J.: Key Concepts for New Service Development. In: The Service Industries Journal (1996), Heft 2, S. 140–164.

Hartmann, E. H. TPM: Effiziente Instandhaltung und Maschinenmanagement Redline Wirtschaft bei verlag moderne industrie, 2001.

Heide, J.B. and John, G.: (1988) The Role of Dependence Balancing in Safeguarding Transaction-Specific Assets in Conventional Channels', Journal of Marketing, Vol. 52, No. 1.

Hentschel, B.: Dienstleistungsqualität aus Kundensicht. Vom merkmals- zum ereignisorientierten Ansatz. Deutscher Universitäts Verlag, Wiesbaden 1992, January, 20–35.

Lorange P. and Roos J.: (1991) "Why some Strategic Alliances Succeed and Others Fail", The Journal of Business Strategy; January/February, 25–30.

Meffert, H. und Bruhn, M.:Dienstleistungsmarketing. Grundlagen – Konzepte – Methoden. Gabler Verlag, Wiesbaden 1995.

Meffert, H.: Marktorientierte Führung von Dienstleistungsunternehmen. In: Die Betriebswirt-schaft 54 (1994) 4, S. 519–541.

Sampson, S. E.: (2000) "Customer-Supplier Duality and Bidirectional Supply Chains in Service Organizations," International Journal of Service Industry Management Vol. 11, Pages 348–364.

Schomburg, E.: Entwicklung eines betriebstypologischen Instrumentariums zur systematischen Ermittlung der Anforderungen an EDV-gestütze Produktionsplanungs- und –steuerungssysteme im Maschinenbau. Dissertation. RWTH Aachen 1980.

Schönsleben, P.: Integrales Logistikmanagement – Planung und Steuerung von umfassenden Geschäftsprozessen.2. AuflageSpringer Verlag, Berlin 2000.

Schönsleben, P.: "Integral Logistics Management – Planning and Control of Comprehensive Business Processes", CRC Press / St-Lucie Press, 2000.

Schönsleben, P. und Hieber, R.: Supply Chain Management Software: Welche Erwartungshaltung ist gegenüber der neuen Generation von Planungssoftware angebracht?, io management, Nr. 1/2 2000.

Sontow, K.: Entwicklung einer Vorgehensweise zur Planung eines potentialorientieren Dienstleistungsprogramms für kleine und mittelständische Unternehmen des Maschinen- und Anlagenbaus. Sonderdruck FIR, Aachen 1997.

Spekman R.E., Forbes T.M. III, Isabella L. A. and MacAvoy T.C.: Alliance Management: A View From the Past and a Look to the Future, Marketing Science Institute, December, 1995.

Spekman, R.E. and D. Strauss: (1986) "An Exploratory Investigation of a Buyer's Concern for Factors Affecting more Co-operative Buyer-Seller Relationships", Industrial Marketing and Purchasing, Vol. 1, No. 3, 26–43.

van Hoek R.: "The Role of Third-Party Logistics Providers in Mass Customization," The International Journal of Logistics Management, Vol. 11, No. 1 (2000), pp. 37–46.

van Hoek, R., Weken, H.A.M.: (1998) "The Impact of Modular Production on the Dynamics of Supply Chains," The International Journal of Logistics Management, Vol. 9, No. 2, pp. 35–50.

van Hoek, R.: (1998) "Re-configuring the Supply Chain to Implement Postponed Manufacturing," The International Journal of Logistics Management, Vol. 9, No. 1, pp. 95–10.

www.supply-chain.org

Youngdahl, W. E. and Loomba, A. P. S.: (2000) "Service-Driven Global Supply Chain Management." International Journal of Service Industry Management, Vol. 11, No. 4, pp. 329–347.

Zsidisin, G. A., Minjoon J. and Adams L. L.: (2000) "The Relationship Between Information Technology and Service Quality in the Dual-Direction Supply Chain: A Case Study Approach," International Journal of Service Industry Management, Special Edition on Supply Chain Management in Service Operations, Vol. 11, No. 4, pp. 312–328.

*Advanced Manufacturing – An ICT and Systems Perspective – Taisch,*
*Thoben & Montorio (eds)*
© *2007 Taylor & Francis Group, London, ISBN 978-0-415-42912-2*

# Challenges in industrial service business development

Vesa Salminen[1] & Petri Kalliokoski[2]
[1] *Lappeenranta University of Technology*
[2] *VTT Technical Research Centre of Finland*

ABSTRACT:  Industrial service business is a fast-growing area in engineering and manufacturing industry. Many companies have tried to develop industrial services to create new business with customers but many of them have failed. Often customers have not valued the proposed service models because of the lack of added value compared to current co-operation between supplier and customer. This paper presents the findings of a BestServ feasibility study that was aimed at recognising the current status, development needs and future challenges of industrial service business in Finnish manufacturing industry.

*Keywords*:  Industrial services, knowledge intensive business, complexity, dynamic transition, value proposition, value network.

## 1  INTRODUCTION

The business environment is influenced by variety of economic and dynamic trends according to which companies have to consolidate on a global scale. Companies can at the same time be driven by technological and business innovations, all kinds of deregulation, customer requirements, and other elements. All these factors and trends add to the complexity of solutions development and make fast new product and service introductions even more important and challenging. Companies that are able to manage these challenges effectively will be the ones that succeed in dynamic markets in which internal and external service providers will execute development activities. Management will be done according to value creation and networking. Clusters of networked organisations are collaborating around a specific technology and are using a common architecture and operational models to deliver independent elements of customer value and solutions that grow with the number of participating organisations.

### 1.1  *Paradigm shift in industry*

The engineering industry is now undergoing a transition from being the product provider to being the provider of customer value and product-related value-added services (Clarke and Clegg 1998). Enterprises have proclaimed and tried to undergo this transition but have failed in reality in several respects. Some of the product-related services are partly implemented with technological solutions, but most of the industrial services are only pilot schemes. The challenge is to identify customer's critical processes and develop services to support these processes.

This transition from *ownership* to *access* and the potential sustainable growth of the business lies with the creation and capture of these services. Business related to industrial services tends to grow out of a commodity trap. This transition can be termed the *framework of value transition*. This framework covers the complete transition of the industry from *parts supplier* to *value provider* (Figure 1). The figure illustrates the main element that an enterprise or value network of enterprises needs to become an integrated product-service provider (Tushman and Anderson 1997; Bainbridge 1996).

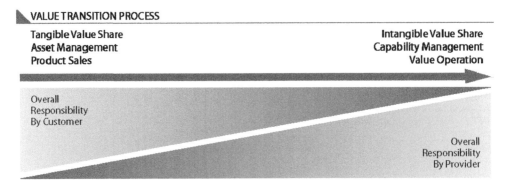

Figure 1. The framework of value transition.

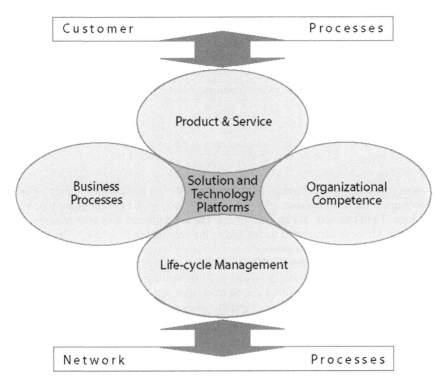

Figure 2. Integrated products and services management.

This value transition means that traditional products are changing into solutions covering products and services. Typically, customers have no capability to use these solutions without solution-provider services. The transition from products to solutions creates the basis for new business and co-operation models among networked companies. Figure 2 illustrates the model of integrated products and services management.

It is estimated in many studies that the volume of intangible elements in products will grow significantly. Figure 3 depicts an estimate of mechanical industry breakthrough development. The estimate is based on the BestServ feasibility study (Kalliokoski et al. 2003) and other reviews of different product and service concepts. This analysis has concentrated only on the concept level, not on the business level.

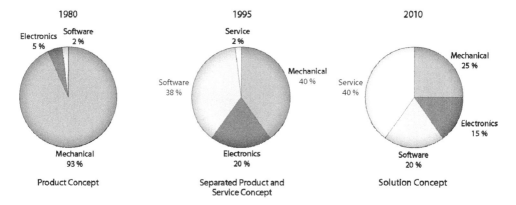

Figure 3.    Breakthrough development vision of the manufacturing industry – concept paradigm.

## 1.2    *Structural change of businesses*

There is a great need for most of businesses to develop their product and service management in an open systems architecture. Business concepts are changing towards a knowledge-intensive and value-critical approach over the product lifecycle. Value networks of companies are responsible for solution offerings consisting of service and product elements produced by various enterprises. Management of offering structure and further innovation needs a semantic structure (Pallot et al. 2004). Intelligence is increasingly embedded into products, and product and service modules are evolving in a dynamic relationship through a growing value network of partners. This means that the management and role of service and product architecture are becoming very important when managing increasing complexity. The enterprises of today are looking for methodologies and tools that will help capture aspects of a business and analyse these to identify and compare options for meeting the requirements of the business. This enables the sustainable growth and development of the customer value offered. Complexity is increasing owing to new generations of products, which include embedded intelligence with multiple functions in diverse operating conditions. The implementation of technologies is becoming multidisciplinary and the use of knowledge-bases has rapidly increased. Systems interaction has become complex. Subsystems share and exchange information, communicate, restructure methods and track new tools. Service and product architecture and integrated modular concepts have become increasingly important in managing the whole business (Salminen and Pillai 2001). The core is a company's own structure and practices, consisting of product and service configurations, business processes connected with activity, organisational competence and lifecycle management, and a learning and innovative culture (Prahalad and Ramaswamy 2004); (Achtenhagen et al. 2003), (Grönroos 2000).

The business model is changing completely, while the quantity of services is increasing in the long run. The hardest transition emerges when a company and its value partners are changing on the road to performance partnerships (Figure 4). The various business models of industrial service business when knowledge intensity is increasing are introduced in section 2. There exist several business transition points on that road, and product and service integration creates a continuously changing structure at the heart of the enterprise. The entrepreneur is surrounded with a new business model that offers solutions and other dynamically forged offerings to match the new circumstances. The new business level is achieving the intensity of higher knowledge, and the pattern of interactions among partners becomes service focused.

## 1.3    *Organisational change*

The paradigm shift from *ownership* to *access* also applies to the organisational business transition process that covers several business processes. New dimensions of organisational concepts and

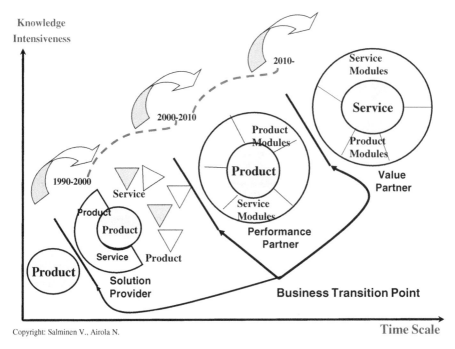

Figure 4. Business transition process from machine supplier towards value partner (Salminen and Pillai 2005).

business models are forming at the moment as a result of this business transition process. Concepts and perspectives on learning and innovation have emerged in connection with discussion about enterprise networks (Hyötyläinen 2000). The result is that product and service development is considered as a new arena for the co-operation of enterprise networks. New organisational forms of networks and information technology platforms are required to realise the potential of the new opportunities (Musgrave and Anniss 1996).

As a result of this development companies are moving closer to their customers, which entails the need for new product concepts and service models. A product requires services that add value to the product (Tomlinson 1997). Industrial services can be seen as the optimisation of customer assets. Service business can be related, for example, to technological co-operation, process improvements, remote diagnostics and financial arrangements. Technology breakthroughs are speeding up the development of product concepts in some areas.

Corporations that are able to manage these challenges effectively will be the ones to succeed in dynamic market places where internal and external service providers will execute the implementation activities. Corporate management will respond to value recognition and networking (Hyötyläinen 2000). Clusters of networked organisations are already collaborating around a specific technology and making use of a common architecture to deliver independent elements of value, which grows with the number of participating organisations (Salminen and Pillai 2003).

## 2  A PRACTICAL INSERT ON THE STATE OF INDUSTRIAL SERVICES

During recent years, there have been feasibility studies on the state of industrial services in Finnish industry. The studies were done by several research groups under one project called BestServ. According to the feasibility studies, a common understanding has been created about what an industrial service is and how it interrelates with normal product businesses (Kalliokoski et al. 2003). Manufacturing companies position themselves differently in terms of customer intimacy

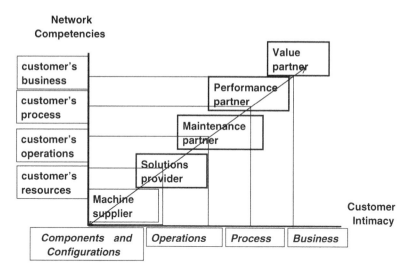

Figure 5.    Business model analysis according to customer intimacy.

through their industrial service offerings and operations. For practical reasons, five different supplier positions or *roles* (Figure 5), relative to the customer, were defined:

- Machine supplier: The focus of the business relationship is on delivering a piece of machinery or equipment that fits the customer's technical specification.
- Solution provider: The focus of business is on delivery of a system, e.g. a production line, which is usually designed for the specific customer's process and comprises a wider scope of supply than just one piece of equipment.
- Maintenance partner: The focus of business expands to also include continued supplier involvement during the continuing lifecycle of the delivery. This role adds contractual aftermarket elements, such as spares and consumables agreements, to the supplier-customer relationship.
- Performance partner: In this role the supplier is closely involved in operating the customer's technical process by taking partial responsibility for the performance of the system, e.g. through availability warranties. This role requires the supplier to maintain at least a minimum of continuous on-site presence. The focus of the customer relationship is on securing the effective operation of the unit or production line.
- Value partner: The supplier is directly involved in the customer's business, e.g. through *operate and maintain* agreements, where the customer pays a pre-determined price for the actual output of the system. Both parties focus on profitable daily operations, and the supplier is responsible for the day-to-day operation of the plant or line.

Each of these five supplier business models has associated with it a specific *mindset*. When a supplier aims to progress from one model to the next, it faces tough challenges, mostly in terms of getting the customer involved in the change and developing its own technical and business competencies such that it is able to advance. The strategic positioning decision between supplier and customer is important and has to be prepared as thoroughly as any other strategic decision.

The first two models focus the supplier's activities on the customer's investment decision and do not concentrate too much on supporting the lifecycle of the customer's process. A solution supplier needs the ability to understand and interpret the customer's actual operations in its offering. A maintenance partner concentrates on professional maintenance management as a continuous process. As a performance partner, the supplier can have a responsibility for the actual daily performance of the customer process. When the supplier is a value partner, it is involved in the customer's value generation, e.g. producing optical cable in a cable factory at a given quality and price. The supplier has to have competence in the customer's business. The level of knowledge and experience is increasing and creates competence for productive communication among partners of a value network.

The BestServ feasibility study showed that industrial services have been seen as a strategic intent to manage global competition and the evolution of current business models. This approach leads to the management of the customer offering through lifecycles, from a solution and a customer perspective.

The main long-term development areas identified were:

- It is difficult to recognise the benefits of industrial services for customers and for all suppliers. The main challenge in this may be the lack of a shared value model of industrial services. The shared value model enables discussion about the potential benefits and values to be captured by the services.
- Industrial services are usually built like extended products based on current product architectures and not on management of customer requirements and values. There is a need for a customer-oriented solution architecture that integrates product and service offering and enables efficient market segment management.
- It is difficult to determine an interoperable structure for industrial services to be integrated with product structure. This complicates the creation of new and innovative business models. Enterprises should have reference business models based on the integrated structure of product related industrial services (earnings logic, business strategy, organisational models, etc.). The efficient development and use of a reference model enables the continuous innovation of an integrated offering over the lifecycle of the customer process and own business model.
- At the moment many industrial services are traditionally oriented, while the need is for knowledge-intensive services (e.g. proactive maintenance, all kind of business consultation). The development of knowledge-intensive services requires a deep understanding of customer processes as well as the development of the service provider's own competencies. These require the reinventing of the customer offering and the related business model.
- Technological solutions (e.g. telecom, automation, and operational systems) are mainly developed to support separate operations and processes. There is a development challenge to manage the integration of separate technological solution. Continuous industrial service development requires parallel development of business architecture and information as well as communication technology architecture.

During the survey it was recognised that most of the companies needed to adapt their business according to the paradigm shift from ownership to access and that potential sustainable growth of business lies in services created and captured. The framework of value transition means a transition from parts or machine supplier to value provider. Then the important element is an adaptive business transition process.

A well understood, structured business model supporting business architecture is a very important strategic tool when business is evolving according to the market requirements. The solution architecture is reliant on business alignment. Solutions consist of integrated service and product elements. When there are reusable elements in the architecture it is easier to build up new ones. Customer and functional requirements over the lifecycle, product and service features, modules, and components and interfaces, build the core structure of an enterprise solution structure.

## 3 DEVELOPMENT FRAMEWORKS ON STRATEGIC MANAGEMENT OF COMPLEXITY

### 3.1 *Business development framework*

The idea of a business development framework is to manage the business model development to promote successful service business concepts and solutions. The purpose of Figure 6 is to depict the process of promoting the right business concept at the right time in the targeted markets. Figure 6

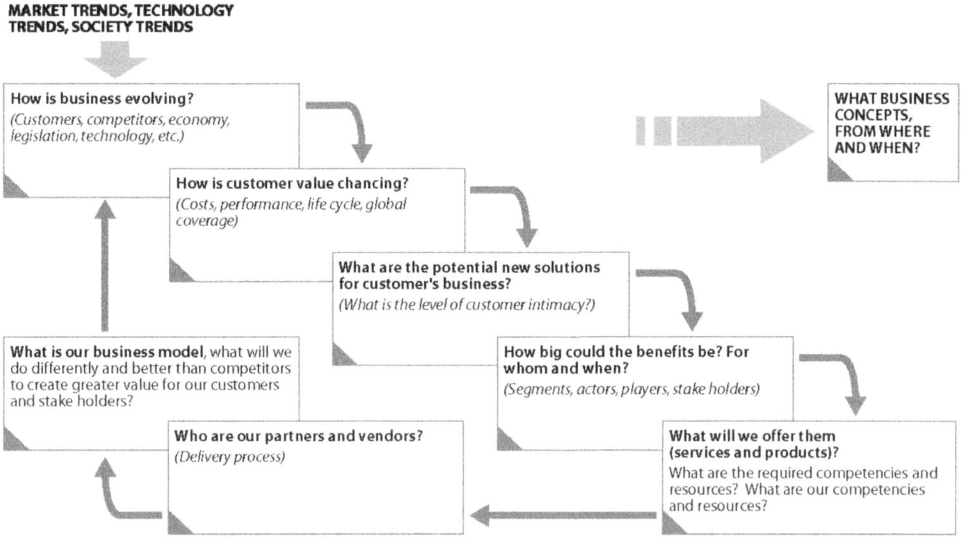

**MARKET TRENDS, TECHNOLOGY TRENDS, SOCIETY TRENDS**

How is business evolving?
*(Customers, competitors, economy, legislation, technology, etc.)*

How is customer value chancing?
*(Costs, performance, life cycle, global coverage)*

What are the potential new solutions for customer's business?
*(What is the level of customer intimacy?)*

What is our business model, what will we do differently and better than competitors to create greater value for our customers and stake holders?

How big could the benefits be? For whom and when?
*(Segments, actors, players, stake holders)*

Who are our partners and vendors?
*(Delivery process)*

What will we offer them (services and products)?
What are the required competencies and resources? What are our competencies and resources?

**WHAT BUSINESS CONCEPTS, FROM WHERE AND WHEN?**

Figure 6.    Business development framework.

consists of different questions, and by answering these questions, companies can focus on and direct the correct development activities. The development framework is a continuous loop and the result of that loop should be new innovative business concepts for managing business.

The main question for service business strategy is the evolution of the customer's business. This means the changes in the customer's values that should be captured with the new service solutions. There are different market trends (e.g. technology, market and society trends) that are changing business. Companies competing within the industrial service business market should track the potential and value of the benefits the customer may get from the services. The main concept of the service offering should be designed based on the expected values. This applies to the service offering concept as well as the offered products. Industrial services distribution networks should also be designed to meet the customer service level and its requirements. Based on these main industrial service business development components, the expected business model and concept can be agreed to promote the designed services and products.

### 3.2    Technology development framework

Based on the designed business concept and model, the technological choices for supporting industrial service business should be addressed. In the past, the development of industrial services has mainly been technology-oriented, without successful business implementation. It had been expected that the technological solutions themselves would provide the added value to the customer. One of the main drivers for developing technological solutions for industrial services should be a business-oriented roadmap that identifies the service business development trends and changed customer expectations and values. Figure 7 illustrates a preliminary framework for managing business-oriented technology development. The technology roadmap process presented is also a continuous loop, but now the first effect of development is coming from the company's own business roadmap.

In developing industrial service business technologies, the challenges are in creating and managing solid ICT architectures and the integration of different kinds of applications and solutions like operational systems (e.g. ERP), field devices, telecommunications and remote systems. The integration and implementation of large ICT architectures is a significant investment. That is why the customer added value should be addressed to avoid risks.

205

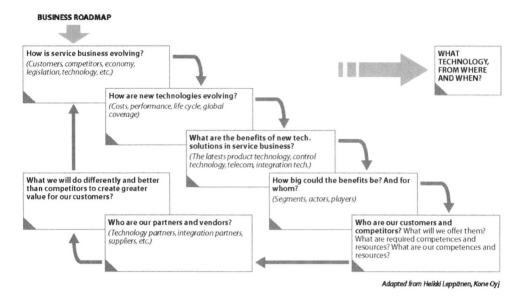

Figure 7. Technology roadmap process.

### 3.3 *Organisational transformation process*

The development of a service business requires people. Organisations and their employees are in a key position to make structural change in business possible. From the organisational standpoint, service business development needs organisational transformation. There are many different attitudes, opinions and cultural aspects that explain why business development is hard to facilitate. An organisation should change to make service business effective and also to turn services into business for a company. Figure 8 presents a framework for the organisational transition process. The framework outlines different tasks that organisations should complete to achieve organisational transition and effective new service business. The first step is to clarify different industrial service reference models. The reference models are needed to clarify how other companies have done, so the first step is similar to benchmarking. There is no sense in creating new models if there are suitable models already being developed. Of course, none of the models are directly suitable, but after analysis there could be common ideas worth adopting for a specific business development. After reference-model analysis, organisations should understand and test customer values. When it is clear what the customer's needs and requirements are, organisations can build their product and service offerings to fulfil customer needs. This step includes segmentation as well as new product and service architecture development. The next step in the organisational transition process is business model implementation, which requires new mindsets and competencies. The next step is the definition of the organisational business transition process. The following step is the development of a collaboration network, where operations with customers and partners should be managed. Business process definitions come after the development of a collaboration network. Then organisations have an almost complete framework for new service business. The last step is architecture creation based on the earlier phases.

By following the above framework, organisations can manage their transition process. The time frame in every case should be clarified: this will depend up on organisational capabilities.

### 3.4 *New industrial service development*

New product development has several methodological process models and tools that are designed for product development from idea to market introduction. Many of these product development

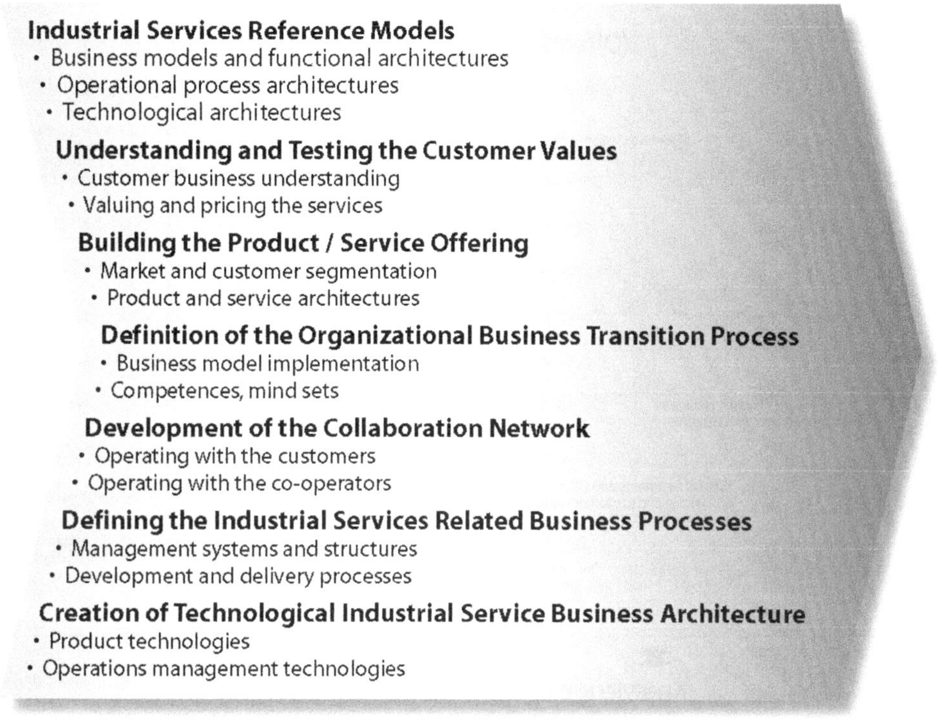

**Industrial Services Reference Models**
- Business models and functional architectures
- Operational process architectures
- Technological architectures

**Understanding and Testing the Customer Values**
- Customer business understanding
- Valuing and pricing the services

**Building the Product / Service Offering**
- Market and customer segmentation
- Product and service architectures

**Definition of the Organizational Business Transition Process**
- Business model implementation
- Competences, mind sets

**Development of the Collaboration Network**
- Operating with the customers
- Operating with the co-operators

**Defining the Industrial Services Related Business Processes**
- Management systems and structures
- Development and delivery processes

**Creation of Technological Industrial Service Business Architecture**
- Product technologies
- Operations management technologies

Figure 8.    Framework for the organisational transition process.

processes and methodologies are based on traditional stage-gate approaches and they are widely implemented in organisations and operational quality systems. Product development in industrial companies is a rather systematic approach to managing cross-functional development processes. Usually they also have some aspects related to services development in support of the products.

During the BestServ project, a preliminary new service development model was developed, based on analysis and discussion with the participating companies. This model focuses on new service development in co-operation with the customer. The model is based on business strategy (product and market management) and gained knowledge and learning. The preliminary model is depicted in Figure 9. The new service development process is based on strategic business management focused on maximising customer value. This is achieved by understanding the customer business logic and operational business processes. The main idea behind the maximisation of customer value is market and business-based customer segmentation with a modular offering structure that combines the products and services required for each customer. The modular design of the offering structure is a continuous process that should focus on solution deliveries. This kind of modular offering can be based on lifecycle innovation of the solution lifecycle management. Solution lifecycle management is based on business intelligence and related knowledge and information.

One of the main components of the model is continuous interaction with the customer about the developed services. Interaction is based on the common value model. The value management process, which is composed on three main phases, can depict the interaction process. The phases are value evaluation/assessment, value creation and value maximisation/delivery. The value creation process can be divided into phases:

- In the first value model phase (value evaluation and assessment), the potential of the new service value is discussed and tested with customers (e.g. leading customer group). As a result of this phase, a value proposition is made to the customer.

207

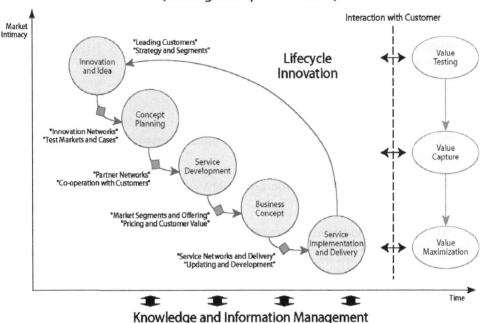

**Strategic Management
(Offering Concept and Markets)**

Figure 9.   New service development process (NSD).

- In the second phase (value capture) the value of the service being developed is captured by the service concept that concretises the value for the customer.
- In the third phase (value maximisation and delivery) the proposed value is delivered to the customer by means of the new and innovative service.


## 4   INDUSTRIAL SERVICE BUSINESS DEVELOPMENT AREAS

According to the industry-driven and oriented feasibility study, the main development areas of industrial service business can be derived from structural change in the businesses of participating companies. Industrial services can be seen as a strategic intent on the part of all the BestServ companies to manage global competition and the evolution of current business models. This approach leads to the management of the customer offering through lifecycles, from a solution and a customer perspective. The main development areas are related to the following generalisation derived from industrial development needs:

- According to the companies studied, argumentation of the benefits of industrial services is difficult for many of the companies. The main challenge for this may be the lack of a shared value model of industrial services. The shared value model enables discussion about the potential benefits and values to be captured by the services.
- Industrial services are usually based on current product architectures and not on comprehensive management of customer needs and values. This kind of customer-oriented approach needs a solution architecture that combines product and service offering and enables efficient market and segment management.
- Industrial services are difficult to structure and manage by companies, which complicates the creation of new and innovative business models. This means that companies should have reference

business models for industrial services (earnings logic, business strategy, organisational models etc.). The efficient development and use of IS reference models enables the continuous innovation of industrial services.

- At the moment, many industrial services are traditionally oriented, while the need is for knowledge-intensive services (e.g. consultation, proactive maintenance etc.). The development of knowledge-intensive services requires a deep understanding of customer processes as well as the development of the service provider's competencies. These require the reinvention of the customer offering and the related business model.
- On the common level, technological solutions (e.g. telecom, automation, and operational systems) are mainly developed to support separate operations and processes. The main development challenge is related to architectural management of technology integration. The business technology architecture for industrial services should also be defined.

The presented development areas together create a need for a solution lifecycle concept that means lifecycle and customer-oriented approaches to managing the customer offering.

All these development areas create a great need for systematic research and development activities, which should be organised to manage the global competition and the creation of innovative business models and customer values. The following sections summarise the main development areas under the project clusters supporting a collaborative approach to research and development.

## 5  MAIN SERVICE BUSINESS LONG-TERM R&D TOPICS

This contribution has concentrated on long-term development activities and system development in the face of a business culture that is changing towards industrial service business. The analysis of the main R&D focus areas from a system perspective is also introduced.

During the survey it was recognised that most of the companies needed to adapt their business according the paradigm shift from ownership to access and that potential sustainable growth of business lies in services created and captured. The framework of value transition means a transition from parts or machine supplier to value provider. Then the important element is an adaptive business transition process. There will be a continuous change in business models, which can be run according to the analysis of the rough value and evolving business models. It is difficult to manage this transition without system understanding and a well-structured customer-value based process for business transition.

The development topics should be understood as a system, with subsystems and individual R&D topics dependent on one another. It is only possible to concentrate on a few topics at the same time in any given business network and individual enterprise. Prioritising according to changing system requirements is essential. It should be remembered that everything is changing as the move towards service-oriented business is made. Results in one development area have an influence on requirements that develop another. Business system understanding should support continuous prioritising.

The main long-term focus areas when developing industrial service business were analysed after the industrial survey. Figure 10 illustrates the R&D topics from a system perspective. The basic idea is that various development areas are dependent on one another.

### 5.1  *Value network management over the solution lifecycle*

There is a great need for a new approach to lifecycle management. This will be based on a value model supported by all stakeholders. Because the main partners in a value network are at some level responsible for customer business, customer value management becomes even more important. Organisational culture becomes more networked and value-oriented. Value has to be created, evaluated, captured and finally maximised and delivered over the lifecycle of a customer application. This cannot be achieved without excellent collaboration network management. Continuous uncertainty should be changed into risk management. Also social capital becomes essential among

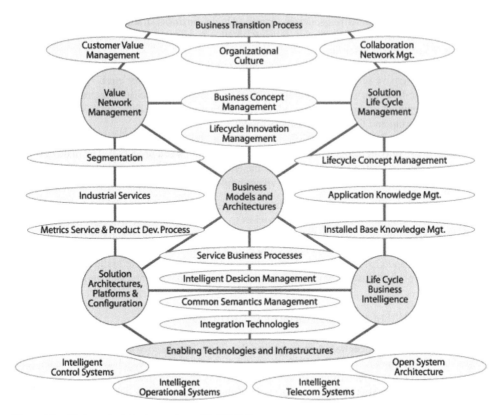

Figure 10.   Main project clusters and specific R&D themes.

other sources of capital. Dynamic relationship management will be needed, especially as some of the business network partners will change over the lifecycle. There will also be certain requirements for new methodologies for life-cycle communication. An installed base application requires continuous maintenance and renewals to maintain application competitiveness in the customer's operational and business process. Collaborative Value Assessment Practice is missing. The challenge is to create and define the customer-oriented value model for developing industrial service implementation.

## 5.2   *Business models and architecture supporting the business transition process*

The business transition process with suitable metrics is the one of the most important development areas. Continuous industrial service development requires parallel development of business architecture and ICT architecture. A well understood and structured business model supporting business architecture is a very important strategic tool when business is evolving according to market requirements. Lifecycle innovation needs new approaches and leads to business-concept management. In future, it will be possible and essential to sell business models based on the available architectural structure of a company. The main problem is that a reference business model and mechanism for the creation of new business model schemas are missing. Business transition is not supported by collaborative change process in a networked environment. The challenge is to ensure the organisational transition process from *product-oriented approach* to *customer-oriented solution approach*, from *ownership* to *access*. The challenge is to create and develop reference models for industrial services. This enables business model evolution and cost efficiency management.

A challenge is also to recognise when it is time to change the business model and to identify which pieces of the old one are usable and which have to be developed further or changed.

### 5.3 Solution architectures, platforms and configurations

Solution architecture is at the heart of business alignment. Solutions consist of service and product elements. When there are reusable elements in the architecture it is easier to build up new ones. Customer and functional requirements, features, modules, components and interfaces build the core structure of an enterprise solution structure. It is important to have an integrated service and product development process to create new offerings according both requirements. Service and product platforms supported by the solution architecture are used in the solution and in the customer configuration. The main challenges are the development and implementation of relevant solution architecture and platform knowledge that enables efficient customer and market management.

### 5.4 Solution lifecycle management and the supporting business intelligence system

A lifecycle business intelligence system is important when the competitiveness of a customer application has been secured. A lifecycle business intelligence concept needs to be created to combine application knowledge and organisational knowledge in the value network for lifecycle information management. An intelligent decision management system will also be needed to support new product and service element development. This is a parallel activity to development competence, service and the product as a whole. Development and other operational processes such as industrial service delivery, business logistics and communication will alter owing to the changes in offerings and targeted market segmentation. Solutions are embedded and completely mechatronic. When an attempt is made to manage a solution lifecycle, the lifecycle business intelligence is usually insufficiently collated and a lifecycle concept to direct the evolution of the customer's operational process does not exist. The challenge is to manage the solution lifecycle and its phases efficiently by providing customer-oriented industrial services. Service metrics is an important means for supporting continuous evolution.

### 5.5 Organisation culture and service competence

Many companies refer to *developing a new mindset* in the organisation, as the main challenge in establishing a new industrial service role (cf. McGraw and MacMillan 2000). The notion of developing a new mindset suggests replacing the old mindset with the new one, e.g. developing from *product driven* to *customer driven*. The risk in this notion is that it leads to undermining the company's strength in its current supplier role. All companies tackle the challenge in their own way.

It is essential when developing business according the related architecture that also information and communication systems are developed at the same time to support the business evolution. Organisational culture and service competence are most important areas to develop at the same time. It is difficult to manage the evolutionary process without changing the organisation and competence structure at the same time.

### 5.6 Enabling technologies and infrastructures

Finally it is essential to develop methodologies, solutions, and customs in information and communication management in a value network. Research and development of enabling technologies and infrastructures is a parallel activity alongside the others presented above. There is a need for new integration technologies to get intelligent control systems and intelligent operational systems as well as intelligent telecom systems and open system architecture to fully support the operational industrial service. There will be a need for new types of remote diagnostics and wireless systems. New types of business hub systems will be developed to support fluent collaboration in a value network. Common semantics management is essential and should be developed for industrial service

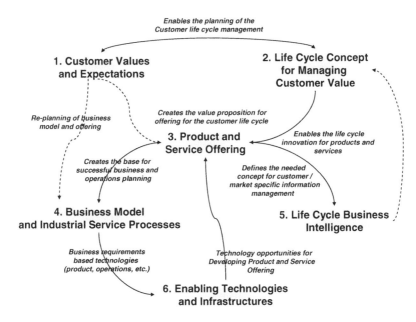

Figure 11.    Customer value-based interrelationship chart for business transition process.

oriented, knowledge-intensive business. The main problem is that enabling technologies and infrastructure of various operational ICT Systems and industrial service architectures are not integrated. The challenge is planning and implementation of industrial service technology architectures and integration of different technologies (product, operations, back-end).

### 5.7    *Continuous business system co-evolution*

In a modern and dynamic business environment, the management of continuous change or co-evolution is essential for competitive edge and sometimes as well for survival. Business is the same type of system as other systems; it has operational subsystems, which function together. They can be collaboratively developed further as a dynamic system by the help of various frameworks. The frameworks are various tools for cross-functional discussion and collaborative working. The *business development framework* can be seen as a preliminary framework for industrial-service business model development (customer value, business model, collaboration network, etc.). The *technology development framework* is based on the business model and it is aimed at developing the technological infrastructures and applications to support industrial service. The *business transition process* from *product supplier* to *value or performance partner* can be seen as a service business development framework. This transition process contains an organisational approach for business model development and implementation. Figure 11 shows in a simple way the introduced customer value-based challenge areas and their relationships as a system model – let it be called the lifecycle service business system.

## 6    CONCLUSIONS

Thirty Finnish companies, all operating in the global market, and several universities and research institutes have been participating in a project and roundtable work called BestServ. It has been an interest group and knowledge community for companies, research organisations and financiers to activate and guide industrial service development. The purpose of this analysis work has been to find out the state of industrial service business globally, in various business sectors and also in individual enterprises. During the roundtable work, several generic frameworks have been created to help enterprises face future challenges. BestServ work has given also guidance and alignment for future research and development challenges.

This paper introduces some generic frameworks created and understanding achieved during that work:

- Short-term development areas and good practices were identified by benchmarking;
- Key long-term research and development areas were defined for further development;
- A preliminary business and technology framework as well as an organisational business transition process was defined for managing industrial service development in an enterprise context;
- The lifecycle innovation process from collaborative idea to implemented service was preliminarily defined for industrial services.

Industrial service development activities should be collaborative, enterprise-driven development and long-term cross-scientific research that combines the various sectors of the relevant research traditions and themes (management, technology, psychology, etc.).

## REFERENCES

Achtenhagen, L., Melin, L., Mullern, T., Learning and continuous change in innovating organizations. Innovative forms of organizing. London: Sage Publications, 2003, p.72–94.

Bainbridge, C., Designing for change – A practical Guide to Business Transformation, John Wiley & Sons, Inc, West Sussex, 1996.

Clarke, T., Clegg, S., Changing Paradigms – The Transformation of Management Knowledge for the 21st Century, Harper Collins Publishers, London, 1998.

Grönroos, C., Service management and marketing. West Sussex: John Wiley & Sons, 2000.

Hyötyläinen, R., Development mechanisms of strategic enterprise networks. Learning and innovation in networks. Espoo: VTT Publications 417, 2000.

Kalliokoski, P., Salminen, V., Andersson, G., Hemilä, J., BestServ. Feasibility Study, Final Report. Teknologiateollisuus, Kerava: Savion Kirjapaino Oy, 2003.

McGrath, R.G., MacMillan, I., The entrepreneurial mindset. Strategies for continuously creating opportunity in an age of uncertainty. Boston. Harvard Business School Press, 2000.

Musgrave, J., Anniss, M., Relationship Dynamics. The Free Press, New York, 1996.

Pallot, M., Salminen, V., Pillai, B., Pawar, K., Business Semantics: The Magic Instrument Enabling Plug & Play Collaboration?, ICE 2004, International Conference on Concurrent Engineering, ICE, Sevilla, June 14–16, 2004.

Pettigrew, A.M., Linking change processes to outcomes. In Beer, M. & Nohria, N. (Eds.), Breaking the code of change. Boston: Harvard Business School Press, 2000, p. 243–265.

Prahalad, C.K., Ramaswamy, V., The future of competition. Co-creating unique value with customers. Boston: Harvard Business School Press, 2004.

Salminen, V., Pillai, B., Strategic Management of Adaptive, Distributed Product Development of a Mechatronic Product. In Arai E., Arai T. and Takano M (Eds.), Human Friendly Mechatronics. Elsevier, 2001, p. 377–382.

Salminen, V., Pillai, B., Methodology on Product Lifecycle Challenge Management for Virtual Enterprises. 4th IFIP Working Conference on Virtual Enterprises, PRO-VE'03, Lugano, Switzerland, 29–31.10.2003.

Salminen, V., Pillai, B., Integration of Products and Services – Towards System and Performance Partner . International Conference on Engineering Design, ICED 05, Melbourne, August 15–18, 2005.

Tomlinson, M., The Contribution of Services to Manufacturing Industry, CRIC Discussion Paper No.5, 1997.

Tushman, M., Anderson, P., Managing strategic innovation and change. New York: Oxford University Press, 1997.

*Part VII*
*Human factors and education in*
*manufacturing*

To survive in the face of worldwide competition, companies are forced to produce ever-increasing volumes at the same costs. Meanwhile the variation in products increases, shorter delivery times are demanded and quality should be high as well. This pressure on companies can be solved partly by the optimal use of the human performance. It has been demonstrated that a motivated and healthy workforce contributes significantly to innovation and productivity. Especially in Europe there is a great potential in the workforce to increase innovation and to be more competitive, because the level of education is relatively high compared to other areas in the world and there is no machine as flexible as a human. For the coming years the challenge is to use this innovation potential in the competition with other parts of the world.

The first paper in Part VII contains a contribution focussing on measures to increase productivity and employability by paying attention to the workforce and the environment in which they work. It shows that flexibility is an important future theme and that the workforce is crucial as they are the most flexible *machine* and because ideas to become more flexible are generated by humans. Six themes that need attention in the future are proposed by Vink and Stahre: (1) Humans will be a core resource in coping with the demand for flexibility; (2) Humans have the ability to cope with complex tasks; (3) Inclusive design to cope with special groups (e.g. the elderly); (4) Systems should support self-control by humans; (5) Time aspects; (6) Towards service manufacturing.

The second contribution written by Rolstadås provides a new approach for education and training of engineers to better comply with the ever-changing needs of a more global environment for manufacturing operations. Industrial manufacturing is no longer home-based; it operates in a global market. Digital business has become a strategy to survive. The extended enterprise is being implemented. Components and even products are made where conditions are most favourable. However, engineering education does not completely reflect today's needs of an industry that faces problems of an integrative nature across the traditional disciplines, such as: working globally in a multicultural environment; working in interdisciplinary, multi-skilled teams; and working with digital tools for communication. This circumstance calls for a new university curriculum reflecting technology and management. The main achievements of the IMS project GEM (Global Education in Manufacturing) are: the definition of the industrial needs for a new type of curriculum in manufacturing industry and the development of a framework of a curriculum in manufacturing strategy accessible over the Internet and allowing appropriate flexibility for implementation according to local tradition and legislation.

As products become more complex, it is often necessary for various enterprises with certain key competencies to collaborate to produce a product. The resulting specialisation leads to an increased demand for knowledge. Thus inter-organisational learning gains importance. As a consequence of the stated developments, the way of working and thus the educational requirements have changed as well. Performance skills about the processes and challenges within inter-organisational product development and learning are becoming vital for efficient participation in future working scenarios. A promising approach to increase the efficiency of learning is the application of simulation games. In the final paper in this part of the book Thoben and Schwesig describe the development and the initial testing of a web-based group simulation game that has been built to address new educational demands.

*Advanced Manufacturing – An ICT and Systems Perspective – Taisch,*
*Thoben & Montorio (eds)*
*© 2007 Taylor & Francis Group, London, ISBN 978-0-415-42912-2*

# Human factors research to increase manufacturing productivity and innovation

Peter Vink
*TNO/Delft University of Technology, The Netherlands*

Johan Stahre
*Chalmers University, Göteborg, Sweden*

ABSTRACT: It is essential that Europe have a flourishing manufacturing industry to keep employment and to be economically healthy. However, the wages of citizens in the European Union are higher than in other regions of the world and the European Union is not able to compete on high volumes with much manual work. In this paper a project is described in which 20 experts in human factors defined the most promising way of using the European workforce in manufacturing. Based on this, the research needs for the coming 10 years with regard to human factors in manufacturing, are described.

To identify what research is needed to have an optimal functioning staff in future manufacturing, literature on the important current problems was gathered and discussed with experts and manufacturing companies. Based on these data in the coming years research attention is needed on:

– optimal functioning of humans;
– participatory simulation; and
– systems that support flexibility in manufacturing.

*Keywords*: Manufacturing, human factors, future, flexibility

## 1 PRESSURE ON MANUFACTURING

To survive in the face of worldwide competition, companies are forced to produce ever-increasing volumes at the same costs. Meanwhile the variation in products increases, shorter delivery times are demanded and quality should be high as well (Eijnatten 2002). This pressure on companies can be solved partly by the optimal use of human performance. It has been demonstrated that a motivated and healthy workforce contributes significantly to innovation and productivity (Eijnatten 2002). Some authors state that the motivated workforce is one of the secrets why Toyota had its highest profits ever (Yamada 2001). Research has shown the positive effects of paying attention to the human factor. Companies that place workers at the core of their strategies produce higher long-term returns on investments than their industry peers. A study among 702 firms showed that better human resource attention is associated with an increase in shareholder wealth of 40,000 Euro per employee (Pfeffer and Veiga 1999). Better ergonomics could increase productivity up to 46% (Rhijn et al. 2005), reduce sick leave costs by 18% (Koningsveld et al. 2004) and reduce the number of mistakes by 37% (Koningsveld et al. 2004).

However, a healthy workforce is still a problem in the European Union (cf. Figure 1). Problems in the future will be fatigue, stress and neck/shoulder problems (Smulders 2002), because of more repetitive work and the increased use of information technologies. Smulders (2002) findings are

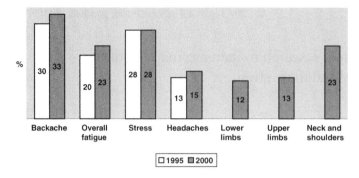

Figure 1. Percentage of 21,500 European Union workers having complaints (Merllié and Paoli, 2000).

based on research in the Netherlands, but the same trends are found in the data produced in 2003 by the European Foundation for the Improvement of Living and Working Conditions. Thus, care for health in manufacturing is also needed. In this paper the future of manufacturing work with special attention to the human factor is described as well as the consequences for R&D in the coming 10 years. This future is based on literature study and discussions with human factor experts in 10 European countries and also discussions with manufacturing companies. In this project a vision on future needs with respect to humans in manufacturing was made. Hereafter a discussion among human factor specialists and manufacturing companies on main themes was held. First the method will be described and how the vision was defined and then the vision will be described, followed by the discussion outcomes.

## 2  METHOD TO DEFINE THE VISION

Twenty experienced European researchers and consultants in the field of ergonomics and human factors in assembly, participated in several sessions where the vision and the roadmap were defined. In the first session 20 pictures (see for some examples Figure 2) were shown of assembly work situations to obtain a common view of *manufacturing*. After these pictures three groups were formed to define how the ideal future would look like in 10 years from now from a healthy human perspective. Two groups discussed *organisation and change (supply)* and one group discussed *development and workstations (design)*. They were asked to take into consideration:

- increasing implementation of digital ways of working;
- more care about musculoskeletal loading;
- more automation, less manual handling and as a result more control tasks;
- emotional wellbeing;
- balance between to much load and too little; and
- fun in work.

The three groups presented their conclusions to the whole group. Hereafter a lunch break was scheduled to create some distance to the work. After lunch every participant was asked to mention the main issue that was still in mind. Based on this, a view of the manufacturing circumstances 10 years from now was made.

Then a discussion was planned about the knowledge that is available and the knowledge that is missing and which needs to be developed for improvements in the future.

After this discussion all the needs were written down on a flipchart and the group voted for the most important needs. These needs for development were the base for the roadmap that fills up the gap in knowledge. This roadmap was defined through a group discussion.

218

Figure 2. Pictures shown to obtain a general view of the work.

More than a year later the results were presented in two sessions to a group of 30 human factors experts. Also, two sessions were organised with specialists and management from manufacturing companies. In these four sessions a discussion was held on the significant issues and themes that were found already, additions were made if new information became available from projects or conferences and the groups were asked to focus more on important themes. Based on the above-mentioned results a vision was defined on themes that needed attention in the future. Several themes were defined and discussed with regard to their priority.

Among the additional information were the result from the project *ManuFuture* and the European Union Lisbon conference: *Facing the challenge: The Lisbon strategy for growth and employment*. Important issues for the future mentioned in this conference were:

1. education is crucial in economic growth;
2. in the current knowledge economy it is inevitable to use knowledge of personnel in production; and
3. economic growth should be sustainable, which means that humans should be used in the long term, which means inclusive design of work: the awareness in work design on not excluding certain groups will be necessary in future.

## 3 RESULTS 1: THE FIRST SESSION

*Ideal future*: Several important themes came up regarding human performance in the future. An important issue mentioned by the groups was that more attention should be paid to an optimal loading of the workforce. Under loading is not beneficial for the company (unproductive) and the worker (no challenge) (cf. Figure 3).

Overload could have the consequence of stress and musculoskeletal injuries (worker), which leads to mistakes in the production work and drop out of workers (cf. Figure 3). The challenge for the future is to arrive at *circumstances in production lines*, in the lowest part of the curve. This is another direction than the current trend in which the workplace is being optimised in such a way, that people are seated or perform the same repetitive work all day.

Other issues mentioned in the discussion are summarised in Table 1. It was mentioned that work would be more decentralised. Employees will have more access to information and more power to innovate and adapt their work. Also, regarding the organisation, these will be dynamic networks instead of hierarchical structures.

*Available and Missing Knowledge*: To arrive at healthy manufacturing 10 years from now, typified by the issues described above, the group discussed the knowledge that is available and the knowledge that is missing and needs to be developed (cf. Table 2). Knowledge that is missing depends strongly on the industry, country and group within a company.

Some knowledge is now available about participatory design, ideal design processes, simulation and integrating information systems, overload and taking differences of culture into account.

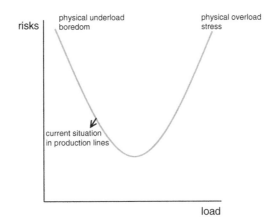

risks | physical underload boredom ... physical overload stress

current situation in production lines

load

Figure 3.    In future the risks should be reduced to increase performance of the company. This means that employees work at the optimum between overload and under load.

Table 1.    Issues mentioned in the sessions on developments in the future.

| Organisational/chain developments influencing human operations | Workplace developments influencing human operations |
| --- | --- |
| Flexibility: volume by technology, variation by people | Manual handling will remain in precision and flexible systems |
| Supportive systems accessible by the workforce (transparency) | More variation, rotation of tasks, teamwork |
| Networks changing dynamically | Computer interfaces self-explanatory |
| Design and simulation possible for a whole workday | Workplaces should allow motion/movement |
| Ideal decision making between human and technology | More participatory design |
| Design of supportive systems together with end-users | Fun/pleasant work environment/aesthetics |
| Design a line including continuous improvement systems | Mobile technology |

Table 2.    Available and missing knowledge.

| What is known | What is missing |
| --- | --- |
| A lot about participatory design: cases, theories, experience | How to get more innovation in flexible organisations? |
| A lot about integrating information systems and sharing information | How to disseminate on good use? |
| A lot about overload (physical, mental) | Promotion of health effects of optimal loading (see Figure 3) |
| A lot about static loads and systems that stay in one condition | Effects of changing environments, allocation of functions |
| How to simulate small tasks | Simulation of teamwork in flexible assembly environments |
| We have a lot of experts | Overcoming barriers among disciplines |
| About teamwork (traditional) | Knowledge on new teamwork (virtual, communities, practice, interdisciplinary) |
| About differences among cultures | What works well in different cultures |

However, this knowledge is only available to a select group of experts and some companies. The challenge for these issues is to diffuse this knowledge to other countries. A very fruitful way of knowledge exchange was seen the Netherlands and Sweden, in nationally funded projects. An improvement was implemented in one company and a group of other companies followed the work and then, at the end of the project, visited the company where the improvements had been implemented.

Table 3.   Topics to be developed in the future.

Development of simulation systems for small tasks that predict all day effects
Development of knowledge to find the individual optimum between over- and under load
Development of simulation tools to test organisation, technology and human aspects for flexible jobs
Design systems for dynamic allocation of functions in flexible organisations
Design computer-aided collaborative self-design and self-control systems to increase flexibility
Systems that support self-control organisations with simulation tools enabling continuous improvement
User-friendliness improvements of the current simulation tools
Design of decision support systems for situations involving dynamic allocation

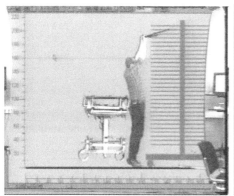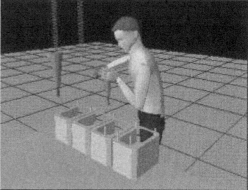

Figure 4.   Simulation of new assembly workstations is needed in the future to increase flexibility and self-control. Left is mixed reality (real person in drawing of a new assembly line); right is a computer simulation.

New knowledge to be developed concerns:

- How do employees learn to find their place in a network organisation. It is important that no time is lost in learning the systems and it is important that employees feel connected to a network. Day doses load effects of variation in tasks. From specific interventions it is known what the effects are. For instance, a better assembly workstation design results in higher productivity and better health (Vink 2005). However, the effect of more varying tasks and different work paces is unknown. Task rotation can even result in more complaints. It is still unknown what dynamics humans need. How can more movement be stimulated?
- The factors that will increase human health are unknown. A lot has been written to reduce complaints, but very little is known about increasing health. What do employees need to *feel good* at work?

*Vision.* Based on *what is missing*, the group made a first version of a roadmap. Important techniques and methodologies to develop *what is now missing* are (cf. Table 3) new simulation systems (cf. Figure 4) that integrate all day effects, variation in tasks and cultures and stimulate optimal human movement. Also, more systems that stimulate innovation at the workstation itself are needed that are easy to use and transparent regarding the organisation and network. Participatory design approaches should also be applied in companies where it is not known yet and the systems should be available in future. More knowledge should also be developed regarding the optimum point between under and overload.

An essential element in the roadmap towards ideal healthy human manufacturing (which is intelligent) is the possibility of simulating the new circumstances and test the effects to stimulate self-design and innovation, check the effects before building and have a high productivity.

221

| | 0–5 years | 0–10 years | 10–15 years | 15–20 years | 20–25 years |
|---|---|---|---|---|---|
| knowledge development | -guidelines for under/over load <br> -employee innovation in flex networks | -all day effects and flexibility <br> -tools for flex-networking | | | |
| tests of knowledge | - participatory design (incl. cultural effects) | -guidelines for under/over load <br> -employee innovation in flexible networks | -all day effects and flexibility <br> -tools for flex-networking | | |
| knowledge transfer to companies | | - participatory design (incl. cultural effects) | -simulation tools for under/overload and innovation in flexible networks | -all day effects and flexibility <br> -tools for flexible networking | |
| application | | | - participatory design (incl. cultural effects) | -simulation tools for under/overload and innovation in flex networks | -all day effects and flexibility <br> -tools for flexible networking |

Figure 5.   A roadmap based on the discussions with human factor experts. After knowledge development, the knowledge is applied in research tests in companies. After this, knowledge is transferred to companies in demonstrations followed by applications.

*Roadmap.* In Figure 5 a roadmap is defined based on the discussions. The coming 5 years knowledge will be developed regarding guidelines for under load and overload (what is the optimum?) and self-control systems that stimulate innovation ion flexible organisations. Based on this knowledge research will be done in and with companies to test effects in 5–10 years from now. It will be demonstrated and applied in the following decennium. Regarding participatory design the process will be faster, because knowledge is available. And regarding tools for networking and all day effects in flexible factories the process will be later, because more time for knowledge development is needed.

## 4   RESULTS 2: MAIN FOCUS

A year after defining the roadmap, the group of human factor experts decided that more focus was needed. The discussed issues were still valid and for future research a more elaborated description of the main themes was seen as necessary. Therefore, a set-up was made of themes and description of the content (see section 5). However, a main theme was missing according to the group. The issues were complicated and for communication it was better to define one theme. Several themes were mentioned, but after voting almost 80% of the group voted for the theme flexibility. For Europe it is essential to have a flexible manufacturing system. In general, high volumes with less intelligence can be produced better outside Europe or in automated factories.

Manufacturing companies are faced with an increase in demand for product variation and product volume variation. This means that human operators, planning and control, and maintenance work, are also changing. Systems as well as workstations should support this need for flexibility. This means that system should be easy to operate, support operators with knowledge on the rest of the factory, and enable quick knowledge on supply information, production systems and operators.

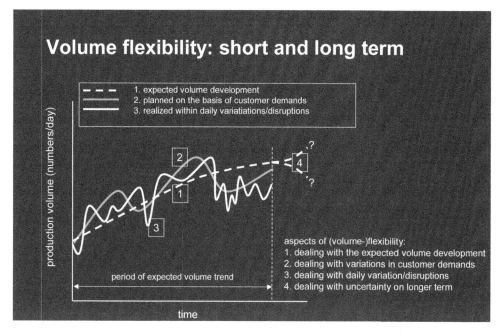

Figure 6. Apart from the increasing types of products, also the variation in volume demand grows. In this figure the volume flexibility demand of one of the participating companies is shown.

Workstations and production lines should support this as well by being adjustable to the subject, adjustable to variations in products and adjustable to speed of work. More knowledge is needed in this area, especially at operation level close to the human factor. On vision and conceptual level a lot is known or preached already.

The discussions with representatives from manufacturing industries arrived at the same main theme: flexibility.

## 5  FUTURE THEMES

Below are the themes described that need further research to support the innovation and competitiveness of manufacturing in Europe.

- *Theme 1 – Humans will be a core resource in coping with the demand for flexibility*

As is described above, in the future humans will still be involved in manufacturing (Vink 2005). Humans are likely to remain the most flexible and innovative manufacturing resource in the foreseeable future. Work will shift from doing the assembly operations to designing, supervisory control, and maintenance (Eijnatten 2002). Worldwide competition forces companies to produce ever-increasing volumes at the same costs. Meanwhile, variation in products increases, shorter delivery times are demanded and high quality is an important requirement (an example of volume flexibility is shown in Figure 6). In the worldwide competition of manufacturing industries, manufacturers are challenged to improve their flow of assembly orders. A current tendency in this respect is known as lean manufacturing (Docherty et al. 2002), which refers to a process where the numbers of unnecessary steps, waste and error are kept to a minimum. Lean manufacturing may well improve efficiency, although others stress the detrimental effects on the operators owing to an intensification of the work (Lansbergis et al. 1999; Vahtera et al. 1997). In addition to the need for more efficiency, there is a clear need for more flexibility. One aspect of this flexibility concerns the ability to deal with the fluctuation in the volume demand throughout the year.

It is well recognised that all this requires a healthy and motivated workforce. Nowadays, productivity, quality, worker's motivation and health are considered the key elements for successful manufacturing. However, in many manufacturing companies, the production lines are mainly designed from a production perspective, whereas the human factors are addressed only slightly. A way to reduce pressure on the companies is to make optimal use of human performance. It has been demonstrated that a motivated and healthy workforce contributes significantly to innovation and productivity (Eijnatten 2002). The motivated workforce may be one of the main secrets why Toyota had its highest profits ever in 2003 and 2004. Motivating a highly skilled workforce and keeping employment are main challenges for Europe.

In general, in the future large low cost volumes will not be produced in Europe. Innovative and varying products are likely to be manufactured in Europe. Flexibility in terms of products and volume is a key issue. This means that the number of employees will vary per month, week or even day. The number of working hours will vary, as will the supply of products. This will have consequences, for example, on workstation layouts. There is a need for new knowledge of the *ideal flexible workforce*, knowledge on varying supplier needs, and end-assembly layout implications. In the decades to come, the level of mechanisation and the use of robots will further increase. This may increase production output as well as reduce labour costs, which will help make Europe more competitive against Asian countries. On the other hand, there is a well-recognised risk of losing flexibility through automation. Thus, the process of automation should be critically considered, keeping in mind that *the human is the most flexible robot*. To keep employment within Europe it is essential to enlarge and disseminate this knowledge among manufacturing companies.

- *Theme 2 – Humans have the ability to cope with complex tasks*

Production will be more complex regarding technology and customisation, but also in the supply chain. More countries will be involved, more specialists for a specific part of a product, and more variation in products means more suppliers. This makes the supply chains increasingly complex. To handle this complexity customers have to be able to build deep supplier relationships. Such strong manufacturer-supplier partnerships are one reason why Japanese car manufacturers such as Honda and Toyota have the fastest product development process or why they were able to bring down manufacturing costs (Liker and Choi 2004). In the future, companies need employees that have knowledge of the best way to cope with such complex supply networks and how to build and strengthen customer-supplier relationships. Human factors as perspective taking and collaborative planning are likely to play a leading role in future supply chains (Parker and Axtell 2001; Windischer and Grote 2003). Also, knowledge is needed about the supportive systems, where employees can perform optimally. This varies from adjustments like designing new workstations where many parts from different suppliers fit to designing new ICT to support supply chain management. However, this field that needs further research.

- *Theme 3 – Inclusive design to cope with special groups (e.g. the elderly)*

In the workforce of the future more elderly people will be active, but also more cultural differences will be introduced. Work should be designed in such a way that no groups are excluded without having a good reason. ICT can support a much higher level of productivity and wellbeing for the large number of elderly people that will be part of the future workforce. However, systems must be developed with the end-users in mind. Best cases and *design for all* methods (also mentioned as inclusive design in the literature) are needed to inform companies and help them anticipate the growing number of older employees.

- *Theme 4 – Systems should support self-control by humans*

High productivity and innovation can only be achieved if people are in control through full use of ICT and advanced manufacturing technology (Vink et al. 2004). The success of manufacturing companies relies on anticipating customer demands leading to manufacturing customisation and flexibility. However, this is only possible if all levels of the workforce (i.e. top management,

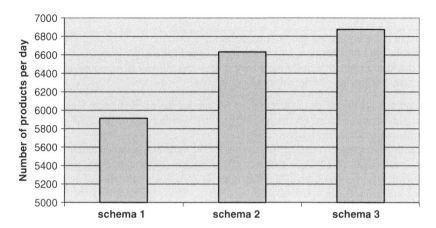

Figure 7. Effect on productivity of different work-rest schemata (time aspects) at Philips DAP. On the same line the productivity can increase 17% by choosing the ideal work-rest schedule. More knowledge on the theoretical background is needed in this area to support EU companies with increasing productivity.

middle management, and operators) are able to understand and make optimal use of ICT systems. Productive manufacturing ICT must therefore be thoroughly tested by end-users on all levels before use. ICT on an even higher level of efficiency should be developed by optimisation of the division of tasks between technology and humans (e.g. the optimal level of automation).

- *Theme 5 – Time aspects*

Nowadays, attention for human factors goes beyond the ergonomic design of workplaces, and the prevention of heavy lifting, pushing or pulling. Rather than the intensity of the workload, it is the time pattern of the load during the day that needs consideration. Aspects of this time pattern or *rhythm*, like work rest schedules, task variation, cycle time, and work pace may clearly affect the worker, mentally and physically (cf. Figure 7). A whole day effect is still unknown (Vink et al. 2004). Meanwhile, these aspects will affect production and also the quality of the production. The effects of time aspects on the worker and on the production need further quantitative study before they can be incorporated in the existing engineering tools.

- *Theme 6 – Towards service manufacturing*

The large advantage of manufacturing in Europe is that much knowledge is added. This is a good ground for not only delivering a product, but to combine it with service. This means that manufacturing improvement and innovation should be combined with delivering services. Much is gained when a good system layout is found for this work. But the options are abundant: Parallel or line, short or long cycles, logistics, fixtures, positioners, and workstation layouts.

There is no clear indication for choosing a specific system. Increasingly, companies choose a line-layout, because of anticipated *decreasing learning time*, and *simplifying material supply* (*less double handling of kits and assembling* and *preventing losses during shift changes*). Much knowledge is available, but dispersed among universities and companies. Also, new knowledge regarding flexible and service oriented manufacturing systems is needed.

6 DISCUSSION

The results described above are *coloured*, because the participants are experts in human factors and the involved companies interested in human factors. This means that the vision of management, designers and engineers might be somewhat underexposed.

However, there is not so much debate about the four main topics:

- the human is the crucial element in innovation;
- look for the optimum between under and overload of humans;
- to get the ideal production line a participatory design with visualisation by simulation; and
- use human intelligence for flexibility of the factory.

Other authors also mention the strength of the innovation of the human factor. It is shown by several scientific studies (e.g. Pfeffer and Veiga 1999; Koningsveld et al. 2004), but also in more political discussions, like the Lisbon conference.

Other authors mentioned the issue of striving for the optimum between under load and overload and (e.g. Smulders et al. 2002; Neerincx 2003). Also, on self-organisation, several authors have mentioned the same trends (Pinto 2003). The need for knowledge on all day effects of combinations of loading and subjects is not described very often. However, the need for this knowledge is demanded by the fact that the optimum between over- and under-load should be known in the future for different populations (including differences in culture).

On participatory design with the help of simulation, the positive effects are often described (see the work of The European Foundation for the Improvement of Living and Working Conditions in the introduction). Also, the positive effect of the combination of participation and simulation has been described before (Eijnatten 2002).

Also, flexibility is discussed often as an important European topic. On flexibility increase a European program was set up within Framework Programme 6 in 2004 to financially support the increase in knowledge in this area. Now several projects are running on this theme.


ACKNOWLEDGEMENT

The authors wish to acknowledge the European Commission for their support. They also wish to acknowledge their gratitude and appreciation to the companies and experts participating in the project.


REFERENCES

Docherty P, Forslin J, Shani AB. Creating sustainable work systems – Emerging perspectives and practice. London: Routledge, 2002.

Eijnatten FM van, Ed. Intelligent manufacturing through participation: A Participative Simulation environment for Integral Manufacturing enterprise renewal. Hoofddorp: The PSIM Consortium, an IMS project, TNO Arbeid, 2002.

European Foundation for the Improvement of Living and Working Conditions. Communiqué July/August. Dublin: EFILWC, 1999:2.

Koningsveld EAP, Houtman ILD. The cost of poor working conditions. Dublin: European Foundation for the Improvement of Living and Working Conditions, 2004. www.eurofound.eu.int/ewco/2004/12/NL0412NU01.htm.

Lansbergis PA, Schnall P, Cahill J. The impact of lean production and related new systems of work organisation on worker health. J Occupational Health Psychology 1999;42:108–130.

Liker JK, Choi TY. Building deep supplier relationships. Harvard Business Review 2004;12:104–113.

Merllié M, Paoli P. Working conditions in the European Union. Dublin: European Foundation for the Improvement of Living and Working Conditions, 2003.

Neerincx MA. Cognitive task load design. In: Holnagel E, Ed/ Handbook of cognitive task design. New York: Lawrence Erlbaum Associates, 2003:Chapter 13.

Parker SK, Axtell CM. Seeing another viewpoint: antecedents and outcomes of employee perspective taking. Academy of Management Journal 2001;44(6):1085–1100.

Jeffrey Pfeffer S, Veiga JF Putting people first for organizational success. The Academy of Management Executive; May 1999; 13, 2; pp. 37–48

Pinto J. Automation unplugged. New York: ISA, 2003.

Rhijn, GW van, Looze, MP de, Vink P. Improving Flexibility and Human Factors in Lean Manufacturing at Philips Shavers. IMS Forum 2004, Como, Italy, 18 May 2004

Smulders PGW, Houtman ILD, Klein Hesselink DJ. Trends in work 2002. (In Dutch). Alphen a/d Rijn: Kluwer, 2001.

Vahtera J. Effect of organizational downsizing on health of employees. The Lancet 1997; 350:1124–1128.

Vink, P. (ed) Comfort and Design: principles and good practice, CRC Press, Boca Raton, 2005.

Vink P, Stahre J, Rhijn G van, Christmansson M. A roadmap to intelligent manufacturing with healthy humans in 2020. In: Taisch M, Filos E, Garello P et al., Eds. International IMS Forum 2004: Global Challenges in Manufacturing. Milano: Politecnico di Milano, 2004:1400–1407.

Windischer A, Grote G. Success Factors for Collaborative Planning. In: Seuring S, Müller M, Goldbach M, Schneidewind U, Eds. Strategy and Organization in Supply Chain. Heidelberg: Physica-Verlag, 2003:131–146.

Yamada S. Challenges in dealing with human factors issues in manufacturing activities. In: Arisawa H, Ed. ER 2001, proceedings of the international workshops. Yokohama: Yokohama University, 2001:10-29.Tom: Paper formatting: theory and practice. Mc Graw Hill, New York, 1993.

*Advanced Manufacturing – An ICT and Systems Perspective – Taisch,*
*Thoben & Montorio (eds)*
© *2007 Taylor & Francis Group, London, ISBN 978-0-415-42912-2*

# Global education in manufacturing

Asbjørn Rolstadås

*Department of Production and Quality Engineering; Norwegian University of Science and Technology,*
*Trondheim, Norway*

ABSTRACT:   Over the last decade industrial manufacturing has undergone a significant change: it is no longer home-based; it operates in a global market; digital business has become a strategy to survive; the extended enterprise is being implemented; and components and even products are made where conditions are most favourable. However, engineering education does not completely reflect today's needs of an industry that faces problems of integrative nature across the traditional disciplines, such as: working globally in a multicultural environment; working in interdisciplinary, multi-skilled teams and working with digital tools for communication. This circumstance calls for a new university curriculum reflecting technology and management. The GEM (Global Education in Manufacturing) project, funded under the IMS initiative, was completed in November 2004. The main achievements of GEM are: The definition of the industrial needs for a new type of curriculum in manufacturing industry and the development of a framework of a curriculum in manufacturing strategy accessible over the Internet and allowing appropriate flexibility for implementation according to local tradition and legislation.

*Keywords*:   Manufacturing strategy, engineering education, curriculum, extended enterprise.

## 1   INTRODUCTION

At the IMS Steering Committee meeting in California, USA, in May 2003, Dr. Thomas J. Duesterberg, president of the Manufacturers Alliance in USA, gave a presentation of the manufacturing industry in the USA and claimed that it is the world engine of growth (Duesterberg 2003). He claimed that the growth in the manufacturing industry is larger than the growth of the non-manufacturing industry. The productivity boom cantered in manufacturing, and products with embedded new technologies come almost exclusively from manufacturing sector. The manufacturing sector also leads in management innovation. Concepts such as *Lean*, *Six Sigma*, *JIT*, *Supply Chain Integration*, etc. all have their roots in manufacturing.

In Europe, the European Commission has also focused on manufacturing. An expert group discussed research and educational issues in European manufacturing and presented its conclusions at a conference in Milan in December 2003 (*ManuFuture* 2003). At this conference five drivers for innovation in manufacturing were identified. One of these drivers is the provision for better education and training schemes in Europe.

The importance of the manufacturing industry is recognised throughout the world. The IMS research co-operation is a proof that progress in manufacturing is important and that international collaboration on research and education may bring new innovations in the industry.

Over the last 40–50 years the manufacturing industry has undergone a significant change. This is clearly demonstrated in the shift of the manufacturing research agenda. It is possible to recognise four generations of research focus areas:

- Machine focus;
- Factory focus;

Extended product life cycle

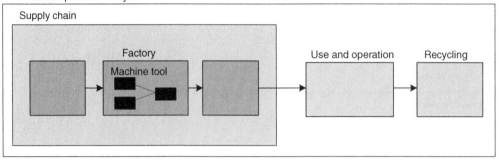

Figure 1.   Manufacturing research generations.

- Supply chain focus;
- Lifecycle focus.

In the machine focus, research activities were targeted at a single process element, for example, around the cutting process in a single machine tool. Improved cutting data, cutting conditions, cutting materials, tools and capability of the machine tool are all typical examples of problems areas that were improved significantly during this generation.

The second generation changed the focus from the single process element to the whole factory. Research focus was on industrial automation, robotics, CNC, flexible manufacturing, operations and process planning (such as the APT language) and in the end on CAD/CAM, CIM and intelligent manufacturing.

The third generation shifted the focus to the whole supply chain. In addition to the pure manufacturing technology, the research now also had a dimension on management and economics. One of the main enablers of this research is the availability of digital solutions for planning, control, management, and business operation. The extended enterprise is a topic that has received much attention in this respect (Browne and Jagdev 1998).

The fourth generation maintains the focus on the supply chain, but extends this into a full lifecycle. The focus shifts to a certain extent from production to the product. The concept of extended products has been launched covering the whole lifecycle from delivered product through use and maintenance to disassembly and recycling. In Europe a number of research projects in this field have been funded by the European Commission. Some examples are the APM, AEOLOS and SMARTISAN projects. A new IMS and European Union project on smart embedded devices has already been (PROMISE) launched in this field.

These four generations and their focus are shown in Figure 1. The generations indicate the focus of new research. The newest generations represent growth areas. This does not mean that research in the other generations is unimportant. On the contrary, all generations still require research as basis for industrial improvement.

For the manufacturing industry productivity and competitive advantage has always been a significant issue. Models for measurement of productivity within an enterprise have been developed (Porter 1980, Sink and Tuttle 1989, Bredrup 1995) and the concept of benchmarking (Andersen and Pettersen 1994) has found wide application.

New business operational strategies have been developed following the research focus generations described above. A possible classification of such strategies into four focus areas has previously been published by the author (Rolstadås 2000):

- Cost          to reduce operational costs;
- Market        to increase market share;
- Organisation  to develop a more effective organisation;

- Business       to find new and improved way of business operation.

The change in the manufacturing industry has also led to a change in need for competence in the engineering field. The industry today focuses on circumstances characterised by:

- Working with digital tools for communication;
- Working in a multicultural environment;
- Working in interdisciplinary, multi-skilled teams;
- Sharing of work tasks on a global and around the clock basis;
- Working in a virtual environment.

This circumstance calls for a new university curriculum reflecting technology and management. The IMS GEM project was launched to develop specifications for such a curriculum (Moseng and Rolstadås 2002).

## 2   THE GEM PROJECT

The GEM IMS project was endorsed in January 2002 and was finished in November 2004. The European part was sponsored by the European Commission (Framework Programme 5). The project had 27 partners from 20 different countries in Europe, North America, Asia and Australia. In addition 75 companies supported the project as co-operation partners of the universities involved. The main objectives of GEM were to:

- *Define and understand the needs of the manufacturing industry for training and education in manufacturing strategy on a global basis to comply with the concept of digital business and extended products.*
- *Develop detailed specifications for a manufacturing strategy curriculum focusing on manufacturing and business administration topics.*

The innovation of GEM is twofold:

- A new, first ever curriculum for manufacturing strategy at masters level to be delivered internationally to meet the future needs of the digital business and extended products and to be accepted as a world standard;
- New pedagogic approach and delivery by ICT to meet the needs of on-the-job training and education and to combine theory and practice.

The main achievements of GEM are:

- The definition of the industrial needs for a new type of curriculum in manufacturing industry;
- The development of a framework of a curriculum in manufacturing strategy accessible over the Internet and allowing appropriate flexibility for implementation according to local tradition and legislation;
- The curriculum implementation strategy.

A full curriculum has been specified in detail covering seven important knowledge areas. It is documented on the web page of the project and is free for all to pick up to develop an education in manufacturing strategy. The GEM curriculum is being implemented in Japan and Korea. Australia will be implementing parts of it. It is unlikely at current that it will be implemented in the USA. In Europe some of the partner universities have formed an alliance and will be implementing this type of education within existing degree structures and programs at their universities. The work plan contained eight work packages and was organised in three project phases as shown in Table 1.

Table 1. Work plan structure.

| Phase | Phase title | WP | WP title |
|-------|-------------|-----|----------|
| 1 | Definition of the industry's needs | 1 | Extended enterprise training needs |
| 2 | Development of draft curriculum | 2 | Training delivery mechanism |
|   |   | 3 | Draft curriculum |
| 3 | Verification and development of final curriculum | 4 | Develop demonstrator, evaluate concept |
|   |   | 5 | Revise curriculum |
|   |   | 6 | Exploitation and dissemination |

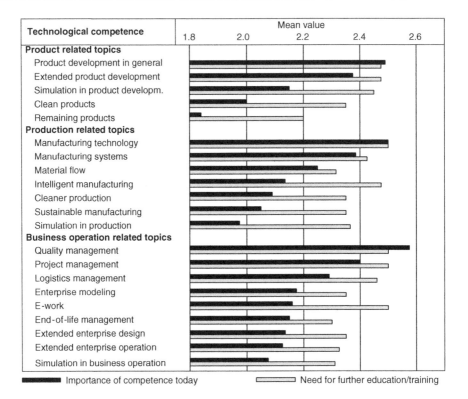

Figure 2. Industrial needs for technological competence.

## 3 THE INDUSTRIAL NEED FOR A NEW MANUFACTURING STRATEGY CURRICULUM

To solicit the needs of the international manufacturing industry for a curriculum in manufacturing strategy, a questionnaire was developed and distributed by the GEM partners to industry in all the participating IMS regions. A total of 556 responses from 22 countries were received and analysed (Kvernberg Andersen et al. 2003).

The competence need was split in three areas (that were again subdivided):

- Technological competence;
- Humanistic competence;
- Business competence.

The results are presented as a number of graphs (Figures 2–4).

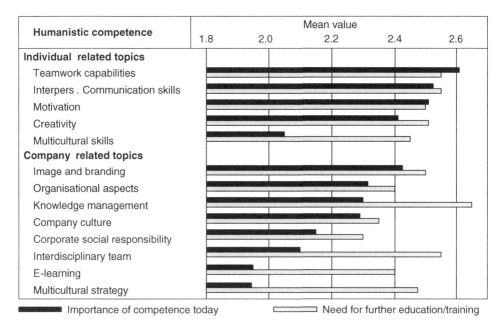

Figure 3.    Industrial needs for humanistic competence.

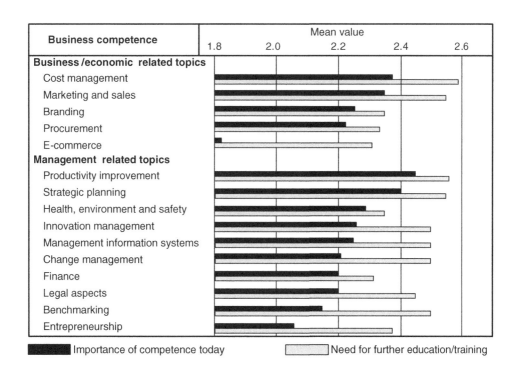

Figure 4.    Industrial needs for business competence.

233

Table 2. GEM knowledge areas.

| Knowledge area | | Description |
| --- | --- | --- |
| A | Development of extended products | The development of a combination of a physical product and associated services/enhancements that improve marketability. |
| B | Digital business along the supply chain | Information on how a business can use e-commerce and related technologies and processes to develop, expand or enhance its business activities along the facilities and functions involved in producing and delivering a product or service. |
| C | End-of-life planning and operation | Techniques on how to develop methodologies and tools to support the end-of-life routing/processing decision based on economic, environmental and societal criteria. |
| D | Business operation and competitive strategy | Explanation of how organisations function and interact with competitors and their marketplace, and deliver performance over time. |
| E | Intelligent manufacturing processes | Elaboration of techniques applicable for handling complex production working in an uncertain, changing environment, with special emphasis on artificial intelligence and machine learning approaches. |
| F | Intelligent manufacturing systems design | Tools on how to model the skills and knowledge of manufacturing experts so that intelligent equipment and machines can produce products with little or no human intervention. |
| G | Enterprise and product modelling and simulation | Information on how to develop and use computational representations of the structure, activities, processes, information, resources, people, behaviour, goals and constraints of a business or a product. |

A survey over the Internet (O'Sullivan et al. 2002) on curricula for manufacturing revealed that there are no existing curricula fulfilling these needs fully. The survey was executed by all project partners, who searched and analysed master programs from their own geographical region. 300 relevant programs were found. Two ranking criteria were applied:

- Whether the masters in manufacturing program was available in English;
- How relevant the program was according to the aims of GEM.

The programs were ranked on a scale from 1 to 4, with 1 as the most relevant. 40 programs obtained a rank of 1. They were further analysed at a course level. This revealed that some of the GEM knowledge areas were inadequately covered. The areas less adequately covered were C (end-of-life planning and operation) and G (product modelling and simulation).

## 4 THE CURRICULUM FRAMEWORK

The GEM framework identifies seven core knowledge areas within any new manufacturing curriculum, all of which reflect the current and future needs of the manufacturing industry. Table 2 shows an overview of the knowledge areas.

Each area focuses on aspects of manufacturing that are future oriented and promote a paradigm shift from traditional manufacturing to digital business. Each area comprises courses and individual subjects (equivalent to an individual lecture). Each course is described and structured in a number of subjects and with a recommendation for their delivery mechanism. Textbooks are recommended and assignments suggested. The specifications should be sufficiently detailed for a university to start delivering lectures.

For the specification of the curriculum, a skeletal framework has been developed (see Figure 5). The framework has a number of elements. Students will enter a particular programme with a bachelor's degree and may or may not be induced into the programme or university through a series of induction, bridging or capstone courses or workshops. The details of these activities are solely

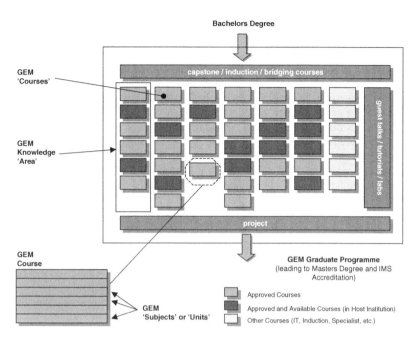

Figure 5.   The GEM framework skeleton.

| Partner | Day | Topic | Type | Nationality |
|---|---|---|---|---|
| CIMRU | 1 | Digital Business along the Supply Chain | E-Learning Course | |
| EPFL KAIST | 2 | Advanced Modelling Methods and tools for E-manufacturing | E-Learning Course | |
| ◙ NTNU | 3 | Productivity Management | E-Learning Course | |
| | 4 | Strategy in Production Management | E-Learning Course | |
| BIBA | 5 | Development of Extended Products | Web based Group Simulation game | |

Figure 6.   Preliminary course structure within the GEM demonstrator.

the responsibility of the university concerned. Students will then have available to them, a number of courses. It is not anticipated that a particular university will offer all defined courses. On the contrary, universities will develop courses based on their core competencies on campus. However, the GEM framework will be available to help educators to identify and specify courses of interest.

The education and training uses modern e-learning based on multimedia techniques. This allows for students to attend independent of time and place. This is crucial for delivering continued education to people working in industry.

5   THE DEMONSTRATOR

To demonstrate the feasibility of the curriculum, a five-day course was developed. The content structure of the e-learning demonstrator is illustrated in Figure 6.

The objectives of the course *Digital Business along the Supply Chain* are to examine manufacturing supply chains; understand the role of business strategies and techniques for digital business; identify new technologies and their influence on e-business; and determine the validity of digital business and why it is increasingly being integrated into the new business model.

The course *Advanced Modelling methods and tools for e-manufacturing* deals first with the definitions of Petri nets and the use of this technique to make useful calculations. Then it overviews the objectives and the process of new product development and then introduces digital engineering tools for this, such as digital mock-up and virtual manufacturing.

The course *Productivity Management in production* focuses on the main areas of productivity management, performance measurement and performance improvement. Performance is generally accepted to cover a wider range of aspects of an organisation – from the old productivity to the ability for innovation, attracting the best employees, maintaining an environmentally sound outfit, or doing business in an ethical manner.

This course *Strategy in production management* focuses on the strategic management of technology and innovation in established firms. The conceptual framework of the course is an evolutionary process perspective on technology strategy and innovation.

The objective of the course *Development of Extended Products* is that the participants should develop an understanding about extended products to experience the related success factors and challenges for their development. The course is delivered as an educational game. Having been assigned to one of nine roles, the players undergo a virtual inter-organisational product development scenario that is implemented in a web-based group simulation game to improve their communication skills as well as their collaborative skills.

The demonstrator has been delivered worldwide to 487 users (Schwesig et al. 2005). In general, the users evaluated the courses content quite good and would have appreciated a greater variety of interactive elements. They often appreciated the courses instructional structure, especially the frequency of interactions. According to the screen design, the simplicity and cleanliness of the design has been appreciated most, while the quality of media elements and graphics has been evaluated controversially. The greatest part of the users appreciated the quality the graphics and the value added by them, while the quality of video/ audio elements has been rated worse.

The participants of the simulation game mostly appreciated the user interfaces structure, but criticised the limited functionality of the chat function. Most of the players were able to identify the problems and challenges of distributed inter-organisational working and then to develop strategies to cope with these problems. They were able to understand the main characteristics of extended products and the success factors for the development of extended products.

## 6 THE IMPLEMENTATION PLAN

GEM has been a complex project with many partners spanning different cultures. However a viable network of universities was developed, especially in Europe where also the integration of the new Member States of the European Union has been successful.

In Australia the curriculum has been taken up at the University of Melbourne. In Japan and Korea the curriculum is being implemented. In the US there is interest in future co-operation to implement the curriculum. In Europe the GEM alliance has been formed with the ambition of co-operation to implement the curriculum.

For The GEM Alliance a memorandum of understanding has been signed by NTNU, University of Bremen, EPFL and AUA. Some important points in this agreement are:

• Within its current degree structure, each party will offer an option for students to select a GEM curriculum as a specialisation.
• A student is enrolled at one – and only one – university (hereafter referred to as host or hosting university). This university has the sole responsibility for the student in all legal terms. The

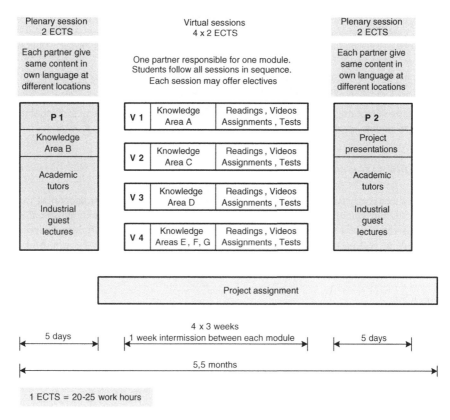

Figure 7.    Model for training at engineer update level.

student will receive all degrees, diplomas, and certificates from the host university, unless otherwise agreed.

- Each party will offer at least 7.5 ECTS credits.
- Each course is concluded with an examination. The courses will be prepared for distance education in such a way that students from the other parties may follow the education at their university.
- All lectures, textbooks, assignments, and examinations will be in the English language.
- All knowledge areas should be included in the curriculum.

For education and training of people in industry, a strategy for implementing industrial training has been developed. Some highlights of this strategy are:

- Interested partners will form an alliance or some other form of partnership to be regulated through a memorandum of understanding or an agreement;
- The partners operate a common training program such that learning objectives and skills attained are the same across all partners;
- Training should be offered at two levels: *Engineer update level* as well as *Executive level*.

The training will be based on e-learning following the *hybrid model* (Hussein and Rolstadås 2002):

- Two plenary sessions: one at the beginning and one at the end;
- A number of virtual sessions in between: each session to be carried out within a given time window;

- Plenary sessions to provide learning through interaction among students and by extensive use of industrial experts;
- Virtual sessions to include varied activities such as readings, video lectures, slide-shows, assignments and tests;
- An optional project assignment extending over the whole learning period to be executed in small groups and to be based on real life problems in the students' companies.

Figure 7 shows an example on how the hybrid model can be used to implement industrial training at the engineer update level.

## ACKNOWLEDGEMENTS

The work has been partly funded by the European Commission through IST Project GEM Europe – Global Education in Manufacturing, IST 2001-32059. The authors wish to acknowledge the European Commission for their support. They also wish to acknowledge their gratitude and appreciation to all the GEM project partners for their contribution during the development of various ideas and concepts presented in this paper.

## REFERENCES

Andersen, B., Pettersen, P.G. (1994): The Basis of Benchmarking: What, how, when and why. Proceedings for 1994 Pacific Conference on Manufacturing. Djakarta, Indonesia.

Bredrup, H. (1995): Performance Measurement in a Changing Competitive Industrial Environment: Breaking the Financial Paradigm, University of Trondheim.

Duesterberg, T. J. (2003): U.S. Manufacturing: The Engine for Growth in a Global Economy, presentation at the Intelligent Manufacturing Systems ISC 16 Meeting, Monterey, California, USA.

Hussein, B., Rolstadås, A. (2002): Hybrid Learning in Project Management – Potentials and Challenges, PMI Research Conference, Seattle, USA

Jagdev, H. S., Browne, J. (1998): The Extended Enterprise – A Context for Manufacturing, International Journal of Production Planning and Control, Volume 9 number 3.

Kvernberg Andersen, T., Gjerstad, T. B., Madsen, B. E., Moseng, B., Rolstadås, A. (2003): Training and Education Needs for Manufacturing Strategy, GEM-EUROPE project report, Trondheim, Norway.

ManuFacture 2003: European Manufacturing of the Future: Role of research and education for European leadership, Milan, Italy 2003.

Moseng, B., Rolstadås, A. (2002): Global Education in Manufacturing – GEM, CIRP International Manufacturing Education Conference, CIMEC 2002, Twente, The Netherlands.

O'Sullivan, D., Precuo, L. E., Duffy, P., van Dongen, S., Guochao, X. (2002): Survey of existing Manufacturing Curricula, GEM-EUROPE project report, Galway, Ireland.

Porter, M.E. (1980): Competitive Strategy: Techniques for Analyzing Industries and Competitors, Free Press, New York.

Rolstadås, A. (2000): Business Operation by Projects, IFIP World Computer Congress, Beijing, August 2000.

Schwesig, M., Rolstadås, A., Thoben, K.-D. (2005): An E-learning experiment in manufacturing strategy, IFIP World Congress on Computers in Education, Cape Town, July 2005.

Sink, S., Tuttle, T. (1989): Planning and Measurement in your Organization of the Future, Industrial Engineering and Management Press, Norcross.

*Advanced Manufacturing – An ICT and Systems Perspective – Taisch,*
*Thoben & Montorio (eds)*
*© 2007 Taylor & Francis Group, London, ISBN 978-0-415-42912-2*

# Simulation gaming to train engineers to efficiently collaborate in manufacturing networks

Klaus D. Thoben & Max Schwesig
*Bremen Institute of Industrial Technology and Applied Work Science (BIBA), University of Bremen,*
*Hochschulring, Bremen, Germany*

ABSTRACT: A simulation has been created that combines aspects of organisational learning and knowledge management with a manufacturing scenario in and among organisations. It aims on mediating communication skills, collaboration skills as well as the proper use of information and communication technologies within authentic (distributed) manufacturing scenarios. The gaming approach used in the simulation game *COSIGA* was combined with identified processes and challenges in organisational learning, principles of distributed manufacturing and the lifecycle of a type of a manufacturing network. These elements then formed an approach that has been described in two scenarios. These scenarios deal with organisational learning and product development in one company and distributed development and manufacturing of an extended product and inter-organisational learning within a manufacturing network of companies. An initial version of the simulation was evaluated and validated in four gaming sessions with 36 students and industry representatives. The learning outcome was assessed by applying a questionnaire based survey as well as interviews. After having played the game, the players were asked to assess their communication and collaboration skills.

*Keywords*: Simulation gaming, organisational learning, distributed manufacturing.

## 1 INTRODUCTION

To successfully respond to global competition, manufacturing companies have to differentiate themselves from competitors by, for example, targeting customer's needs through extending their physical products with additional features such as services or other characteristics. Development and manufacturing of these complex products require the combination of certain key competencies that can often only be accomplished in manufacturing networks. The resulting specialisation within companies accelerates innovation speed and thus increases the demand for knowledge. As half-life of knowledge is continuously decreasing, organisational capacity for learning has been identified as one of the key abilities for manufacturing organisations to survive. As a consequence of the high knowledge intensity within product development and manufacturing, manufacturing engineers are in particular are affected by these learning and working processes.

As a result of these developments, the way of working and thus the educational requirements that engineers face have changed as well. Communication skills (Probst and Büchel 1994) as well as social competencies and the knowledge and skills that determine how relationships with others are handled (Smith et al. 2001) are becoming important to effectively exchange information within and among companies.

Appropriate tools to mediate such skills are simulations. One type of simulation is simulation games. They provide significant educational benefits as they allow students to experience the complexity of cause and effect relationships and consequences of decisions in a sufficient practical manner, while significantly reducing time, money and adverse real-world consequences (Oh et al. 2001). Such learning experiences enable learners to cope with real problems and authentic circumstances that are close to reality (Kriz 2001). Accordingly, this paper describes the development and the initial testing of a web-based group simulation game that has been built to address these new educational demands by mediating the required skills in an authentic manufacturing environment.

Existing simulation games in the field of organisational learning, and the closely related knowledge management (Büchel and Probst 2002) such as the simulation game *KM Quest* (KITS 2004) and *Wissensmanagement Planspiel für soziale Organisationen* (BBJ Consult 2002), emphasise the mediation of basic principles of knowledge management as well as the implementation and intervention related aspects.

Existing simulation games in the field of manufacturing like *COSIGA* (Hoheisel et al. 2001), *City Car Simulation* (Goffin and Mitchell 2002) and *GLOTRAIN* (Windhoff 2001) focus on the mediation of certain approaches or emphasise the important success factors in product development or even distributed manufacturing. So far, (inter-) organisational learning, company collaboration, and manufacturing have not been considered in a single gaming approach. To address the mentioned educational requirements, a web enabled group simulation has been developed, addressing processes and challenges in inter-organisational working and learning.

## 2 RESEARCH APPROACH

As the initial evaluation and validation of the COSIGA simulation was very encouraging (Riedel et al. 2001), this four-step product development and manufacturing process was used as a basis for the design of the new simulation. Since this new game emphasises the experience of intra-organisational and inter-organisational learning, it was necessary to identify key processes and challenges within intra-organisational and inter-organisational learning. To simulate inter-organisational manufacturing, it is important to identify organisational principles of distributed manufacturing as well as the lifecycle of a manufacturing network of companies. Together, these design elements were used to shape the simulation game.

### 2.1 *Processes and challenges of organisational learning*

Organisational learning is interpreted from a multi-level perspective, consisting of the individual, group, organisational and inter-organisational level (Inkpen and Crossan 1995, Nonaka 1994), since this perspective enables the main levels of action within an enterprise to be considered. According to this perspective, the following working definitions of the different learning activities were created:

Individual level learning focuses on individual knowledge acquisition without further social interaction. Group level learning happens, if more that one individual consciously or unconsciously acquires knowledge interactively. Organisational level learning focuses on perspective taking among groups in a company (Sumner 1999). As organisations are typically composed of multiple interacting communities, each with highly specialised knowledge, skills and technologies, knowledge intensive firms require these diverse communities to bridge their differences to create a new-shared perspective on an organisational level. Inter-organisational learning happens in two ways: either through the transfer of existing knowledge from one organisation to another, or through the creation of new knowledge (Larsson 1998). Group level learning, organisational level learning, and inter-organisational learning are all affected by *people barriers* like proprietary thinking and scepticism towards the sharing of knowledge and various fears (Barson 2000). Additionally, inter-organisational learning is affected by organisational boundaries like space, time, (cultural) diversity, structure and distribution of knowledge and results (Bosch-Sijtsema 2001).

Organisational level learning is especially difficult to simulate, as it requires the professional and technical identification of the players with their departments' specialised backgrounds. As this effect cannot be reached in a multiple hour simulation, individual, group, and inter-organisational level learning in the simulation has been considered.

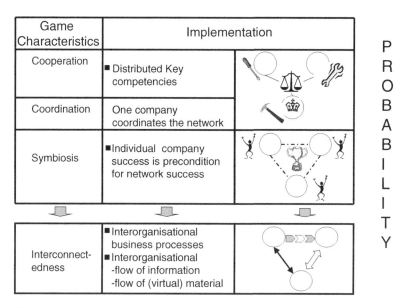

| Game Characteristics | Implementation | |
|---|---|---|
| Cooperation | ▪ Distributed Key competencies | |
| Coordination | One company coordinates the network | |
| Symbiosis | ▪ Individual company success is precondition for network success | |
| Interconnect-edness | ▪ Interorganisational business processes<br>▪ Interorganisational<br>-flow of information<br>-flow of (virtual) material | |

P R O B A B I L I T Y

Figure 1. Characteristics of distributed manufacturing and their implementation.

## 2.2 *Principles of distributed manufacturing*

According to Windhoff, game relevant characteristics of distributed manufacturing are co-operation, interconnectedness, symbiosis and probability (Windhoff 2001, p. 31). As key competencies and knowledge are distributed, companies have to co-operate to combine them to be successful. Owing to this co-operation, the different systems are interconnected, so that especially inter-organisational business processes and the related material and information flows have to be considered in a simulation. As the companies are standing in symbiosis to other companies, the success of an individual company depends on the performance of the other partners. On the other hand, the optimisation of one company does not automatically lead to the optimisation of the net-work. Additionally, the behaviour of the system cannot be predicted precisely, which might lead to complex non-transparent working circumstances. Figure 1 illustrates the mentioned characteristics of distributed manufacturing and their implementation.

## 2.3 *The lifecycle of a manufacturing network*

As manufacturing networks are dynamic and constantly changing, temporary consortia of enter-prises that strategically join skills and resources to better respond to business opportunities are gaining importance. The lifecycle of such virtual manufacturing organisations consists of four phases: creation, operation, evolution and dissolution (Camarinha-Matos and Afsarmanesh 1999). Within the creation phase, the virtual organisation is planned; partners are selected according to their competencies to then proceed with the negotiation of the contract, which is usually based on the co-ordinators call for tender and the other partners' related offers. On the basis of these offers, the contracts are completed and form the basis for the collaboration. During the operation of the virtual organisation, distributed business processes have to be co-ordinated. The evolution and dissolution of the virtual organisation form an additional cycle that comprises the occasional replacement of partners. After having performed the joint effort, the virtual organisation usually dissolves. Figure 2 illustrates the lifecycle of such a virtual organisation.

As the simulation model is rigid and the inclusion of additional partnering organisations would increase the model's complexity, the evolutionary aspect of virtual organisations is not considered in the simulation. Since the simulation does not aim to simulate the complete lifecycle of a virtual

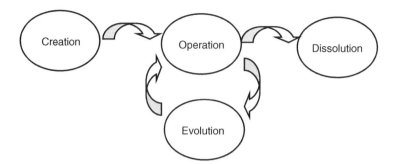

Figure 2. The lifecycle of a virtual organisation (Camarinha-Matos and Afsarmanesh, 1999).

organisation, the dissolution of the virtual organisation is not considered either. Thus, the simulation considers the creation and operation of the virtual organisation.

## 3  THE SIMULATION GAME

The processes of organisational learning, the principles of distributed manufacturing as well as the lifecycle of a manufacturing network are used to form a gaming approach that has been implemented in two scenarios. The simulation model has been described in greater detail within previous papers (Schwesigand Thoben 2004, Thoben and Schwesig 2004).

### 3.1  Level 1 of the simulation

In the first level, nine players have to specify, design and manufacture a Jet Ski in one company. They act as employees of an organisation that covers the basic economic functions. Each department is responsible for the successful completion of at least one step within the product development and manufacturing process. The players are divided into three groups resembling the departments of the company. These three groups are distributed in disperse locations. Still, the players have the opportunity to visit one another to enable the players to experience working in a typical co-located working environment. The players use an individual web interface to accomplish their given tasks within the simulated product development and manufacturing process. They have the opportunity to communicate using the built in chat function as well as the telephone.

To simulate organisational knowledge exchange, essential information that is required to accomplish the product development and manufacturing process is distributed unequally, so that the players have to co-operate and to communicate to be successful. Following their role descriptions, some players act in a non-collaborative way to form *people barriers*. The information-seeking players have to convince their non-collaborative colleagues to share their knowledge. To support this process, they can apply *trust-enhancing measures* implemented in the simulation such as gifts or simply negotiate by applying negotiation tactics that have been previously taught. The layout of the first level is illustrated in Figure 3.

After the simulation the players come together to reflect on what has happened and why certain events took place. Within this debriefing phase, they identify problems and initiating events that occurred in the areas of communication, collaboration, and, of course, trust. Together, they develop problem solutions to improve their communicative and co-operative skills.

### 3.2  Level 2 of the simulation

Within the second level, the players use the acquired knowledge and skills about handling intra-organisational people barriers in the inter-organisational contract negotiations to then specify, design and manufacture an extended product inter-organisationally: a cell phone enriched by certain

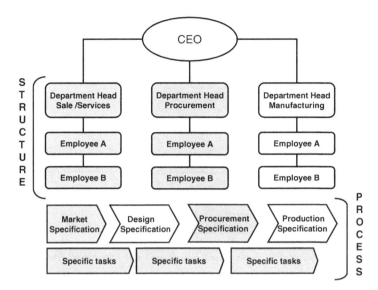

Figure 3.  Simulated organisational structure and operational processes in level 1.

services. While the simulated service-company takes leadership in the consortium and develops services, the two simulated manufacturing companies develop and produce the cell phone in a generic approach. Again, the players are divided into three groups resembling the particular companies in the production network. These three groups are again dispersed in different locations to enable the players to experience working in distributed environments. Again, the players use individual web interfaces to accomplish their given tasks within the simulated product development process. To simulate inter-organisational learning related challenges, the constant flow of information is affected by the simulated organisational boundaries, space, time, diverse cultures and structure. To successfully accomplish the scenario, the players are required to find appropriate solutions to overcome the barriers. These challenges require intense communication and collaboration among the simulated organisations. The layout of the second level is illustrated in Figure 4.

After the simulation session, the players come together again to reflect on what has happened and why certain events took place. Within this debriefing phase, they identify problems and initiating events that occurred in the areas of inter-organisational knowledge exchange and related boundaries. Together, they develop strategies how to overcome such boundaries to acquire performance skills in inter-organisational working and learning.

### 3.3  The web-based user interface

The graphical user interface was developed according to current usability guidelines DIN EN ISO 9241-10 and the usability framework model (Bevan 1991), while considering the particularities of web usability such as: the lack of standardised design guidelines that lead to a confusing mix of design elements; the heterogeneity of user groups and their related demands; the variety handling approaches as well as various browser platforms that do interpret program code differently (Heinsenn 2003, Keeker 1997). After having logged into the simulation by using individually assigned login data, the players enter their individual web interface.

The web screen design is divided into five basic parts. The top part provides explanatory, role specific and dynamic information about the simulation and the sharing of documents. The left part provides explanatory information about the different users or companies or both. The right part provides the different documents of the product development process, the specific options related to the product development as well as trust enhancing measures. The lower part displays the chat as well as dynamic information about the current status. The central part of the screen is always

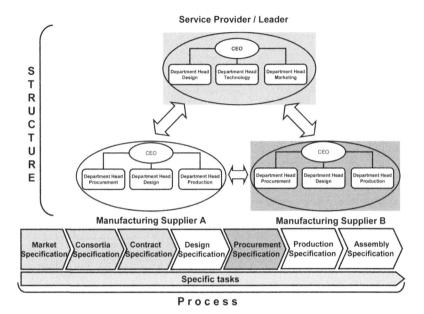

Figure 4. Simulated organisational structure and operational processes in level 2.

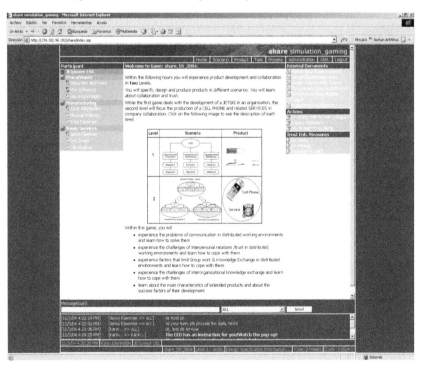

Figure 5. Screenshot of the web based user interface of the simulation game.

used to display the selected information. This measure should support the user's orientation within the interface. Following the Microsoft Office software, specific self-explaining icons should help the user to identify the particular elements to ease the navigation. Figure 5 illustrates a screenshot of the web-based user interface of the simulation.

244

## 4  CONCLUSIONS

A simulation has been created that combines aspects of organisational learning and knowledge management with a manufacturing scenario in and among organisations. It aims on mediating communication skills, collaboration skills as well as the proper use of information and communication technologies within authentic (distributed) manufacturing scenarios.

The gaming approach used in the simulation game *COSIGA* has been combined with identified processes and challenges in organisational learning, principles of distributed manufacturing and the lifecycle of a type of a manufacturing network. These elements then formed an approach that has been described in two scenarios. These scenarios deal with organisational learning and product development in one company and distributed development and manufacturing of an extended product and inter-organisational learning within a manufacturing network of companies. An initial version of the simulation was evaluated and validated in four gaming sessions with 36 students and industry representatives. The learning outcome was assessed by applying a questionnaire based survey as well as interviews. After having played the game, the players were asked to assess their communication and collaboration skills:

- Communication skills in distributed environments
  Most of the players were able to identify problems of communication in distributed environments and to find strategies to cope with these problems. For instance, the players recognised the fact that misunderstandings are more difficult to solve in such environments and suggested approaching the other partner more openly to discuss potential problems honestly.
- Collaboration skills
  Most managed to identify the challenges within interpersonal relations and trust and found strategies to cope with these problems. The players identified the increased difficulty to build trust without non-verbal communication as a problem and found that increased tolerance and increased trust at the beginning of a distributed working relation could be solution strategies. Almost all the players identified the non-collaborative attitude of the players as an element that limits knowledge exchange among individuals. They also identified strategies how to cope with circumstances like the trial to understand the perspective of the non-collaborative person, and then to convince them.
- Collaboration skills in distributed environments
  The players were able to identify challenges within the inter-organisational knowledge exchange such as time zone conflicts and different cultural backgrounds. They proposed solutions strategies such as goal-oriented attitude and strong tolerance towards the habits and values of other cultures. Most of the players believed that their sensitivity towards other company cultures improved.

To apply this gaming approach within different settings, it is now planned to develop an adjustable simulation that is able to simulate different kinds of vertical and horizontal collaborations. A suitable system architecture is under development. Apart from this, modern wireless technologies will be integrated to enable an easy implementation in working and learning environments and to present future ubiquitous learning environments for engineers.

## REFERENCES

Barson, R.; Foster, G.; Struck, T.; Ratchev, S.; Pawar, K.; Weber, F.; Wunram, M. (2000). Inter- and Intra-Organisational Barriers to Sharing Knowledge in the Extended Supply Chain. Proceedings of eBusiness and eWork Conference 2000; The Key Action II Annual Conference, Madrid

BBJ Consult AG (Eds.), (2002). Wissensmanagement-Planspiel für soziale Organisationen. Abschlussdokumentation. München

Bevan, N.; Kirakowski, J.; Maissel, J. (1991): "What is usability?", in Proceedings of the 4th International Conference on HCI, Stuttgart, September, webpage: http://www.usability.serco.com/papers/whatis92.pdf

Bosch-Sijtsema, P. (2001). Knowledge development in a Virtual organisation: An Information Processing Perspective, Licentiate dissertation from Lund University, KFS AB Lund, Sweden

Büchel, B.; Probst, G.(2002). From Organizational Learning to Knowledge Management. Genf. 2000. URL: http//hec.info.unige.ch/recherches_publications16.11.2003

Camarinha-Matos, M.; Afsarmanesh, H. (1999). The virtual enterprise concept In: L.M. Camarinha-Matos and H. Afsarmanesh, Editors, Infrastructures for Virtual Enterprises—Networking Industrial Enterprises, Kluwer Academic Publishers, pp. 3–14

Goffin, K.; Mitchell, R.(2002). "Teaching Innovation and New Product Development using the "City Car" Simulation" 13th Annual Meeting of the Production and Operations Management Society, San Francisco 5th–8th April

Heinsenn, S.; Vogt, P. (2003): "Usability praktisch umsetzen", Hanser Verlag, 2003

Hoheisel, Jens; Thoben, Klaus-Dieter; Echelmeyer, Wolfgang: Almost Like Real Life – An Experience Based Learning Concept for Engineers. In: New Engineering Competencies – Changing the Paradigm", Proceedings of the SEFI, Annual Conference, Copenhagen 12–14 September 2001, pp. 99ff and CD-ROM

Inkpen, A.; Crossan, M. (1995). Believing is Seeing: Joint Ventures and Organizational Learning. In: Journal of Management Studies 32:5 / 1995, pp. 595–617

Keeker, K. (1997): "Improving Web Site Usability and Appeal", online unter http://msdn.microsoft.com/library/default.asp?url=/library/en-us/dnsiteplan/html/improvingsiteusa.asp , 24. Juli 1997

KITS (2004). Internetseiten des KITS Projektes über das Planspiel KM-Quest. WWW-Seite. http.//kits.edte.utwente.nl/kmquest/index.html.08.05.2004

Kriz, W. (2001). How to facilitate the debrief of simulations/games for effective learning. LMU München. In: Psychologische Beratung und Intervention, PAB 2001/2, München

Larsson, R.; Bengtsson, L.; Henriksson, K.; Sparks, J. (1998). The Interorganizational Learning Dilemma. Collective Knowledge Development in Strategic Alliances. Special issue. Managing Partnerships and Strategic Alliances. Organization Science Vol. 9, pp. 285–306

Meier, C.; Herrmann, D.; Hüneke, K. (2001). Medien- und Kommunikationskompetenz – Schlüsselqualifikation für die Zusammenarbeit auf Distanz. In. Wirtschaftspsychologie – Aktuell. Vol. 4, 2001. pp. 12–20

Nonaka, I. (1994). A dynamic Theory of Organizational Knowledge Creation, in. Organization Science, Vol. 5 , pp. 14–37

Oh, E.; van der Hoek, A.(2001). Adapting Game Technology to Support Individual and Organizational Learning in: Proceedings of the 13th International Conference on Software Engineering and Knowledge Engineering, Buenos Aires, Argentina, June 2001

Probst, G.; Büchel, B. S. Organisationales Lernen. Wettbewerbsvorteil der Zukunft. Wiesbaden 1994

Riedel, J.; Pawar, K. S.; Barson, R. (2001). Academic and Industrial User Needs for a Concurrent Engineering Simulation Game. In. Concurrent Engineering Research and Applications, Vol.9 (3), pp.223–237

Schwesig, M., Thoben, K.-D. (2004). Developing a web based group simulation game to simulate organisational and inter-organisational learning in production networks. In: Proceedings of the 8th international workshop of the IFIP WG 5.7 special interest group (SIG) on Experiential Interactive Learning in Industrial Management, Wageningen

Smith, P.G.; Blanck E.L. (2001). From experience: Leading dispersed teams. Journal of Product Innovation Management. Vol. 19 (4), pp. 294–304

Sumner, T.; Domingue, J.; Zdrahal, Z.; Millican, A.; Murray, J.(1999). Moving from On-the-Job Training towards Organisational Learning. Proceedings of the 12th Banff Knowledge Acquisition Workshop, Banff, Alberta, Canada, pp.16–22

Thoben, K.-D.; Schwesig, M. (2004). Web enabled simulation gaming to mediate performance skills in interorganisational learning to engineers. In: "Integrating Human Aspects in Production Management". Kluwer Academic Publishers. New York, London

Windhoff, G. (2001). Planspiele für die verteilte Produktion. Entwicklung und Einsatz von Trainingsmodulen für das aktive Erleben charakteristischer Arbeitssituationen in arbeitsteiligen, verteilten Produktionssystemen auf Basis der Planspielmethodik. Dissertation Bremen

*Part VIII*
*Collaborative engineering*

Constantly changing customer demands and an intense global competitive environment imposes the compelling need for creativity, innovation, as well as enhanced inter-personal productivity. All these facets are an extension to the principles of concurrent engineering. On one hand, customers are increasingly requesting additional services associated with the supply of products. On the other hand there is an increased tendency to form new types of partnerships, alliances and organisational arrangements to reflect changing customer requirements. All these issues need to be considered throughout the whole product design and development process. Hence the future of collaborative engineering lies within the context of integrated design of customisable products, comprehensive services, and flexible organisations where individuals, groups, businesses and communities collaborate together within loosely or tightly coupled networks through the use of online shared workspaces.

The virtualisation and the distribution of product development activities have lead to new challenges for organisation and technology in European companies. Global competitiveness can only be assured through the development of effective and lasting strategies for creating and managing innovation. The IMS-NoE Special Interest Group on *Co-operative and Virtual Engineering* analysed problems of co-operative engineering as well as methods and tools of virtual engineering of extended products. Based on these analyses a multi-stakeholder roadmap is proposed by Goossenaerts et al. In the first contribution, that articulates public and private sector roles in coping with future engineering challenges. The public sector role targets the creation of a knowledge intensive global business eco-system conducive for balanced private sector innovation and sustainable growth. The private sector roles evolve under strategies that implement proven co-operative and virtual engineering practices with a focus on value creation.

The second contribution from Eschenbaecher et al. deals with the management of dynamic virtual organisations and presents results from a collaborative engineering case. Industrial production is now changing dramatically because competition is not anymore taking place among companies but more among virtual organisations, forcing each network player to operate more efficiently than its competitors. Many management challenges in planning, co-ordinating, and supervising such dynamic operating virtual organisations have not been addressed so far. The reason is that these dynamic virtual organisations, which can also be considered as collaborative networks, represent an evolution of enterprise development, a topic dealt with by the ECOLEAD project. These management challenges have to be addressed by powerful methods to leverage success of virtual organisations. The identification of challenges and management requirements in six case studies shows the tremendous practical needs, methods, concepts and tools to support various European SME networks. These SME-based virtual organisations clearly need governmental directives guiding them though their collaborative processes. For one case study concerning collaborative engineering, the management challenges are discussed in detail. The paper concludes with a list of management aspects that must be addressed by research activities.

One of the main results of CE-NET (Concurrent Engineering Network) was the roadmap for the Collaborative Enterprise 2010. This roadmap is presented in the third contribution. The term collaborative enterprise captures the shift from the existing organisational approaches to more dynamic and re-configurable peer-to-peer networks and from ownership to sustainable services. Several complementary perspectives, namely social, customer, business, workplace, technology, and legal have been concurrently considered to build up the collaborative enterprise vision and roadmap. The roadmapping process and the relationship among today's circumstances, the CE vision, gaps, research needs and solutions are explained. State-of-the-art, vision, gaps, needs and the required solutions, in the short, medium and long-term, to achieve the vision for product, service and organisation development of collaborative enterprising 2010 are outlined.

*Advanced Manufacturing – An ICT and Systems Perspective – Taisch,*
*Thoben & Montorio (eds)*
© *2007 Taylor & Francis Group, London, ISBN 978-0-415-42912-2*

# Co-operative and virtual engineering: A multi-stakeholder roadmap

Jan Goossenaerts[1], Christian Brecher[2], Frank Possel-Dölken[2] & Keith Popplewell[3]

[1]*Eindhoven University of Technology, Paviljoen MB Eindhoven, Netherlands*
[2]*Laboratory for Machine Tools and Production Engineering (WZL), RWTH Aachen University, Aachen, Germany*
[3]*Coventry University, Manufacturing and Systems Design Research Group, Coventry, UK*

ABSTRACT: The virtualisation and the distribution of product development activities have lead to new challenges for organisation and technology in European companies. Global competitiveness can only be assured by the development of effective and lasting strategies for creating and managing innovation. The IMS-NoE Special Interest Group 6 has analysed problems of co-operative engineering as well as methods and tools of virtual engineering of extended products. Based on these analyses a multi-stakeholder roadmap is proposed that articulates public and private sector roles in coping with future engineering challenges. The public sector role targets the creation of a knowledge intensive global business eco-system conducive for balanced private sector innovation and sustainable growth. The private sector roles evolve under strategies that implement proven co-operative and virtual engineering practices with a focus on value creation.

*Keywords*: Co-operative Engineering, Virtual Engineering, Product Development, Public Goods, Institutions.

## 1 INTRODUCTION

To increase value for stakeholders, to shorten *time-to-market*, to handle increasing complexity of products, and to lower the development costs, novel approaches for co-operative engineering are needed. Within this context innovative and rapid product development is a key issue. Two important topics are the *co-operative engineering* (CE) and the *virtual engineering* (VE). While CE is mostly an organisational issue, VE focuses more on the technological infrastructure that enables and supports CE and the lifecycles of extended products.

Figure 1 shows the Three-Cycle Model of Product Development, which identifies the significant issues of holistic product development, namely *strategic product planning, virtual product development*, and *virtual production system development*. All these issues have an organisational perspective that must be addressed by CE approaches as well as a technical perspective, which is addressed by VE.

Because product stakeholders also affect the product lifecycle via regulations, transportation, marketing, usage, repair and upgrade, take-back and recycling and disposal, the organisational aspects extend into the social domain: *the socio-industrial eco-system*. The *multi-stakeholder product development and life* paradigm requires intimate information sharing among all product and production stakeholders. Hence also the technological aspects extend into the social domain: the *engineering information infrastructure* (Kimura 2005).

In the extended context for product development very strong social demands and constraints and environmental considerations are directing manufacturing activities and product use into more resource-saving and environmentally benign manners. At the same time manufacturing industry

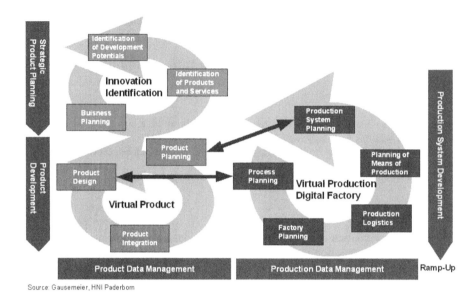

Figure 1.   The Three-Cycle Model of Product Development (Gausemeier 2004, p. 100).

must be competitive to survive in severe global markets. Information technology is clearly an enabler to accommodate both requirements, and to lead to a new manufacturing paradigm: from product manufacturing to function or service manufacturing (Kimura 2005).

From this perspective, the *product development and use community* must adopt a practice of co-operative engineering of extended products. With a focus on total benefit and cost of ownership and socio-environmental impacts, this community must achieve a high maturity level in obtaining and structuring *knowledge* from external and internal sources.A powerful environment bundling VE-methods, tools and infrastructure for the multi-stakeholder product development process must enable co-operative engineering. The deployment of VE-methods and tools in a socio-industrial global community goes hand in hand with enhanced co-operative and virtual engineering maturity.

## 2   VISION AND TRENDS FOR FUTURE ENGINEERING

The vision for the future is that *an excellent level of holistic harmonisation and fit of technologies, organisational concepts, and company and market culture, is achieved society-wid*e. The *improvement of the state of manufacturing industries as a whole*, envisioned in the IMS program (Yoshikawa 1994), includes industry's ability to respond to global challenges and to contribute to realising public policies.

Regarding the development and use of extended products this vision requires the integration of organisational approaches, technical solutions and company culture into an innovation and development environment that is equitable, target and customer-focussed, high performance, and leading in quality.

Several trends support the vision. In co-operative engineering they include:

- Product development as integrated, formalised enterprise business process;
- Networked, team-oriented, inter-departmental and inter-company process collaboration;
- Product lifecycle management;
- Set-up of virtual enterprises of small and medium-size companies (SMEs);
- Increased systematic innovation and knowledge management.

Supporting trends in virtual engineering include:

- Development Application Integration (DAI) and Enterprise Application Integration (EAI) based on portal or peer-to-peer technologies;
- Engineering Object Management and new approaches for distributed data and knowledge models;
- Development of aligned market, industry and company specific standards for data exchange among different applications.

The social domain also sees new institutional practices that aim to respond to challenges of the technology intensive network economy:

- The new approach to technical harmonisation and standardisation (Bilalis and Herbert 2003);
- The drafting of new intellectual property schemes that aim for a new balance between public benefits from creation and private rewards for creators (Creative Commons, Lessig 2001).

To achieve the vision, a roadmap must also leverage the current state of practice. The most important tools for product development are the software applications that facilitate the entire development process. Nearly all documents generated as part of the product development and project management are created and managed electronically. Robert Bosch GmbH, for instance, in 2002, applied thirteen different types of software applications, among them CAD, FEM, simulation, document management, multimedia, and office communication (Eversheim 2002).

To cope with the heterogeneous system environment and their data exchange problems companies today strive to set up internal standards and interfaces to integrate the different software solutions that are part of the product development process. This does not only apply to large companies, such as automotive industry, but also to SMEs (e.g. in tool and die industry and in sheet metal processing) that face the challenges of customer-oriented job shop manufacturing with increasingly short delivery times. In the tool and die industry, the CAD-CAM-NC chain is the important optimisation target for reducing lead times. Many of the successful tool and die companies have managed to develop their own CAD-CAM-NC environments that consist of individual commercial software systems with company-specifically integration. This technical trend is accompanied by an extended management approach, *Product Lifecycle Management* (PLM) that puts a holistic view on all lifecycle phases of products from creation to disposal. For job shop producers this means, for instance, integrating sales business processes with product development processes. In mass production, companies strive for lifecycle spanning, IT-supported business processes from marketing, through development and production, to sales and service (cf. Figure 2).

The machine tool industry illustrates the importance of integrated development processes. Machine tools are among the most complex mechatronic products and face the highest requirements concerning mechanical design, usage of latest electronic equipment and advanced software solutions for control, operator support and maintenance. Machine tools are also subject to requirements that are of public interest, such as the protection of health and safety of users, and the protection of the environment. However, machine tool builders in Europe are mostly SMEs with limited financial and personnel resources. To keep up with the growing customer requirements and competition from overseas, the machine tool industries as well as related solution providers together with research institutions, such as the WZL, have invested heavily into researching and exploiting the potentials of the *Virtual Machine Tool* (VMT).

In the first step, the VMT vision and concept calls for a coupled design environment for machine tools and related components that is based on commercial CAE applications but allows for bi-directional exchange of product models among the development phases, design, component design and calculation, FEM, analytic mechanical analysis with multi-body simulation, and performance evaluation and optimisation by matching calculation results with measured machine tool behaviour (cf. Figure 3). In the second stage this approach is further extended to the mechatronic development domains, such as the design of electronic circuits with E-CAD systems, the controls

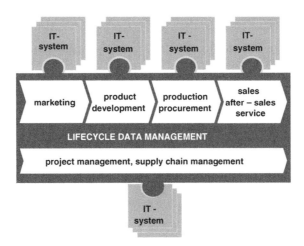

Figure 2.    Product Lifecycle Data Management (source: BMW, WZL).

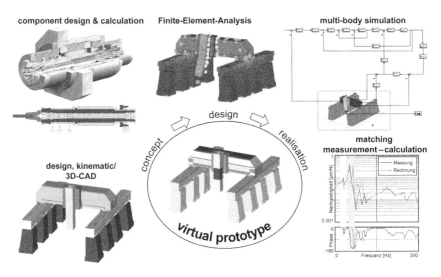

Figure 3.    Virtual machine tool design (source: WZL).

architecture development, the controls programming and the control system simulation. Furthermore current research activities focus on the coupling of machine tool simulation and process simulation. The third stage of the VMT concept has a long-term perspective and calls for integrated development environments that offer holistic support for mechatronic design. These environments require integrated data models, which so far are not available. However, the closer collaboration of CAE software vendors with automation providers and machine tool builders (also several recent acquisition activities in the CAE business) offer the perspective of more integrated development environments that will build on extended Product Data Management systems.

The high complexity especially of mechatronic products like machine tools requires domain-specific product models suited to the various development teams and stakeholders involved. From the long-term perspective new approaches for integrating the various domain-specific product models need to be integrated as part of a DAI platform. Agent-oriented approaches provide the necessary means to set up information environments where the objects of a unified product model

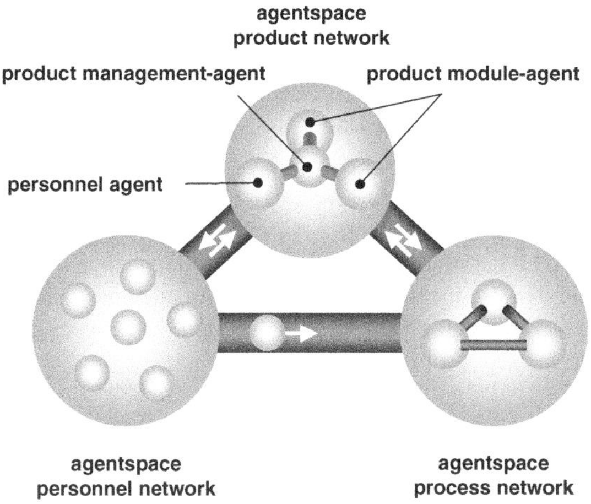

Figure 4. Multi-agent approach for Engineering Object Management (source: WZL).

(UPM) can be flexibly linked and jointly administered while different stakeholders access only domain-specific views on to the UPM (cf. Figure 4). At WZL such an agent-based PDM platform has been developed (Klement 2005) that can consistently manage product data objects from different engineering domains as well as project management. Various CAE applications can connect to the platform via their Application Programming Interfaces (API), load their internal data objects into the generic data model of the agent platform, and link to data objects of other CAE applications. The CAE application thus becomes a tool that provides a specific view on to the overall, agent-managed product and project data structure and that offers domain-specific functionality for manipulating selected data objects.

## 3 GAPS

In industry today, there exist many pockets and chains of excellence. Multiple gaps must be bridged to achieve a society-wide excellent level of holistic harmonisation and fit of technologies, organisational concepts and company and market culture. Those objectives cut across all multi-stakeholder product development and lifecycle activities. Raising the levels of co-operative engineering maturity to quantitatively managed and optimising is a socio-technical challenge for a *global* IT-reliant work system. (According to Alter (2003) a *work system* is a system in which human participants or machines or both perform work using information, technology, and other resources to produce products or services or both for internal or external customers. *IT-reliant work systems* are work systems whose efficient or effective operation, or both, depends on the use of IT).

The vision points in a direction, and proven ICT-based solutions and experimental institutions demonstrate that opportunities, gaps and problems still exist in practices, knowledge and tools, as well as in the institutional infrastructure.

### 3.1 *Co-operative engineering*

It is mainly in the co-operative engineering sphere that the dynamism in the IT sector does not match well with the long investment horizons in some industries, such as process industries, or the limited learning means of others, such as SMEs. A process industry sector study into the adoption of product data standards, undertaken by USPI NL and the Information Systems Department of the Eindhoven University of Technology, illustrates some aspects of the resulting *in-need* state. Problems that emerge in the network of plant lifecycle stakeholders (Dreverman 2005) include:

253

- *Data interpretation differences among* engineering, procurement and construction (EPC) *contractors and plant owners*: The engineering data used by EPC contractors and equipment vendors differs greatly from the operational data used by the plant owners. Plant owners need a fraction of this engineering data. Plant owners impose compliance to their own standards, but have decreasing knowledge of which data they need. EPC contractors on the other hand often work strictly to contract specifications and do not pro-actively participate in formulating improved specifications. As a result plant owners have much unnecessary and redundant data.
- *Fear of revealing critical information to competitors*: EPC contractors and also equipment vendors are reluctant to co-operate owing to fear of revealing critical information to competitors.
- *Short learning curve*: The frequency of projects is low; a plant is never built twice. A slight change of parameters results in a greatly different design. Even when an exactly identical plant is built in sequence, the second plant differs significantly owing to rapid progression of the technology used.
- *Presence of a multitude of standards creating a need for multiple mappings*: The plant owners may impose the use of their own standards and software solutions on the EPC contractors. As plant owners traditionally did engineering in-house many have developed in-house standards. For this reason EPC contractors have to maintain multiple software systems or translate the designs afterwards. In both cases this means higher costs. Equipment vendors differ significantly regarding their data readiness. Bigger vendors may be able to deliver in the required format, but many small suppliers use very old systems or sometimes only are capable of delivering paper drawings. Furthermore, equipment vendors also use various catalogues and libraries, making this circumstances more complex. As a result mapping of all these different standards and systems is slow and difficult.
- *Differences in style and taste of engineers*: In global multi-office engineering it is very hard to co-operate in design. This because of the differences in style and taste of engineers. Aligning these different styles may cost more than the advantages of co-operation. For this reason global engineering is mainly done on unit and module level.

The organisational aspects extend beyond the supply chain:

- Virtual engineering is a technology that exhibits strong *network effects*. These are complementary relationships in value creation among adopters of a common standard (Farrell and Klemperer 2003). For such technologies, lock-in circumstances are common and the lack of expectation sharing among stakeholders may be just one component of adoption inertia (Au and Kauffman 2005)
- The lack of *systematic approaches to and support tools for the capture, evolution and re-use of knowledge* applicable in the support of collaboration among departments or companies. Generic knowledge of the process of collaboration can be identified as applicable to almost all industrial sectors, as can sector specific knowledge dependent on market and technological factors (Popplewell and Harding 2004). Companies or virtual enterprises generate knowledge of their own collaboration issues, which if captured and maintained forms a growing body of enterprise experience that improves decision-making and hence all measures of performance (i.e. financial, environmental, societal, etc.). Such knowledge at all levels is dynamic and expanding, and support tools must be able to use a wide range of techniques for knowledge capture, discovery and retrieval.

## 3.2 *Virtual engineering*

The significant gaps in virtual engineering can be summarised as follows:

- *Insufficient integrated models for product lifecycles*: So far no holistic reference models for different application domains that describe domain-specific product lifecycles are available.
- *Lack of methodological and technical support for integrated product, project and production information management*: Companies still lack the holistic understanding of their networked

product development processes, their customer-focussed project management, as well as their production and operations management that finally results in inefficient business processes.

- *Lack of formalised product lifecycle workflows*: The implementation of VE methods must be embedded into a sophisticated organisational approach that specifies and standardises company-specific workflows of how to proceed in the development of products and processes.
- *Heterogeneous system environments*: Today's VE applications lack interfaces for facilitating and supporting integrated information technology. This impedes deployment and implementation of VE approaches.
- *Innovation barrier required investments*: Currently used CAE applications are large-scale, grown engineering environments that need to be fundamentally adapted / changed to support integrated product lifecycles.
- *Gap between technical opportunities and available experience on the user side*: The high technical competencies of VE solution providers often face little experience and strong reservations against VE technologies by far too many product developers.
- Required system complexity for holistic VE solutions would possibly overstrain the capabilities of users, provided that *software intelligence* does not significantly increase.
- *Computing power* is still insufficient *for integrated simulation environments*: This concerns particularly the simulation of complex automation and control solutions with respect to real-time behaviour.
- Effort for *setting up virtual worlds*: The design of virtual models of machinery or entire production systems requires extensive efforts. The success and benefits of simulation experiments on the other hand can often not be predicted for sure.
- Lack in integrated simulation environments for production systems and *integration of* these *virtual worlds into standardised development workflows*: The commercial market for integrated simulation environments is just beginning to emerge.

### 3.3 *Institutional gaps in the knowledge economy*

Free riding and under-supply plague knowledge that becomes a pure and global public good. An actor that lacks the knowledge used to create value in a modern economy suffers from an idea gap (Romer 1993). Whereas the partial and temporary monopoly power that is granted by patents, or by control of a large fraction of the market, lock-in or network externalities, may act as motivators for intentional efforts to produce and transmit knowledge (Romer 1993, 2003), there is as yet *no consensus on the basic institutional infrastructure for market exchanges when knowledge is the good exchanged.*

The Global Public Goods Task Force (2005) writes: "*Most governments are heavily involved in regulating the production and dissemination of knowledge, using two instruments – (i) intellectual property laws that protect the rights of patent and copyright holders, and (ii) support for the common knowledge platform via funding of research in specific areas and protection of common use rights. However, the balance between the two kinds of systems has been tilting increasingly towards private intellectual property during the past 20 years, reflecting the pursuit of national commercial interests in the absence of international standards. The European Union has adopted a highly restrictive directive on the protection of databases; the United States has expanded the scope of patentable research findings, including what otherwise would be considered discoveries of nature rather than inventions; and other developed nations have moved to protect business software, plant and animal varieties, genetic sequences, and biotechnological research tools. The bottom line is a major contraction of the common knowledge platform, which is prejudicial to public interests.*"

For today's innovators the relentless expansion of the current legal apparatus may be generating a highly fragmented and complex system of rights whose management incurs high transaction costs, with the effect of discouraging those types of creative activities that cannot afford these new costs (Ramello 2005). *Tragedy of the anticommons* (Heller, 1998) is the name of the institutional failure that is looming with the trend to stronger patent systems. The Open Source Movement (Lerner and Tirole 2001) and the Creative Commons licensing contracts (Lessig 2001) are emerging as practices at the other extreme. Neither of these extremes seems to establish the appropriate balance between the public benefit from creation and the private reward for the creator.


## 4  A BROAD ROADMAP FOR CO-OPERATIVE AND VIRTUAL ENGINEERING

The gap analysis shows that improving the state of *manufacturing industries as a whole* is beyond the means of the manufacturing community alone. The connected nature of the global fabric is widely recognised. To overcome the inability to make engineering and manufacturing part of concerted practices that *respond to global and local challenges*, it is necessary to expose the industry's embedding in the global community, as an engine of growth, and as a beneficiary of right-sized institutions and new technologies.


### 4.1  *Global context*

Over the past decade, while globalisation has been studied as a driver for competitiveness, the international community has also articulated desirable outcomes, including social and environmental, and it has achieved consensus about global development goals, such as the Millennium Development Goals (Sachs et al. 2005, and http://www.developmentgoals.org/), and environmental targets, such as the Kyoto Protocol. Reporting frames such as that of the Global Reporting Initiative help organisations to report on environmental and social outcomes in addition to profits or losses.

The GRI was launched in 1997 as a joint initiative of the U.S. non-governmental organisation Coalition for Environmentally Responsible Economies (CERES) and United Nations Environment Programme with the goal of enhancing the quality, rigour, and utility of sustainability reporting. The initiative has enjoyed the support and engagement of representatives from business, non-profit advocacy groups, accounting bodies, investor organisations, trade unions, and many more. Together, these different constituencies have worked to build a consensus around a set of reporting guidelines with the aim of achieving worldwide acceptance (Global Reporting Initiative (2002), Sustainability Reporting Guidelines (www.globalreporting.org)).

Suddenly the pre-competitive and post-competitive phases of the knowledge production process (Yoshikawa 1994) can be addressed in a more mature, socio-technical global environment. A new performance paradigm is being shaped. It recognises the broad context within which production capabilities develop, as well as their enabling role in achieving development goals. The result-focussed management of knowledge in the pre- and post-competitive phases of product lifecycles is a significant challenge, for which Kimura (2005) lists critical issues. Co-operative and virtual engineering require institutions and practices regarding knowledge and idea flows that cannot escape the public-private context.


### 4.2  *A multi-stakeholder roadmap*

In a global co-operative context, a roadmap must identify tasks for a large number of actor-stakeholders, who together can achieve the scaling of excellence from their current pockets to society wide. Society is subject to the developments in and the limits of its natural and physical environment: the baseline and vulnerability context. Within this context, the excellent society should achieve bold development goals, as expressed in the Millennium Development Goals and the Kyoto Protocol. This society also is a community of value exchanging actors moving into increasingly advanced uses of content, ICT and application software to achieve outcomes that contribute to a total system of environment, economy, society and culture (Monnai et al. 2005).

Gaps are diverse, and conflicting trends exist. Tasks to overcome gaps must be allocated to the natural stakeholders, i.e. those responsible for the assets involved. Where conflicting trends inhibit total welfare, decision frames must be established first, before courses of action are decided. For the purpose of this broad roadmap, the stakeholders will be subdivided into two categories: those pursuing public benefit and those pursuing private reward.

### 4.3 *Public sector stakeholders and their roles and tools*

This category of stakeholders includes the public sector, academia and research institutes, standards bodies, etc. Often, the asset concerned is knowledge or institutional infrastructure. Knowledge has one of the oldest traditions of international co-operation. Stakeholders in the public sector must re-articulate their purposes and their role models taking into consideration the following guidelines.

- *Focus on Balancing Public and Private Interests.* Given the international interest, there needs to be a transparent international dialogue – and process with a view to achieving a consensus – on the appropriate balance between private intellectual property and knowledge in the public domain (Global Public Goods 2005).
- *International Framework to Promote Access to Data.* An emerging complex cyber infrastructure is rapidly increasing our ability to produce, manage and use data. As research becomes increasingly global, data-intensive, and multi-facetted, it is imperative to address national and international data access and sharing issues systematically in a policy arena that transcends national jurisdictions. Open access to publicly funded data provides greater returns from the public investment in research, generates wealth through downstream commercialisation of outputs, and provides decision makers with facts needed to address complex, often transnational problems (Arzberger et al. 2004).
- *Academia and Research Institutes* must ensure that their activities, in particular research target finding, publication of results, and dissemination approaches, are informed also by enhanced public sector policies that balance public and private interests, and ensure global access to data for greater returns from the public investment.

The *Global* Public Sector must give adequate attention to the provision of knowledge as a global public good, and to the removal of hurdles that prevent its flow. With a strategic horizon, the Global Programs of the World Bank (World Bank Operations Evaluation Dept. 2004) may offer one possible instrument to address this challenge. The proposals of the International Task Force on Global Public Goods describe a roadmap with far-reaching changes. Thus, the international community must architect and deploy an improved knowledge exchange infrastructure that is as pervasive as the international financial architecture, and plays a special role as a conduit that lets productive ideas flow across the globe. Without this, change processes of most private sector stakeholders will continue to be slow and laborious at best, and infeasible in most cases.

Immediate action is possible for the principles regarding the establishment of access regimes for digital research data from public funding, to which thirty-four OECD member governments declared their commitment. Contracts regarding publicly funded research, for instance in the proposed Seventh Framework Program, could include clauses that implement these principles, for all contractors in line with their focus on public or private utility (thus reducing the contractual incompleteness that now reigns most knowledge exchanges, and encourages free-riding and rent-seeking aims). Simultaneously, supporting infrastructure services should be identified, or be implemented where gaps exist. An example in this area is the *Global Bio-diversity Information Facility* (GBIF). GBIF is an interoperable network of bio-diversity databases and information technology tools that enables users to navigate and put to use the world's vast quantities of bio-diversity information to produce national economic, environmental and social benefits (http://www.gbif.org/).

For the private sector stakeholders, the transformations of the knowledge exchange infrastructure and the articulation of private and public phases in the product knowledge lifecycle should have the following implications:

- *From shallow contracting and contractual incompleteness to knowledge aware and triple bottom line contracting*: supported by a more capable market, and institutional infrastructure, knowledge spaces, social and environmental values will become part of dedicated clauses in the future contracts that will bind the stakeholders in product lifecycles. This will reduce risks of rents in contractual relations, stimulate local content in global supply chains and lead to an improved socio-environmental impact of products.

- *From private matters most to balanced public-private partnerships*: whenever knowledge on a product is captured by any stakeholder, this must be done early and without loss, to allow for further sharing beyond the private boundary, and with transparent reflection of its hosting frame of reference in the knowledge evolution.

A second area for immediate action is the application to the area of interoperability of enterprise software applications of the new approach to technical harmonisation and standardisation (Bilalis and Herbert 2003). The lack of interoperability of software for product development has long been addressed by standards alone (e.g. STEP), without the support by Directives. Yet, there is a sound argumentation for institutional intervention where market failures support rent seeking by software vendors, especially since user-owned data can be separated from application-added data through a proper architectural style (Goossenaerts 2004). Dul et al. (2004) illustrates the new approach in the ergonomics area: several CEN standards are related to legislation in European Directives. These place generally formulated essential requirements on, for instance, safety, health, or environment. Linked to these Directives, European standards are developed that give detailed requirements. Any company that meets these standards is assumed to meet the general requirements set in the Directives. Thus implementing the standards is an effective and efficient way to meet the legal requirements.

### 4.4  *Private sector growth enabled by knowledge infrastructure and institutions*

A study of the consulting firms Roland Berger and Frost and Sullivan forecasts an average annual increase of the total turnover for software and services for the Digital Factory of approximately 35% in the next 5 years. This shows the dynamic changes that are beginning to take place in today's production companies. However, there is no single threaded path into the future. Instead there are manifold issues that need to evolve and finally merge to establish a culture and technological environment for efficient networked product lifecycles. Changes in practices are required; many of these changes involve network effects. Enhanced knowledge infrastructures and institutions will accelerate the growth of co-operative and virtual engineering maturity of the market players.

- *From information sharing to knowledge sharing*: While progress has been made in supporting the sharing of knowledge among VE partners, this has been based mainly on request and response, under human or occasionally automated control. Where automation is possible it relies on shared information models. Recent research in system interoperability must be applied to increase the level of automation of the sharing of information and, through ontology and epistemology based methodologies (Abramov et al. 2005), to ensure common understanding of shared information – the partners sharing information must have a common understanding of the meaning and implications of the information.
- *From workflow management towards intelligent collaboration*: Current VE collaboration depends upon a combination of workflow management procedures and the informal skills of a project manager to insure timely communication and co-operation among VE partners who are geographically and organisationally distributed, and who have differing areas of technical discipline, expertise and experience. Without such management, VE partners can and will make decisions which, while optimal within their own discipline or plant, conflict with other partners' decisions, resulting in sub-optimality in terms of the global VE, with potentially very expensive results. Workflow management can identify only the predictable points where synchronous decision-making is necessary, and project managers' experience can only have limited predictive ability. Intelligent knowledge-based agents embedded in collaborative systems, already demonstrated in some scenarios, must be exploited to embody and apply as wide a range as possible of knowledge of collaboration, the relevant industrial sector, the VE and the partners, to identify and resolve potential decision conflicts at the earliest possible point, before avoidable expense is incurred.
- *From knowledge capture to knowledge discovery*: The volume of knowledge potentially available to support collaborative engineering is too great to be adequately embedded in support agents

through conventional elicitation and encoding. Knowledge discovery methods (e.g. data-mining) offer the ability to derive new understanding from current and historical information systems, to be maintained in structured hybrid knowledge-bases, greatly extending the capabilities of knowledge-based support agents.

- *From static co-operation towards dynamic enterprise networks*: While the awareness of the importance of co-operation is increasing, companies need to establish processes and culture that will spur and enable dynamic co-operation with suppliers, customers, or even competitors if required. This concept of so-called virtual enterprises is vital for European SMEs to compete with large companies from overseas while keeping their independence. The virtual enterprise concept is thus the only alternative to mergers and take-overs if SMEs are to stay ahead in terms of innovation. The key feature of virtual enterprises is a temporary, project-oriented co-operation of companies, which is aimed at handling projects that none of the partners are able to do alone, see (Zheng et al. 2002).

- *From simultaneous engineering to systemic, networked engineering approaches*: The concept of simultaneous engineering can be considered as state-of-the-art in most advanced engineering firms. The next step to go will be networked engineering approaches that link the expertise and knowledge of past engineering projects with new engineering tasks across inter-departmental and inter-company boundaries. Furthermore, participating people need to be involved in a process of lifelong learning that must help them to gain experience in neighbouring areas of work so that systemic thinking in engineering can evolve over time. This is indeed a key to cultivate lasting innovation capabilities.

- *From case-based integration towards integrated PLM*: Integration activities in business information management have increased and concern Enterprise Application Integration (EAI) as well as vertical integration on the level of Manufacturing Execution Systems (MES). Also in product development, integrated (vendor-specific) software environments emerge. One of the important tasks for information technology for the next 10 years will be the development of solutions to integrate the increasing number of *integrated islands* into a flexible framework for integrated PLM where different stakeholders and application domains engage in information networks.

- *From workflow management to intelligent workflow generation*: As pointed out before, work-flow management is a key issue in realising integrated PLM. The short-term task in this domain is the development of reference workflows or workflow patterns. These need to be extended by methods and tools for the automated, context-sensitive generation of workflows, i.e. based on semantic networks and domain ontology. Thus, staff from engineering to production can be supported and guided in their daily work to save time and costs (see Weck et al. 2004).

- *From virtual processes to virtual process chains*: While simulation activities today are usually centred around specific engineering or application domains, these approaches need to be brought together to optimise not just the development of single manufacturing steps, but to focus on entire process chains. Many of the individual steps in such process chains are linked by manifold con-straints. To achieve and guarantee lead times, product quality, production costs, etc. there is a need to analyse the production processes beginning with the very raw material for the different component threads that finally merge into a single product, such as a machine tool or automobile.

- *From virtual reality via augmented reality to the real world*: As discussed before, simulation technologies strive to deliver reliable simulation analyses. To achieve this goal, simulation systems for factory process flow simulation, or simulation of automated manufacturing and assembly systems, step by step, become similar to the information and computer systems used to plan and control real-world production. Thus, the differences between a simulation application and its real-world counterpart fall apart, so that the same application could be used for both purposes. In the future, it might be possible to design anew factory shopfloor and related operations in a digital factory that is later used for production planning and control as well. In the same way the same software might be used to simulate machine controllers and to control the real-world machine tool. The development of production systems and its components might thus evolve as a path from the virtual world to the real world that is supported by unique multi-purpose information systems. In between the virtual and the real world, there is VE information to support the real world operations in production, which can be supported and handled by Augmented Reality (AR) technologies. AR technologies thus help to enrich the information collected from the real world with additional information and knowledge provided by a VE environment.

- *From reference modelling to multi-dimensional modelling frameworks*: The successful implementation of integrated PLM requires integrated models of products, required processes, and related stakeholders. Today's modelling approaches are not yet sufficient for any such complex modelling effort. However, there is a continuous trend that leads from modelling via reference modelling to meta-modelling that rather describes the structure and rules for application-specific models. With a long-term perspective the meta-modelling approaches will lead to multi-dimensional modelling concepts where different meta-models as well as application specific model implementations can be integrated. This can be realised based on integrated modelling frameworks that supply tools for automatically managing, synchronising and controlling models. This kind of modelling framework will further be the germ cell for the development of integrated PLM information technologies.

- *From product data management to intelligent agent-based engineering management*: From the technological perspective today's approaches for product data management or integrated CAE solutions will only work satisfactorily if they are vendor-specific. However, the entire enterprise information infrastructure will never be a one-vendor solution. Thus, there is a need for sophisticated approaches that enable the networking of different workspaces from governance and management level right to actuator/sensor level in a machine tool, from suppliers to customers across multiple stages in a supply network, and from product development to service and back. Available EAI solutions, like IBM's WebSphere or SAP's Portal solution support the development of application specific views on conventional distributed and database oriented information system environments. However, they do not deliver satisfactory solutions for the integration on the data level itself. There is a need to develop new IT concepts beyond traditional database-centred approaches that allow the distribution of networked engineering and business objects. Agent technology will most likely play an important role in this issue and can be considered as a key enabler in designing such distributed information environments.

### 4.5 *Results-based monitoring and evaluation*

On the challenges for society and organisations, Zall Kusek and Rist (2004) write: "*With the advent of globalisation, there are growing pressures on governments and organisations around the world to be more responsive to the demands of internal and external stakeholders for good governance, accountability and transparency, greater development effectiveness, and delivery of tangible results. As demands for greater accountability and real results have increased, there is an attendant need for enhanced results-based monitoring and evaluation of policies, programmes, and projects.*"

Private and public stakeholders must take aligned steps to co-operatively deliver extended products that create sustainable prosperity. Virtual engineering technologies promise to make these steps affordable, but only if key participants purposely embark on a complex performance improvement program: societal and organisational. The different scope of public and private sector roles must be supported by dedicated decision and reporting frames. The decision and reporting frames of the Millennium Project and the Global Reporting Initiative must be further refined to serve as a yardstick for progress and a resource for planning.

ACKNOWLEDGEMENT

The authors thank all SIG 6 members and all participants to the Idea Factory sessions in Como, May 2004, and Toronto, October 2004, for sharing views and knowledge. The authors also wish to acknowledge the European Commission for their support to the IMS Network of Excellence within which the roadmap was developed.

# REFERENCES

Abramov, V.A.; J.B.M. Goossenaerts; P. De Wilde; L. Correia: Ontological stratification in an ecology of infohabitants. In: E. Arai, J. Goossenaerts, F. Kimura, K. Shirase (eds) Knowledge and Skill Chains in Engineering and Manufacturing: Information Infrastructure in the Era of Global Communications, Springer, 2005, pp 101–109.

Alter, S.: 18 reasons why IT-reliant work-systems should replace the "IT-artifact" as the core subject matter of the IS field. Communications of the Association for Information Systems, Vol. 12, 2003, pp. 366–395.

Arzberger, P.; P. Schroeder; A. Beaulieu; G. Bowker; K. Casey; L. Laaksonen; D. Moorman; P. Uhlir; P. Wouters: An International Framework to Promote Access to Data (Policy Forum), Science, Vol 303, 19 March 2004.

Au, Joris A.; R.J. Kauffman: Rational Expectations, optimal control and information technology adoption, Information Systems and e-business management, 2005, pp 347–370.

Bilalis, Z.; Herbert, D.: (IT) Standardisation from a European Point of View. J. of IT Standards and Standardization Research, 1(1), Jan–Mar 2003.

Brecher, C. et al.: Effiziente Entwicklung von Werkzeugmaschinen – Mit virtuellen Prototypen direkt zum marktfähigen Produkt. In: Eversheim, W.; Klocke, K.; Pfeifer, T., Weck, M. (Editors), Wettbewerbsfaktor Produktionstechnik. Aachener Perspektiven. Shaker, 2002, pp. 157–190

Cooper, R. G.: Top oder Flop in der Produktentwicklung – Erfolgsstrategien: von der Idee zum Launch. 1. Aufl. Wiley-VCH, 2002.

Dreverman, M.: Adoption of Product model data standards in the Process Industry, MSc thesis, USPI-NL and Dept. of Technology Management, Eindhoven University of Technology, 2005.

Dul, J.; H. d. Vries; S. Verschoof; W. Eveleens; A. Feilzer: Combining economic and social goals in the design of production systems by using ergonomics standards. Computers & Industrial Engineering, vol. 47, 2004, pp. 207–222.

Eversheim, W. et al.: Mit e-Engineering zum $i^3$-Engineering (With e-Engineering to $i^3$-Engineering). In: Eversheim, W.; Klocke, K.; Pfeifer, T., Weck, M. (eds): Wettbewerbsfaktor Produktionstechnik. Aachener Perspektiven. Shaker, 2002, pp. 127–155

Eversheim, W.; Schuh, G.: Integrierte Produkt- und Prozessgestaltung (Integrated Product and Process Design). Springer, 2004.

Farrell, J.; P. Klemperer: Co-ordination and Lock-In: Competition with Switching Costs and Network Effects. In: R. Schmalensee and R. Willig, Handbook of Industrial Organization 3, North Holland, Amsterdam, 2003.

Gausemeier, J.: Strategische Produkt- und Technologieplanung – systematische Entwicklung von Produkt- und Produktionssystemkonzeptionen (Strategic Product and Technology Planning). In: Proceedings of the XI. Internationales Produktionstechnisches Kolloquium PTK 2004. Fraunhofer IPK, Berlin 2004, p. 99–109.

Global Public Goods, Secretariat of the International Task Force on: Meeting Global Challenges: International Co-operation in the National Interest – Towards an Action Plan for Increasing the Provision and Impact of Global Public Goods, www.gpgtaskforce.org, 2005.

Goossenaerts, J.B.M.: Interoperability in the Model Accelerated Society. In: P. Cunningham and M. Cunningham (2004) eAdoption and the Knowledge Economy: Issues, Applications, Case Studies. IOS Press Amsterdam, pp. 225–232.

Heller, M.: The Tragedy of the AntiCommons: Property in the Transition from Marx to Market, Harvard Law Review, 111, 1998, pp 621–688.

Kimura, F.: Engineering Information Infrastructure for Product Lifecycle Management. In: E. Arai, J. Goossenaerts, F. Kimura, K. Shirase (Editors), Knowledge and Skill Chains in Engineering and Manufacturing: Information Infrastructure in the Era of Global Communications, Springer, 2005.

Klement, R.: Agentenbasiertes Produktdatenmanagement (Agent-based Product Data Management). Ph.D. thesis, Aachen University, 2005

Lerner, J; J. Tirole: The Open Source Movement: Key Research Questions, European Economic Review, 45, 2001, pp 819–826.

Lessig, L.: The Future of Ideas; The fate of commons in a connected world. Random House, New York, 2001.

McCarthy, I.; K. Ridgway; M. Leseure; Fieller, N.: Organisational diversity, evolution and cladistic classifications, Omega, The International Journal of Management Science, 28, 2000, pages 77–95.

Monnai, T.; E. Arai, J. Oda, T. Tomiyama, A. Hotta: A proposal of Design Vision for Artifact Design and Production in the 21st Century, A Perspective from the Design Engineering Section of the National Committee for Artifact Design and Production, Science Council of Japan, 2004.

Popplewell, K.; Harding, J.A.: Impact of Simulation and Moderation Through the Virtual Enterprise Life-Cycle. In: Mertins, K. and Rabe, M. (eds), Experiences from the Future: New Methods and Applications in Simulation for Production and Logistics, Berlin, 2004, pp 299–308.

Ramello, G.B.: Intellectual Property and the Markets of Ideas, Review of Network Economics, Vol.4(2), 2005, pp 161–180.

Romer, P.: Idea gaps and object gaps in economic development, Journal of Monetary Economics, 32, 1993.

Romer, P. (interviewed by J. Kurtzman): The Knowledge Economy, in: C.W. Holsapple (ed) Handbook on Knowledge Management 1: Knowledge Matters, Springer, 2003.

Sachs et al.: The UN Millennium Project, Investing in Development, A Practical Plan to Achieve the Millennium Development Goals, The United Nations Development Program, 2005.

Weck, M.; Hoymann, H.; Lescher, M.: Effizienz und Flexiblität beim mobilen Einsatz von AR im Service (Efficiency and Flexibility for Mobile Application of AR in Service Scenarios). In: wt Werkstattstechnik, 94 (2004) 5, pp 242–246

World Bank Operations Evaluation Dept.: Addressing the Challenges of Globalisation; An Independent Evaluation of the World Bank's Approach to Global Programs, 1st edition, Dec. 2004. http://www.worldbank.org/oed

Yoshikawa, H.: Intelligent Manufacturing Systems: Technical Co-operation that Transcends Cultural Differences. In: H. Yoshikawa and J. Goossenaerts, eds. Information Infrastructure Systems for Manufacturing. Elsevier North Holland, Amsterdam, 1994. IFIP Transaction B-14.

Zall Kusek, J.; R.C. Rist, Ten Steps to a Results- Based Monitoring and Evaluation System; A Handbook for Development Practitioners, The International Bank for Reconstruction and Development / The World Bank, Washington D.C., 2004.

Zheng, L.; Possel-Dölken, F.: Strategic Production Networks. Springer, Berlin 2002.

*Advanced Manufacturing – An ICT and Systems Perspective – Taisch,*
*Thoben & Montorio (eds)*
© *2007 Taylor & Francis Group, London, ISBN 978-0-415-42912-2*

# Management of dynamic virtual organisations: Results from a collaborative engineering case

Jens Eschenbächer[1], Falk Graser[1], Klaus Dieter Thoben[1] & Berthold Tiefensee[2]

[1] *BIBA, University of Bremen, Bremen, Germany*
[2] *CeBeNetwork Engineering & IT GmbH, Bremen, Germany*

ABSTRACT: Industrial production is now changing dramatically because competition is no longer just taking place among companies, but more among virtual organisations, forcing each network player to operate more efficiently than its competitors. Many management challenges in planning, co-ordinating, and supervising such dynamic operating virtual organisations have not been addressed so far. The reason is that these dynamic virtual organisations, which can also be considered as collaborative networks, represent an evolution of enterprise development which will be addressed in the ECOLEAD project It is believed that these management challenges must be addressed by powerful methods to leverage success of virtual organisations. The identification of challenges and management requirements in six case studies shows the tremendous practical needs for methods, concepts and tools to support the various European SME networks. SME-based virtual organisations clearly need governmental directives guiding them though their collaborative processes. For one case study about collaborative engineering, the management challenges will be discussed in detail. The paper concludes with a list of management aspects that must be addressed by research activities.

*Keywords*: Dynamic Virtual Organisations, Management approaches, Collaborative engineering.

## 1 INTRODUCTION

Efficient approaches for the dynamic virtual organisation management (DVO Management) are needed to respond effectively to the increasing market requirements for new extended products (Thoben and Eschenbaecher 2003). Indeed, most of today's DVO are not using any kind of explicit management approach to better supervise, co-ordinate, and monitor their collaborative processes (Karvonen et al. 2004). Issues like resource allocation and synchronisation, task distribution, co-ordination principles, incentive systems, information flows, etc., have an impact on the management of a DVO. The challenge is to develop a better understanding of the relationship between management approach and the fulfilment of the tasks given to the DVO in the co-operation phase (Wittenberg 2004). This relationship should give the basis for the measurement of the management performance and also support possible actions for enhancement of the management.

The management of DVO has been neglected throughout the past ten years. Most researchers thought that management procedures and technologies could easily be transferred from single enterprises towards networks. It can be concluded that this belief was misleading for several reasons (for instance, problems in harmonising different enterprise cultures, processes, and IT infrastructures; resource allocation among and across the various partners, etc.). This raises several research questions that need to be addressed (Karvonen et al. 2005).

Consequently, supporting the management of DVO is a key research question. In a first set of studies, the industrial state-of-the-art in DVO management has been analysed. The conclusion from this analysis is that collaboration mechanisms developed for supply chains are related to a rather hierarchically structured thinking, thus disqualifying them for transfer to a democratic DVO. Development of a DVO management system should be bound to the following research questions:

- Is it possible to derive generic requirements for a DVO management system?
- What are the main management challenges and needs for a DVO?
- Which players should be included in a DVO management system?
- What conclusions can be stated from a detailed case study on collaborative networks?

In the context of this paper, research has been undertaken using two approaches:

- Desk-top research and consolidation of the results through discussions within the ECOLEAD consortium;
- Field research in through six case studies.

In the first section of this paper, the ECOLEAD project and respective research challenges for managing DVOs are introduced. Following this, the results of the six case studies related to the subject of management challenges are presented. In one of these case studies a detailed analysis on collaborative engineering challenges is presented. Finally some conclusions summarise the paper.


## 2   MANAGEMENT CHALLENGES FOR DVO

ECOLEAD is an European Commission funded Integrated Project that aims at creating foundations and mechanisms for establishing the most advanced collaborative and network-based industry society in Europe (Camarinha-Matos et al. 2005). The fundamental assumption in ECOLEAD is that a substantial increase in materialising networked collaborative business ecosystems requires a comprehensive holistic approach. Given the complexity of the area and the multiple inter-dependencies among the involved business entities, social actors, and technological approaches, substantial breakthroughs cannot be achieved through incremental innovation in isolated areas. Figure 1 shows all research areas that will be addressed in ECOLEAD.

Summarising the ECOLEAD IP structure highlights three vertical and two horizontal core research items. The vertical ones represent three manifestations of the underlying class of DVOs, whereas the horizontal ones represent measures that support integration of all three VO manifestations (VO breeding environments (VBE), dynamic VOs (DVO), and professional virtual communities (PVC)) into common reference models.

Therefore, ECOLEAD addresses the three most fundamental and inter-related focus areas that are the basis for dynamic and sustainable networked organisations:

- Breeding environments (Afsarmanesh 2005): A VO breeding environment is a pool of organisations and their related supporting institutions that have the potential and the intention to co-operate with one another, through the establishment of a *baseline* long-term co-operation

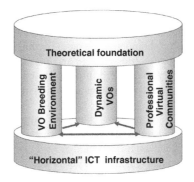

Figure 1.   ECOLEAD research areas (ECOLEAD 2004).

264

agreement. Co-operation within a VBE is built on common trust, compatible value-systems, and baseline ICT standards among the partnering organisations.

- Dynamic virtual organisations (Karvonen et al. 2005): DVOs are temporary alliances of organisations joining competencies and resources to better respond to business opportunities and produce value-added services and products. DVOs are strongly supported by computer networks. A VO can quickly be created out of a VBE. Within the VBE context, partners out of the VBE context are rapidly assembled into a single business entity called a VO, enabling the collaboration among enterprises and individuals to respond to a specific business opportunity.
- Professional virtual communities (Crave and Ladame 2005): A PVC comprises the concepts a virtual community and a professional community. Virtual communities are defined as social networks of individuals, who use information and communication technologies (ICT) to mediate their relationships. Professional communities provide environments for professionals to share their individual knowledge such as similar working cultures, problem perceptions, problem-solving techniques, professional values, and professional behaviour. PVCs cannot be dissociated from the underlying business ecosystem of the society, owing to their contractual links (social-bounds) with all the consequences at the intellectual property and life maintenance levels. PVCs are one of the most relevant elements for keeping the business ecosystem *alive* and for launching and operating dynamic VOs of the future.

The holistic approach is reinforced and sustained on two horizontal activities: the theoretical foundation for collaborative networks and the ICT infrastructure that support the three vertical focus areas. These specific key points provide the basis for a technology-independent and low-cost ICT infrastructure for the establishment of truly dynamic collaborative networks. Additionally two horizontal research areas are addressed:

- Theoretical foundations: Sustainable development of collaborative networked organisations needs to be supported by strong fundamental research leading to the establishment of collaborative networks as a new scientific discipline.
- Horizontal infrastructure – the horizontal ICT infrastructure: Implantation of any form of collaborative network depends on the existence of an ICT infrastructure.

This paper deals with the management issues identified and analysed for DVOs. Figure 2 shows a summary of all DVO issues. This Figure shows a summary of issues that need to be considered

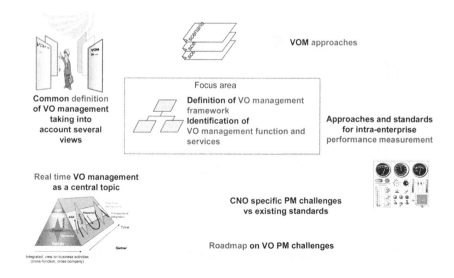

Figure 2.   DVO Management challenges (adapted from Camarinha-Matos 2004).

265

when discussing DVOs. The focus area of this paper is a VO management framework and VO management functions and services as indicated. The box with defined by dashed lines indicated the focus area of this paper. Management requirements and functions deliver a good basis for an understanding of the complexity of management problems in DVOs.

## 3   ANALYSIS OF COLLABORATIVE ENGINEERING CASE STUDIES

This section will discuss the requirements selected from six SME-network case studies about the current usage of DVO management functions and respective requirements. The requirements will be elaborated in six case studies in total: two case studies from Finland; two from Italy; one from Spain; and one from Germany. The results of this study on also include challenges and functions regarding VO management.

The following networks have been analysed:

- CeBeNetwork: Aeronautical network in Bremen, Germany;
- Verkko, Varkaus: Forest industry and automation Systems, Finland;
- ISOIN: SME networks located in Andalusia, Spain;
- KER: Kouvola Engineering Region, SME-Networks Tapio Vankalo, Finland;
- Treviso Technologia: Production Networks in Northern Italy;
- Torino Wireless, Italy: High technology cluster, grouping ICT players in the Piedmont area in a shared system of values, strategies, and actions to increase regional competitiveness through strong integration of research and development (R&D), entrepreneurial skills and venture capital.

Table 1 provides a first, rough summary of some insights gained from the field studies. Additionally it includes three categories:

- Focus VO management: to what degree does the network put emphasis on VO management?
- Preference on methods/tools: The network is interested to understand business models to create VO. A collaborative network concept as investigated within ECOLEAD is still not in use in the considered networks; and
- Preference on tools: To what degree are networks already prepared to use tools?

Table 1 show that all networks are still not in the phase of adopting the concept of the VO. Most of them just made some experience with VO type of constellations. Consequently their preference is on methods and concepts as output from ECOLEAD. This illustrates clearly that democratic-oriented co-operation can still not be considered as state-of-the-art. All networks clearly stated that they prefer to have a simple supplier-customer relation instead of co-operation in which one partner keeps the link to the customer. Research is necessary to provide the necessary business models to break this *resistance against new business strategies*.

The networks have also shown that VO management is an important subject for conducting collaborative projects. Nevertheless most of the networks have not conducted any VO project in

Table 1.   Preferences of Networks (" + " indicates the degree of interest) (ECOLEAD 2005).

| | Focus VO management | Preference on business models/methods/concepts | Preference on tools |
|---|---|---|---|
| Torino Wireless | medium | ++ | + |
| Network A | low | ++ | + |
| CEBENETWork | high | ++ | ++ |
| ISOIN | high | ++ | ++ |
| Kouvola | medium | ++ | + |
| Treviso Technologia | low | ++ | + |

the democratic-oriented sense of ECOLEAD. Indeed only CeBeNetwork and ISOIN have conducted collaborative-engineering oriented co-operative projects. The experience of CeBeNetwork especially reveals the complexity of management challenges that need to be addressed in a DVO. Nevertheless the following aspects can be summarised from a VO management perspective:

- The networks were not very advanced i.e. almost no experience of operational VO was given;
- Mostly VBE requirements were discussed.

The continuing research work will identify further networks to understand if this observation is a general one or if other cases show different results. Table 2 presents a second result that summarises requirements, functions, the source of these, and the priority. These requirements have been elaborated as a result of a common discussion among all networks and can be considered as research in progress. This is not a complete list but rather a collection of ideas.

Table 2 shows altogether seven requirements with respective management functions. These have been ordered to the respective source. Indeed some of the VO management functions can be found in 2–3 networks. The highest priority became contract management as the contractual basis of any co-operation. Governance approaches have also been judged as very important to ensure a common approach among all DVO partners. Finally performance measurement has been given priority.

In the following the CeBeNetwork collaborative engineering case has been selected to provide a deeper insight in the management issues for DVO which need to be addressed.

## 4 CeBeNETWORK CASE COLLABORATIVE ENGINEERING CASE

### 4.1 *Collaborative engineering*

The CeBeNetwork co-operation exists and is proven since November 2002 as an international aerospace subcontractor network of more than 20 co-operation partners with over 20 years of experience in aerospace. It provides competitive services for the aviation sector in France, Germany, UK, Spain and in some low-cost countries and is trying to create new business in automotive and other industries. The co-operation also contributes to space activities: It is among the top 10 of *EADS Space Transportation* preferred engineering supplier list.

The co-operation provides a broad common portfolio of concept, design, prototyping and testing for taking over work packages and responsibilities to realise both, the complete development of projects, from initial concept to series production, as well as the specific execution of individual

Table 2. Summary of VO management (ECOLEAD 2005).

| Virtual organisations | | | Priority |
| --- | --- | --- | --- |
| Requirements | Functions | Source | |
| Good performance of the VO | Time, cost, quality measures | Torino Wireless | 2 |
| Safe contractual basis | Contract management | Network A, CEBENETwork | 1 |
| Feedback about collaboration performance | Performance measurement | CEBENETWork | 3 |
| Profit distribution | Profit & loss management | ISOIN, CEBENETwork | 4 |
| collaboration tools | collaboration platform | Kouvola | 5 |
| Performance of the collaboration | Collaboration management system | Treviso Technologia | 6 |
| Clarification about where and how information of common projects is archived | Project database management | CEBENETWork | 7 |

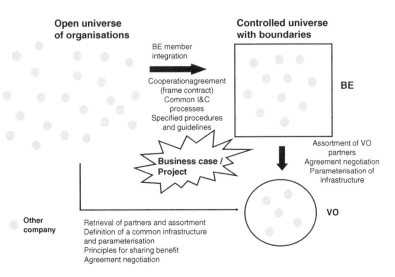

Figure 3.   From VBE to the VO (ECOLEAD 2005).

processes. The integrated portfolio for product engineering consists of IT, onsite experts, design work packages, testing, jigs and tools, and manufacturing.

Based on its concept, the co-operation with five initial partners has already finished 230 Airbus sub-orders. The co-operation ensures quality managed services based on EN 9100 and Airbus Supplier approval to all its partners. The advantages of the co-operation are better performance, better quality, adaptability, reduction in the duration of work time, and last, but not least, only one contact for all services.

The reasons of these advantages result from internal communication among the partners, synergies, and know-how exchange in the co-operation. Additionally members have established compatible solutions together while each partner is specialist in its own domain and is responsible for his domain and core competence.

CeBeNetwork accomplishes the central management of the co-operation obtaining the role of consortium co-ordinator. So it is a highly sophisticated network of professionals specialising in IT and management solutions in the area of product development as well as in full service packages for the entire engineering process in the automotive and railroad industries, mechanical engineering and in the aircraft industry. Co-operation is highly motivated by achieving a high level of customer satisfaction and innovation.

Figure 3 shows the transformation from the open universe to the CeBeNetwork VBE. Out of this VBE a VO is created. The aspects being considered differ substantially from those specified in ECOLEAD. Nevertheless the objective and general mechanisms are the same.

### 4.2   Case specification

One of the main customers of CeBeNetwork is intending to reduce the number of suppliers, and started the call for a supplier consolidation in a particular department (Cabin) in 2003. The suppliers have to deliver the whole spectrum of requested collaborative engineering services. Turnover, as well as the workload for each supplier in this area, will be agreed in advance and means an enormous increase for the selected companies. The customer expects a specification from each company or DVO and its proposed means to address the upcoming requirements. Qualifying employees for prospective work, as well as pre-financing and taking over financial risk for projects are essential tasks. CeBeNetwork as a current supplier in this field needs to manage this challenge to remain in this field and become one of the key suppliers.

CeBeNetwork needs to co-operate with its own companies to become a key supplier. This requires highly motivated partners. The profile of each partner where co-operation is planned in this task is well known by the customer because the companies have delivered this service in former times on their own. The following aspects need to be addressed in reorganising the networks towards a DVO oriented structure:

- CeBeNetwork, as the only interface to the customers procurement department, needs a well-structured and organised network that is a fundamental basis to achieve the own target and become a system supplier. Therefore the collaboration of the existing network needs to be intensified.
- The basis for a detailed overview is to create a database with all relevant information of the involved companies skills, education and capability of employees, available software tools, and experience in the different fields of work. The information collection was one first step in the whole project. The database includes information about applied quality standards, other branches of activity and potential for growth of the partners.
- Processes like the joint offer preparation and the following order management can be improved and accelerated with an extensive informational background. CeBeNetwork will then be able to achieve the rising customer desires in a more efficient way.
- To supervise the performance of running projects, review meetings for each project are agreed before the single contracts are made. Milestones for payments in between are made as well. In these meetings the quality of the delivered work is to be discussed and necessary improvements and re-workings are agreed.
- To get an overview how many hours are needed for the different shares in a project and if the offer planning was right, employees count the number of hours worked.
- With regard to the closer co-operation within the network, the risk-share as well as the profit will be transferred to all participants. For project teams with employees from different companies, solutions must be found how to allocate exactly the share of risk of each party.

To become a *system supplier* each collaborative engineering network must realise the following topics:

- Show the capabilities and competencies to support the customer's competence clusters;
- The way of working as the system supplier for some competence cluster;
- Capabilities for growth and flexibility;
- Capabilities to support this approach with off-shore activities to reduce work package prices.

Figure 4 shows the ECOLEAD demonstration case scenario. In the cloud, a number of new companies (not member of the CeBeNetwork VBE) and new partners (new members of the VBE) are trying to initiate a co-operation. Additionally, freelancers form a professional virtual community, which could be freelancers or programming experts, are selected which together create the CEBNET work collaborative engineering VO. This DVO is co-ordinated by CeBeNetwork as DVO co-ordinator, which creates the connection to the customer.

The grey box indicates that all VO functions identified in Table 2 do play a role in the collaborative engineering case of CeBeNetwork.

### 4.3 *Findings regarding VO management*

As the CeBeNetwork co-operation is a quite young network not many experiences concerning VOs have been made so far. Consequently CeBeNetwork does not have a long experience in organising DVOs. In a questionnaire the following findings provide an indication of the perspective that CeBeNetwork takes towards management issues (ECOLEAD 2005):

- In general, interviewed partners and CeBeNetwork itself agreed on most issues;
- Although participation in a VBE was mainly a customer requirement partners may choose in which network to participate;

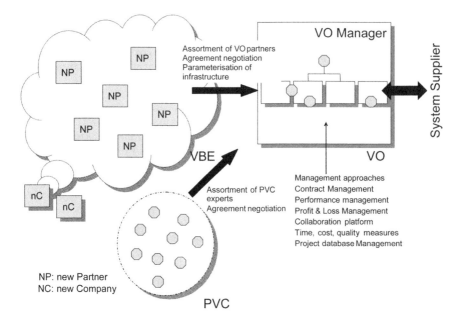

Figure 4. Demonstration case scenario.

- The VO management is now very weakly supported by any specific business model, method or concept; and
- Currently there is a quite positive sentiment within the VBE.

The SME network case has triggered a lot of attention inside the CeBeNetwork co-operation that many management issues can be substantially improved. The following points summarise the results:

- Information collection provides a basis for future joint projects;
- Sharing of risk and profit is only partly specified within a framework contract, and does not cover issues relating to risk sharing within joint common projects;
- Clarification concerning generation of profit;
- Securing that the strategy of the VBE is known to all partners;
- Clarification that main competitors are outside the own VBE;
- Establishment of a list / database about the knowledge and capabilities of partners;
- Establishment of a competence and resource catalogue;
- Establishment of a preparation procedure with regard to a VO;
- Definition of criteria for measurement of the performance during a common project;
- Other performance parameters, e.g. amount of reviews needed, used hours per activity;
- Establishment of a common server and integrated tools with access for all partners; and
- Clarification about where and how information of common projects is archived.

## 5    CONCLUSIONS AND FUTURE RESEARCH

The analysis presented in this paper has revealed that the management of DVOs is in its infancy. Business models, concepts, and methods are not clear to all the selected SME networks. The reason for this is that SMEs are used to managing fairly simple one-to-one business relationships. In a DVO relationships are much more complicated comprising business-to-business relationships and the relationship of the entire network towards its customer. Since the entire network/ DVO is represented externally by its manager, the customer remains invisible for all the other partners. Efficient management of all these relationships will be the most crucial success factor for future DVOs.

The CeBeNetwork collaborative engineering case has proven this general result. The SMEs have to re-configure their business thinking towards more collaborative models. The complexity of collaborative engineering and the intensity of the co-operation provide enormous challenges regarding management needs.

The management skills in DVO have not yet been addressed because such co-operation presents a new and probably significant new business model for most of the companies. Most of the SMEs are used to working with one customer, which means very low complexity. It can be expected that supply chain oriented business models have to be supplemented by business models for DVOs. These business models will be more complex for the individual organisation because point-to-point interaction with customers might be replaced with complex multi-point interaction in large networks. This requires rethinking of current business thinking towards more democratic co-operation.

## ACKNOWLEDGEMENT

The paper is mainly based on work performed in the project ECOLEAD (FP6-IP 506958; www.ecolead.org) funded by the European Commission within the IST-Programme of the 6th Framework Research Programme. The authors would like to acknowledge the ECOLEAD consortium, especially to the participants of WP3 (VO Management).

## REFERENCES

Afsarmanesh, H., Camarinha-Matos, L.M. (2005) A framework for the Management of Virtual Organization Breeding Environments. In: Collaborative Networks and Their Breeding Environment. Edited by: Luis M. Camarinha-Matos, Hamideh Afsarmaneh and Angel Ortiz. Springer 2005, P. 35–49.

Camarinha-Matos, Luis: New collaborative organisations and their Research needs, In: Processes and Foundations for virtual organizations, Kluwer Academic Publishers, Boston/Dordrecht/London, 2004, p. 3–12.

Camarinha-Matos, L.-M., Afsarmanesh, H. and Ollus, M. (2005) ECOLEAD: A holistic approach to Creation and Management of Dynamic Virtual Organisations. In: Collaborative Networks and Their Breeding Environment. Edited by: Luis M. Camarinha-Matos, Hamideh Afsarmaneh and Angel Ortiz. Springer 2005. P 3–17.

Crave, Ladame (2005): Crave, S., Ladame, S.: "Professional Virtual Communities (PVC) inside Networks of Firms", published in Pawar, K.; Weber, F.; Thoben, K.-D.; Katzy, B.: "Proceedings of the 11th International Conference on Concurrent Enterprising", ISBN 0-85358-221-1, Nottingham Business School, Nottingham 2005, pp. 189–192.

ECOLEAD Deliverable D322 (2005) Report on Methodologies, Processes and Services for VO Management.

Hess, T./Wittenberg, S. (2005): IT-gestütztes Netzwerkcontrolling, in: HMD – Praxis der Wirtschaftsinformatik, 41. Jg., Nr. 2, S. 52–62.

Higgins, P., Eschenbaecher, J., Strandhagen, J. O., Horten, A.: An Operations Model for the Extended Enterprise. . In: Proceeding of the 11th International Conference on Concurrent Enterprising. Munich, Germany 20–22 June 2005, S. 259–262.

Karvonen, I., Salkari, I., Ollus, M. Characterizing Virtual Organization and Their Management. In: Collaborative Networks and Their Breeding Environments. Edited by: Luis M. Camarinha-Matos, Hamideh Afsarmaneh und Angel Ortiz . Springer New York 2005, S 194–204.

Scheer, A.-W., Jost, W. (2003) Real-Time Enterprise – Mit beschleuni gten Managementprozessen Zeit und Kosten sparen. Springer Verlag: berlin New York.

Thoben, K.-D., Jagdev, H., Eschenbaecher, J. (2003) Emerging concepts in E-business and Extended Products. In: Gasos, J., Thoben, K.-D. (Eds.): E-Business Applications – Technologies for Tomorrow's Solutions; Advanced Information Processing Series, Springer, 2003 (ISBN 3-540-44384-3).

Thoben, K.-D., Eschenbaecher, J. (2003) Die Erweiterung des Produktbegriffs – Konzept und Praxisbeispiele. In: Industrie Management (19) 16, 4, S. 48–51, 2003.

*Advanced Manufacturing – An ICT and Systems Perspective – Taisch,*
*Thoben & Montorio (eds)*
© *2007 Taylor & Francis Group, London, ISBN 978-0-415-42912-2*

# Integrating the engineering of product, service and organisation within the collaborative enterprise: A roadmap

Klaus-Dieter Thoben[1], Kulwant Pawar[2], Marc Pallot[3] & Roberto Santoro[4]

[1]*BIBA, University of Bremen, Bremen, Germany*
[2]*Centre for Concurrent Enterprising, Nottingham University, Nottingham, UK*
[3]*ESoCE-Net, Paris, France*
[4]*ESoCE-Net, Roma, Italy*

ABSTRACT: One of the main results of CE-NET (Concurrent Engineering Network) is the roadmap for the *collaborative enterprise 2010*. The term collaborative enterprise captures the shift from the existing organisational approaches to more dynamic and re-configurable peer-to-peer networks and from ownership to sustainable services. Several complementary perspectives, namely social, customer, business, workplace, technology, and legal, have been concurrently considered to build up the collaborative enterprise vision and roadmap. The roadmapping process and the relationship among today's circumstances, the CE vision, gaps, research needs and solutions will be explained. State-of-the-art, vision, gaps, needs and the required solutions, in the short, medium and long-term, to achieve the vision for product, service and organisation development of collaborative enterprising 2010 will be outlined.

The full version of the roadmap covering the areas Human Aspects, Business Models, integrated Product-Services-Organisation Development, Information and Communication Technology, and Policy and Regulation, is available at: www.ce-net.org/roadmaps.html.

*Keywords*: Product-Service-Organisation (PSO), Collaborative Enterprise.

## 1 PRODUCT-SERVICE ORGANISATION, COLLABORATION, COLLABORATIVE ENTERPRISE, HUMAN FACTORS ROADMAPPING PROCESS

Several complementary perspectives, namely social, customer, business, workplace, technology, and legal, have been concurrently considered to build up the collaborative enterprise vision and roadmap. Figure 1 gives a detailed view of the roadmapping process characterised by five key areas: Human Aspects that characterise the social perspective; Business Models and Organisations which illustrate the business perspective; Product-Service-Organisation (PSO) Development that focuses on the customer perspective; Information and Communication Technology (ICT) Aspects which characterise the technology perspective; and the Policy and Regulation Aspects which characterise the legal perspective. The workplace perspective is embedded into the human and ICT aspects as well as PSO development. The roadmapping approach describes the state-of-the-art for each area on one side and then develops the vision on the other side, as illustrated in the Figure 1. *Gaps* are then identified which need to be addressed to reach the vision in comparing the state-of-the-art to the vision statements. Through a gap analysis, time specific choices have been developed. From this, research challenges have been derived that can enable the development of solutions. These challenges require further validation to achieve the vision of the collaborative enterprise and product-service-organisation engineering. The next section of this paper, *Roadmap Overview*, explains how this roadmap summary document has been structured, while the third section introduces the collaborative enterprise paradigm. This document synthesises the output of the roadmap task of CE-NET over the last three years. It encapsulates contributions from CENET

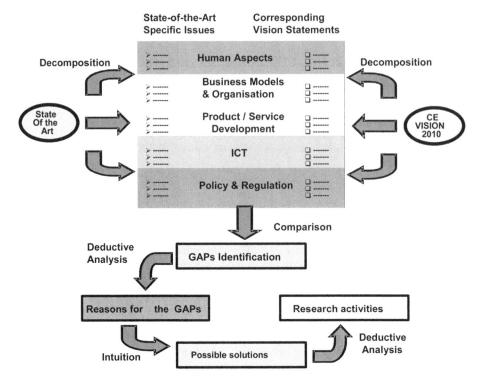

Figure 1. Roadmapping process: overview.

partners engaged in various research projects, as well as contributions from the CE community at large, especially during the ICE conferences and other events.


## 2 ROADMAP OVERVIEW

The domain collaborative enterprise is broken down into five complementary areas that will be considered in the roadmap: Human Aspects; Business Models; Integrated Product-Services-Organisation Development; Information and Communication Technology; and Policy and Regulation. Figures 2 and 3 give an overview of the collaborative enterprise roadmap concept. Figure 2 shows the gaps that were derived by comparing the state-of-the-art with the vision of the future. The future is always difficult to predict; the target vision considered is sometime after 2010.

The foreseen solutions to fill the deduced gaps are categorised into three time spans, namely short term, medium term and long term, as illustrated in Figure 3.

The roadmap document introduces briefly the collaborative enterprise vision and then presents an overview of each area that was investigated in the roadmap process. Each area overview includes two tables that respectively describe elements of the roadmapping stage 1 and then stage 2 as presented in Figure 2 and Figure 3. The first table details the five most important topics in terms of actual position, vision of the future, deduced gaps and then the research that needs to be planned to develop the foreseen solutions.

The second table details the five most important topics in terms of the deduced gaps and foreseen solutions planned to be realised into short, medium and long term implementation steps. After the five roadmapping sections, a specific section is dedicated to research perspectives and challenges which introduce other research scenarios.

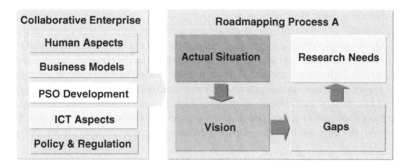

Figure 2. Roadmapping process: stage 1.

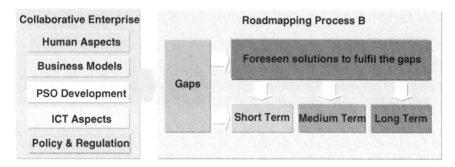

Figure 3. Roadmapping process: stage 2.

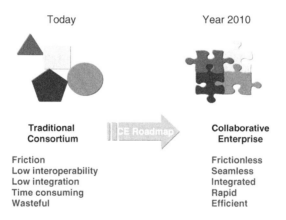

Figure 4. CE vision 2010: industrial innovation transition.

## 3 COLLABORATIVE ENTERPRISE VISION

The collaborative enterprise vision is shown in Figure 4. The prediction is that this will be realised after the year 2010. The vision is to achieve a seamless collaborative enterprise, which is a networked organisation, dynamically formed to create sustainable value through the delivery of integrated Product-Service-Organisation (PSO) configurations providing benefits to customers.

Nowadays, many large companies are going through the transition from traditional organisational arrangement towards more dynamic and re-configurable peer-to-peer network of business organisations. This is already happening in different business sectors. Increasingly products are being provided along with associated services, for example, operational support during the full lifecycle. Customers are shifting the product sales paradigm from direct ownership to sustainable access. This roadmap introduces the innovative new concept of Product-Service-Organisation (PSO). Up to now concurrent engineering has been implemented to integrate the design of product and processes with a lifecycle perspective. Now, the need exists to extend concurrent engineering principles throughout a networked organisation. The concurrent integration of the design of a product, its services as well as the delivery organisation should lead to more sustainable business. This new engineering approach is termed *collaborative enterprising*. Industry is now in a transitional phase (Figure 4); the purpose of this roadmap is to provide guidance and direction to achieve the transition.

This is foreseen to be the next revolution in the domain of enterprise and product engineering. The social, legal, business, customer, workplace, and ICT perspectives are analysed to establish how long-term sustainability can best be achieved.


## 4 PSO DEVELOPMENT

This section first covers the state-of-the-art, vision, gaps and needs of the third roadmap area, Product-Service-Organisation Development. In the following, the gaps and the required solutions, in the short, medium and long-term, to achieve the vision for PSO development of collaborative enterprising 2010 will be outlined.

The full version of the roadmap covering the areas Human Aspects, Business Models, integrated Product-Services-Organisation Development, Information and Communication Technology, and Policy and Regulation is available at: www.ce-net.org/roadmaps.html.


### 4.1 *PSO Development vision and needs*

Product-Service-Organisation Development is tackled by integrating the engineering of product and its related services (the concurrent engineering approach) with associated services and networked organisation design, within a total lifecycle approach to develop a sustainable approach.

There is a lack of a global vision for lifecycle services and organisation design. This is coupled with an unclear path for achieving long-term sustainability in PSO. The existing tools and approaches only address the product. There is a need for practices and tools for PSO development, and design methods for realising customer benefits.

The five most important topics are: integrated PSO, theories for PSO development, transition from ownership to sustainable access, extend concurrent engineering to encompass PSO and concurrent networked organisation design.


### 4.2 *PSO Development: gaps and solutions*

Foreseen solutions to fill the gaps and solutions are presented in the Table 3b. This provides an overview of the five most important topics and solutions categorised according to the short, medium and long-term. To reach the long-term solution of the design and delivery of concurrency across PSO, it is first necessary to extend CE principles to encompass PSO and to eventually arrive at an integrated CE and PSO lifecycle framework. This requires the development of theories and approaches for PSO design, including those for concurrent networked organisation design. These will aid the transition from ownership to sustainable access.

The extension of concurrent engineering principles and development of new practices for PSO lifecycle design and management can be expected in the short-term. A business scenario experimentation test bed using computer-based simulation gaming, sustainability transition approach, and design for customer benefits constitute medium term solutions. Virtual Reality is considered a key component for the development of visionary business innovation and to enable experimentation in providing customer benefits and lifecycle sustainability.

276

Table 1. PSO Development vision, gaps and research needs.

| Actual Situation | Vision | Gap | Research Need |
|---|---|---|---|
| Partial implementation of integrated product and services | Deliver benefits to customers as sustainable product, service & organisation | Lack of global vision and lifecycle services integration to better satisfy customers | Develop business scenario testbeds to experiment and evaluate business innovations |
| Usually product are made to be bought and owned by consumers | Paradigm shift from ownership to sustainable access | Lack of transition for sustainability approach | Develop approaches for measuring the cost of sustainable access |
| CE implementation is mostly restricted to product development | Concurrent design of product, services and organisation lifecycle | Lack of practices and tools to extend CE principles to the whole PSO lifecycle | Define an extended CE framework to. encompass PSO lifecycle |
| Classical product design theories are not anymore applicable to PSO | New knowledge focused methods for prediction and design of usage, benefits & organisation | Lack of design methods focused on customer benefits | Develop new methods, techniques and tools to design for customer benefits |
| Existing engineering approaches are not considering concurrently organisation design | Organisation is designed for sustainability of product and services | Lack of integrated engineering approach covering concurrently organisation design | Develop engineering approaches to integrate concurrent organisation design |

Table 2. PSO Development: gaps and foreseen solutions.

| Gap | Solutions to fill the gaps | | |
|---|---|---|---|
| | Short Term | Medium Term | Long Term |
| Partial implementation of integrated product and services | Provision of services onto existing products | Business scenario testbed to predict customer benefits | VR framework to develop visionary business solutions |
| Usually product are made to be bought and owned by consumers | Increasing provision of sustainability services | Transition approach towards sustainable access to services | Costing framework to evaluate ROI on sustainable access |
| CE implementation is mostly restricted to product development | Extend CE principles to encompass product, services and organisation lifecycle | PSO Lifecycle management | Integrated CE and PSO lifecycle framework |
| Classical product design theories are not anymore applicable to PSO | Collect sustainability practices within different sectors applicable to PSO | Design to customer benefits | VR environment to experiment customer benefits |
| Existing engineering approaches are not considering concurrently organisation design | Collect and integrate existing practices for organisation lifecycle design | Organisation lifecycle management | Design, deliver and manage concurrency during PSO lifecycle |

277

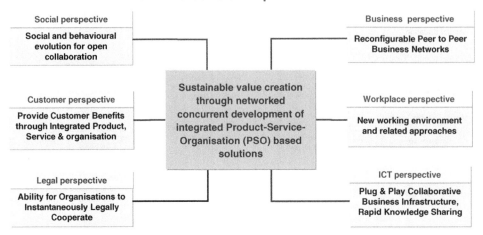

**Collaborative enterprise Vision**

| Social perspective | | Business perspective |
|---|---|---|
| Social and behavioural evolution for open collaboration | | Reconfigurable Peer to Peer Business Networks |

Sustainable value creation through networked concurrent development of integrated Product-Service-Organisation (PSO) based solutions

| Customer perspective | | Workplace perspective |
|---|---|---|
| Provide Customer Benefits through Integrated Product, Service & organisation | | New working environment and related approaches |

| Legal perspective | | ICT perspective |
|---|---|---|
| Ability for Organisations to Instantaneously Legally Cooperate | | Plug & Play Collaborative Business Infrastructure, Rapid Knowledge Sharing |

Figure 5. Research perspectives.

## 5 RESEARCH PERSPECTIVES

Figure 5 shows an overview of the research perspectives that are necessary to implement the research challenges to achieve the CE 2010 Vision. The central rectangle in Figure 5 contains the golden nugget *Sustainable value creation through networked concurrent development of Integrated Product-Service-Organisation (PSO) based solutions*. This vision is to achieve a seamless, collaborative enterprise, dynamically formed to create sustainable value through delivery of integrated Products-Services-Organisation.

The need to integrate multidisciplinary research expertise and resources, which encapsulates the Product-Service-Organisation (PSO) triangle, is crucial to avoid the development of partial or incomplete solutions. CE-NET and its successor ESoCE-NET have recognised this need and propose to capitalise on the decade long experience of the community. Bringing resources to bear through integrating experts from these areas across Europe, and other parts of the world, will help to secure world leadership for European industry. The CE-NET roadmap is a basis for creating integration and coherence of future European research, and from this foundation a systematic and holistic approach to rapidly form collaborative enterprises can be realised.

## 6 RESEARCH CHALLENGES

This section describes the inter-relations and dependencies among the identified research challenges. Figure 6 shows an overview of the eight research challenges, and their interrelations, necessary to achieve the CE 2010 Vision. The diagram should be read left to right, starting with the customer benefits to be delivered to achieve sustainable value. The research challenges are enumerated below.

● **Approaches for better meeting customer value (benefits)**
To define and develop new Products, Services and Organisation (PSO) engineering approaches and to understand, capture, represent, assess and measure customer value. Create a dynamic customer value proposition model and evaluation framework to assess sustainable value. Develop ICT techniques for the capture and formalisation of customer benefits. Develop tools, technologies and methods to translate customer benefits into PSO configuration and functionality. Develop an interactive approach to involve customers in the PSO development process. Understand behavioural, social, business and environmental impact of the PSO transition.

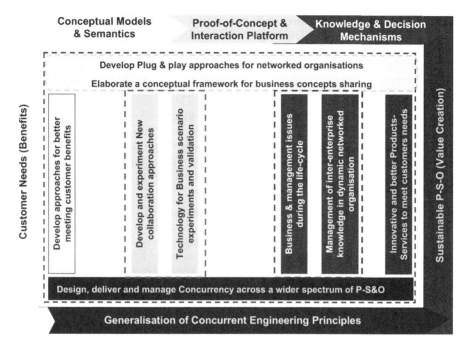

Figure 6.   Inter-relationships among research challenges.

● **Plug and Play (P&P) approaches for networked organisation**
Define a reference model, tackling organisational and legal issues, depicting a range of plug and play business scenarios. This model will encapsulate the dynamic and re-configurable nature of networked organisation for the development of integrated PSO engineering. Develop a conceptual model and ontology for plug and play virtual organisations.

● **Conceptual framework for shared business concepts**
Define a semantic representation for collaboration in PSO engineering in a multicultural and multi-lingual environment. Develop a PSO engineering, semantically agreed-upon description of concepts, their relationships, properties and rules that are related to collaborative work. Develop related business semantics services to access and manipulate concepts along the PSO lifecycle. Experiment with new user interfaces based on semantic representations.

● **Methods and technology for business scenario experimentation**
Experiment and validate different business scenarios to assess the potential benefits for customers and for the value creation process, through the usage and manipulation with a virtual representation. Identify, analyse and specify the requirements for a PSO virtual reality experimentation framework. Develop and elaborate a PSO collaboration virtual reality experimentation platform (VREP) which assembles existing and newly developed tools and components. Develop a VREP builder for generating experimentation platforms for analysing different business scenarios.

● **New collaboration approach and workplace**
Develop and experiment collaborative workspaces (CWS) based on VR technology and semantic meaning approaches. The CWS is intended to support PSO multidisciplinary teams, distributed PSO lifecycle management and collaborative project management. Identify, analyse and specify the requirements for this PSO collaborative workspace. Develop a CWS Builder for PSO Engineering, with different access mechanisms – such as mobile technologies.

## • Innovative and better PSO to meet customer needs

Develop innovative solutions, methods and tools to support sustainable value creation throughout the PSO lifecycle. Develop new PSO definition, modelling, and lifecycle management approaches and tools. Define and design new services and methods for PSO engineering. This typically should include the responsibilities and traceability along the lifecycle of the PSO configuration for better meeting customer benefits.

## • Business and management issues during the P-S-O lifecycle

Envision future business scenarios and identify the gaps between now and the future. Capture the state-of-the-art of collaborative lifecycle PSO development. The challenge is for business modelling and financial investment planning for new business opportunities, in the early phases of the lifecycle, to establish sustainable value creation. Assessment of the impact upon the downstream lifecycle phases is also needed.

## • Inter-enterprise KM in dynamic networked organisations

Envision a networked organisation centric knowledge management (KM), with and underlying capability for IPR conflict resolution to manage knowledge in dynamic inter-enterprise organisations. Develop and experiment with knowledge-based systems and KM services based on semantic meaning approach, and including capturing and reusing knowledge from stakeholders.

ACKNOWLEDGEMENT

This work has been partly funded by the European Commission through IST Project *CE-NET: Concurrent Enterprising Network of Excellence* (No. IST-1999-29107). The authors wish to acknowledge the European Commission for their support. They also wish to acknowledge their gratitude and appreciation to Andrea Bifulco, Ip Shing Fan, Johann Riedel, Rene Stach and all the CE-NET project partners for their contribution during the development of various ideas and concepts presented in this paper. Many thanks also to those who contributed by filling in the vision or research challenge questionnaires during the ICE conferences, and other CE-NET and ESoCE-Net events. The expanded version of this roadmap is available at: www.ce-net.org/roadmaps.html.

*Part IX*
*Supply chain integration*

Supply chain integration has become a vital element in obtaining and maintaining competitive advantage through the creation of value for all participants in the supply chain. Competitive advantages should not only be measured by product design, functionality, costs, and quality, but also by flexibility and lead-times. Today companies need to operate within effective and efficient supply networks that are highly flexible, innovative and dynamic to respond to pressures such as cost reductions, product development, high quality, value adding products, and frequent, on time deliveries.

The contribution from Gulledge et al. discusses different ways of obtaining more or less integrated supply chain solutions in terms of collaboration incentives and applications. The state-of-the-art, the expected future state and the challenges are discussed for the industrial business applications used at present, and a roadmap for supply chain integration is provided. The categories of applications being considered are: customer-oriented applications; internal organisational applications; and supplier orientated applications. The applications selected are Customer Requirement Management (CRM), Material Requirement Planning (MRP) and Enterprise Resource Planning (ERP), Advanced Planning Systems (APS), Vendor Managed Inventory (VMI), Product Lifecycle Management (PLM), and finally, Portals (this includes former expressions such as supplier hubs and electronic markets). The paper concludes that especially the shift from a function-oriented view to a flow-oriented view is a significant challenge regarding industrial applications and company organisation.

The next three papers following the first contribution, all by Hvolby and Steger-Jensen, support the roadmap presented, by discussing selected topics.

The first contribution by Hvolby and Steger-Jensen focuses on the most commonly used application for supply chain integration – Vendor Managed Inventory. The paper combines theory and industrial experience of Vendor Managed Inventory (VMI) based on results from industrial research projects. Other supply chain initiatives are briefly discussed in the introduction. The premises for a profitable VMI implementation at the manufacturer and the supplier(s) are discussed in relation to systems requirements and segmentation.

The third and fourth contributions focus on selected technologies from the enabling layer.

The second paper by Hvolby and Steger-Jensen provides an overview of different web-based integration technologies for e-business solutions. The focus is on supply chain integration and related areas such as integrated application and integration of different enterprise information systems. The web-based integration technologies discussed are Business-to-Business (B2B) and Enterprise Application Integration (EAI). Advantages and disadvantages of these web-based integration technologies are discussed and reviewed from a business perspective.

The third contribution by Hvolby and Steger-Jensen focuses on web services and integration of different enterprise information systems. During the last decade, B2B integration has obtained an extraordinary focus. Most companies have already implemented ERP systems for internal use, and are now ready to implement e-business and Supply Chain Management solutions. B2B integration evolution today has reached a level, which enables implementation of B2B solutions based on Internet, standards, and technologies supported by nearly all the leading hardware and software vendors. B2B integration and connectivity only includes features that are required to interact with external claimants, and will typically not include the deep business process integration that is required when interfacing enterprise systems. This is where the use of web services becomes relevant.

The final contribution written by Meyer provides an overview of the OpenFactory Project (From Enterprise Resource Planning (ERP) to Open Resource Planning (ORP)). Dynamic business networks are a common type of organisation in the mechanical engineering industry. Owing to the complex structure and temporary business relationships of these networks today's ERP solutions do not sufficiently support efficient inter-organisational order processing. This paper describes the conceptual basis of Open Resource Planning (ORP), an innovative approach to realise seamless order management within dynamic business networks. Three important components are outlined: a data and process standard for inter-organisational order management within mechanical engineering industry; a web-based information system; and an innovative Internet business model.

*Advanced Manufacturing – An ICT and Systems Perspective – Taisch,*
*Thoben & Montorio (eds)*
*© 2007 Taylor & Francis Group, London, ISBN 978-0-415-42912-2*

# Supply chain integration: State-of-the-art, trends and challenges

Thomas Gulledge[1], Hans-Henrik Hvolby[2], Con Sheahan[3],
Ray Sommer[1] & Kenn Steger-Jensen[2]
[1]*Public Policy and Enterprise Engineering, George Mason University, Fairfax, Virginia, USA*
[2]*Manufacturing Information Systems Group, Department of Production, Aalborg University, Denmark*
[3]*The Enterprise Engineering Group, University of Limerick, Ireland*

ABSTRACT: This paper discusses different ways of obtaining more or less integrated supply chain solutions in terms of collaboration incentives and applications. The state-of-the-art, the expected future state and the challenges are discussed for the industrial business applications used at present. The paper concludes that the shift from a function-oriented view to a flow-oriented view is a significant challenge regarding industrial applications and company organisation. Further, the specialised systems managing, for example, product lifecycles and vendor inventories, are expected to merge into the large Enterprise Resource Planning applications within the coming years.

*Keywords*: Collaboration technology, Roadmap, Portals, Supply Networks.

## 1 INTRODUCTION

Supply chain integration is a vital element in obtaining and maintaining competitive advantages by creating value for all participants in the supply chain. Competitive advantages should not only be measured by product design, functionality, costs, and quality, but also by flexibility and lead-times. This is much in line with Williams (2001) who conclude that "*companies need to operate within effective and efficient supply networks that are highly flexible, innovative and dynamic to respond to pressures such as cost reductions, product development, high quality, value adding products, and frequent, on time deliveries*".

Since the 1990s many companies have implemented lean ideas in the production to reduce set-up time, stock, work-in-process, etc. But the shift towards customisation has increased the need to control the order through the administrative business processes as more departments and partners are involved in order processing. As a consequence of the growing outsourcing activities to local suppliers and suppliers in low-cost countries, the missing co-ordination and integration is no longer only an internal problem.

Lee and Whang (2001) have defined four key dimensions of supply chain planning: information integration; synchronised planning; workflow co-ordination; and new business models. In Table 1 an adapted version of elements and benefits of the dimensions are listed.

A well co-ordinated and effective support infrastructure in a supply network consisting of individual companies can only be accomplished by allowing individual decision-making. This is why the advanced planning systems (APS) so far only have been successful in large organisations, as they are based on a central decision-making in the supply network.

To allow individual decision-making, each partner needs to have relevant information from the entire supply chain, and the information should be provided in a seamless way using the least resources at exchanging data in the supply chain. This will be difficult to obtain by using the

Table 1. Key elements and benefits of supply chain integration (Adapted from Lee and Whang 2001).

| Dimension | Elements | Benefits |
|---|---|---|
| Information Integration | Information sharing and transparency Direct and real-time accessibility | Reduced bullwhip effect Faster response |
| Workflow Co-ordination | Co-ordinated planning, procurement, order processing Integrated, automated business processes | Optimised capacity utilisation Efficiency and accuracy gains Lower cost Improved service |
| Synchronised Planning | Joint design and engineering changes Mass customisation | Lower cost Earlier time to market Able to penetrate new markets |

present ERP systems, as most of these do not support information sharing among companies in the supply network (Parry 2004).

## 2  SUPPLY CHAIN INTEGRATION ROADMAP

The benefits of supply chain integration have to be obtained by new technologies combined with organisational and behavioural changes. According to Gulledge (2002) some of the challenges related to collaboration and trust are:

- lack of trust and an unwillingness to share supplier data with the separate exchange company;
- unwillingness to outsource supply chain operations to a third party (the exchange company), especially if the third party is also hosting a competitor;
- channel conflict and inability brand products;
- lack of consensus among partners about where functionality should reside;
- high costs of integrating partner back-office systems with the exchange technology;
- inability to recruit and integrate suppliers.

Experiences from a recent research project among Danish SMEs support several of the identified problem areas (Hvolby et al. 2005). This study clearly shows that SMEs are not willing to invest in expensive, integrated IT-solutions. Instead they prefer low-cost and non-integrated solutions. An example of this is a company supplying office equipment on a B2B basis, which invested in a stand-alone web-application for 24-hour sales, with the orders being manually entered into the ERP-system afterwards. For test purposes this kind of procedure is satisfactory, but on a long-term basis this will be far too expensive.

In Figure 1, a supply chain integration roadmap is illustrated. The roadmap is divided into an application layer and an enabling layer. The application layer contains backbone business applications of industrial companies such as Enterprise Resource Planning (ERP), Product Life-cycle Management (PLM) and Vendor Managed Inventory (VMI). The enabling layer contains the methods and technologies used to develop and support the applications such as Supply Chain Operations Reference model (SCOR), Business to Business standards (B2B and Service Oriented Architectures (SOA).

As many applications from different vendors apparently have the same features it becomes relevant to look at the technologies used by the specific vendors to decide which would be the right for a specific company. This could be compared to, for example, the features of a bicycle, which in general seem to be the same for any bike, although the quality of the components used in a cheap bicycle are not the same as those used in a more expensive model.

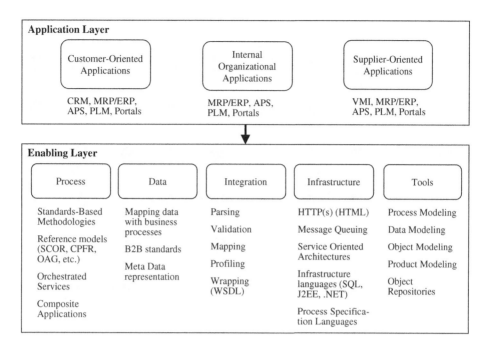

Figure 1. Supply chain integration roadmap organisation.

The application layer is divided into customer-oriented applications, internal organisational applications, and supplier orientated applications. In the following the state-of-the-art, the expected future state, and the challenges will be discussed for each type of application. The applications selected are Customer Requirement Management (CRM), Material Requirement Planning (MRP) and Enterprise Resource Planning (ERP), Advanced Planning Systems (APS), Vendor Managed Inventory (VMI), Product Lifecycle Management (PLM) and finally Portals (this includes former expressions such as supplier hubs and electronic markets).

## 3   CUSTOMER-ORIENTED APPLICATIONS

The keywords in describing the current state-of-the-art for customer-oriented applications are functional orientation and missing integration. The main challenge for ERP systems is to support process orientation especially as regards daily operations (e.g. workflow support) and set-up of appropriate (joint) planning parameters. In general parameter-settings seem to be a significant problem in industry. Even in large companies many examples have been found of contrasting parameters (in-house) leading to inefficient control, increased inventories and loss of money and opportunities.

Workflow support is a strong tool to reduce resource consumption and lead-time. To implement supply chain integration with customers requires improved systems support, cultural changes, and trust building.

Owing to the dominance of ERP systems and the need for integration it is expected that the functionality of the current specialised systems (CRM and PLM) will merge into the ERP-systems (and CAD-systems for the technical functionality of PLM). Regarding APS applications integration is not the main issue, as most APS applications are add-ons to the leading ERP systems such as Oracle and SAP. Instead a more holistic view is needed if the intention is to involve customers (and suppliers) in the planning process.

An overview of state-of-the-art, the expected future state, and challenges are listed in Table 2.

Table 2. State-of-the-art, expected future state and challenges within the customer-oriented applications.

| Components | State-of-the-Art | Expected future state | Challenges |
|---|---|---|---|
| CRM | Separate applications (more or less integrated with other applications) | Integrated with or part of other applications (ERP) | Integrate master profiles |
| MRP/ ERP | Functional oriented implementation Functional-based design | Process oriented implementation Process-based design | Cultural changes needed towards process ownership No process based implementation methodologies |
| APS | Focus on solving co-operate optimisation problems Less integrated in the supply chain | Integrated applications Distributed and Service Oriented Applications (SOA) | Holistic view (less dependant on the hierarchical planning approach) |
| PLM | Separate applications (more or less integrated with other applications) | Integrated with other applications (ERP) | Standards need to mature and converge |
| Portals | Web-based application service Few portal solutions are integrated with backbone (ERP) applications | Process-based integrated applications based on workflow | Security based on federated identities (create user confidence) Rule-based security profiles |

Table 3. State-of-the-art, expected future state and challenges within the internal-organisational oriented applications.

| Components | State-of-the-Art | Expected future state | Challenges |
|---|---|---|---|
| APS | Mathematical capabilities that far exceed the layman's capability to exploit them | New user interfaces will improve the user's ability to manage this complexity | Qualifying internal data Short product lifecycles |
| PLM | Fragmented interfaced components Fragmented structured and unstructured data | Integrated Service Oriented Architectures (SOA) Process-oriented data integration Integrated with other applications (ERP) | Defining the links between processes and data Standards need to mature and converge |
| Portals | Web-based application service | Process-based integrated applications based on workflow | Cultural challenges to a new work paradigm |

## 4 INTERNAL ORGANISATIONAL APPLICATIONS

The applications in this case are more or less the same as for customer-oriented applications, but the view is somewhat different. Regarding APS systems the mathematical capabilities far exceed the capability to exploit them. Only few companies have high quality manufacturing data (set-up and operation times, other process routes, part lists, etc.) and for many companies it does not make sense to run an APS optimisation. The inappropriate counterbalance of late delivery and cost has also previously been discussed (Steger-Jensen and Hvolby 2002). Regarding PLM a significant challenge is for standards to mature and converge. An overview of state-of-the-art, the expected future state and challenges are listed in Table 3.

Table 4. State-of-the-art, expected future state and challenges within the supplier-oriented applications.

| Components | State-of-the-Art | Expected future state | Challenges |
|---|---|---|---|
| VMI | Separate applications (more or less integrated with other applications) | Integrated with or part of other applications (ERP) or based on portals | Reducing inventory in the supply chain instead of putting the pressure on a single supplier<br>Create buyer/supplier confidence |
| APS<br>PLM<br>Portals | These applications are similar to the customer viewpoint – see Table 2 | | |

## 5  SUPPLIER-ORIENTED APPLICATIONS

The applications for suppliers and customers are the same except for CRM and VMI. In this connection VMI covers all initiatives on the supplier side such as Continuous Replenishment, Efficient Customer Response, and Vendor Managed Inventory. CR and ECR are primarily used in the retail industry, whereas VMI is primarily used in the manufacturing industry. The current VMI functionality is expected to merge into future ERP applications. The challenges are once again trust and to establish a joint goal of reducing costs instead of putting the pressure on the single suppliers. The important weakness of VMI lies in the insufficient visibility of the whole supply chain (Barratt and Oliveira 2001). An overview of the state-of-the-art, the expected future state and challenges are listed in Table 4.

## 6  CONCLUSIONS

The main challenge regarding industrial applications and company organisation is a shift from a function-oriented view to a flow-oriented view. Further, the specialised systems handling, for example, product lifecycles and vendor inventories are expected to merge into the large Enterprise Resource Planning applications to fully gain benefits of integration among companies in the supply chain. Further, it is expected that future applications will be server-based and user-interfaces will be web-based to ease maintenance of client installations, ease hardware requirements, and increase user access and mobility.

## ACKNOWLEDGEMENT

The authors wish to acknowledge the European Commission for their support.

## REFERENCES

Barratt, M.A. & Oliveira, A (2001), Exploring the experiences of collaborative planning initiatives, International Journal of Physical Distribution & Logistics Management, Vol. 31 No. 4, 2001

Bicheno, John (2004), The New Lean Toolbox: Towards Fast, Flexible Flow, Picsie Press, England

Gulledge, Thomas (2002), "B2B eMarketplaces and small- and medium-sized enterprises", Computers in Industry, vol. 49

Gulledge, Thomas; Hvolby, Hans-Henrik; Sheahan, Con; Sommer, Ray & Steger-Jensen, Kenn (2004), Supply Chain Integration Roadmap, in proceedings of the second IMS SIG9 workshop, Copenhagen, ISBN 87-91200-42-3

Hvolby, H.-H. & Trienekens, J.H. (2002), "Supply chain planning opportunities for small and medium sized companies", Computers in Industry, Vol. 49, No.1

Lee, H & Whang, S (2001), "E-Business and Supply Chain Integration" Stanford Global Supply Chain Management Forum, Publication no. SGSCMF-W2

O'Brian, K (2003), "Value-Chain Report – Vendor Managed Inventory in Low-Volume Environments", Industry Week Publications (www.industryweek.com accessed 22 July 2004)

Parry, G (2004), ERP; implementation & maintenance in a lean environment, in proceedings of the Logistics Research Networks Conference, Dublin

Pine, B. Joseph; Victor, Bart & Boynton, Andrew C (1993), Making Mass Customization Work. Harward Business Review, September–October

Steger-Jensen, Kenn & Hvolby, Hans–Henrik (2002), "Constraint Based Planning in Advanced Planning and Scheduling Systems". Proceedings of the 7th International Conference on Concurrent Enterprising (ICE), Rome, Italy

Williams, Sharon J (2001), Defining Supply Chains Networks to achieve Best Practice amongst SMEs: A Review of the Pilot Methodology. Proceedings of the Fourth SMESME International Conference, Aalborg, Denmark

*Advanced Manufacturing – An ICT and Systems Perspective – Taisch,*
*Thoben & Montorio (eds)*
*© 2007 Taylor & Francis Group, London, ISBN 978-0-415-42912-2*

# Vendor managed inventory as a supply chain application

Hans-Henrik Hvolby & Kenn Steger-Jensen

*Manufacturing Information Systems Group, Department of Production, Aalborg University, Denmark*

ABSTRACT: This paper combines theory and industrial experience of Vendor Managed Inventory (VMI) based on results from industrial research projects. Supply chain initiatives other than VMI are briefly discussed in the introduction. The premises for a profitable VMI implementation at the manufacturer and the supplier(s) are discussed in relation to systems requirements and segmentation.

*Keywords*: VMI, collaboration, supply networks, case studies, supply chain planning

## 1 INTRODUCTION

Several solutions exist to obtain supply chain integration. Some have been used in industry for decades under other names in less automated solutions. An example is the weekly visit of the supplier of, for example, bolts and nuts which is a manual and local based version of VMI (*candyman*), whereas a wholesaler of, for example, electrical equipment is a predecessor of a supplier hub. The Danish Hi-Fi manufacturer Bang & Olufsen have for many years used the expression *Online Suppliers* for suppliers who do not receive purchase orders, but who instead are responsible for delivery of parts according to Bang & Olufsen's needs.

Over the past decade, VMI has evolved from primarily high-volume retail and OEM applications to a wider use across a variety of industries and operating environments (O'Brian 2003). VMI is slowly becoming more accepted in manufacturing supply chains and an important initiative towards integration in the supply network. Currently, VMI is the preferred supply chain integration initiative among Danish SMEs owing to the low investment and low complexity of the initiative and the quick results.

Most companies have realised that competitive advantages have to be gained by improvements in the supply chain and by improved customer satisfaction. As uncertainties in customer demand along the supply chain usually are buffered in inventories, it is obvious to focus on initiatives supporting reduced inventory cost. In addition many administrative resources are used to handle repetitive and non-value adding activities in connection with the supply of components.

Most authors distinguish several forms of inter-company collaboration. For example Lambert et al. (2000) state that relationships among organisations can range from arms length relationships (consisting of either one-time exchanges or multiple transactions) to vertical integration. Cooper et al. (1997) use the concept of the value tree. Every company has multiple relationships with customers and suppliers. Every single relationship has its own characteristics and may have its own design. This means that even in the realm of one company, different types of vertical inter-enterprise relationships can be distinguished. Thoben and Jagdev (2001) define three types of collaboration between legally independent companies (with increasing level of integration):

- supply chain type of collaboration when nodes in the chain must operate synchronously to meet customer demands;

- extended enterprise type of collaboration when information and decision systems and respective production processes of chain participants are integrated;
- virtual enterprise type of collaboration when loosely related enterprises bundle their competencies with the help of ICT to meet customer demand.

They argue that the concept of supply chains is well established, whereas the emergence of IT has expedited the nature and scope of collaboration to new and higher levels.

## 2   VENDOR MANAGED INVENTORY PHILOSOPHY

Vendor Managed Inventory is a natural step towards supply chain integration. The supplier is entrusted with the responsibility to handle inventory holding at the buyer based on access to plans and forecasts regarding production and sales. This simplifies the supply chain planning process between the buyer and the supplier. On the buyer side the business process flows related to planning and submitting orders and order changes are more or less eliminated. On the supplier side the business process flows related to receiving orders and replying lead-times to customers are also more or less eliminated, and based on a more accurate information of buyer demands it enables the supplier to adjust production and distribution planning in a wider perspective. To fully use this opportunity the supplier needs some guarantee from the buyer that he will purchase a certain amount of items (e.g. equivalent to two months' production) if the buyer decides to terminate the collaboration. Otherwise the supplier is not able to utilise the flexible production conditions on a long-term basis.

According to O'Brian (2003) and Disney and Towill (2003), VMI programs often leads to improved buyer-supplier relationships and a better insight into the spending patterns of buyers. The measurable results of VMI programs include reduced inventories, reduced administrative costs associated with replenishment, reduced inventory transport, and finally reduced obsolescence, enumeration, damage and shrinkage as a result of a closer control of inventories. Further, the supplier is often able to handle inventory ordering and cycle counting at lower costs than the manufacturer. Experience from the retail sector, however, reveals a number of factors that make the efficient operation of VMI difficult. Manufacturers are less eager to work with their competitors or other suppliers to co-ordinate deliveries to the retailer, whereas retailers are unwilling to share all their marketing plans and range strategies with their manufacturers (Blatherwick 1998).

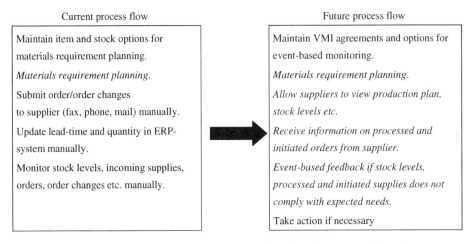

Figure 1.   Possible current and future activities in a typical supply order process. The *italic* processes are carried out by the ERP or VMI application without using manual resources.

When shifting from traditional purchasing to VMI, most companies consider the future ownership of the inventories. Consigned inventory refers to supplier ownership of inventory reserved for future use by a buyer (O'Brian 2003). Inventory is typically, but not always, located at the buyer's facility or at a location accessible to, or controlled by the buyer. Though some suppliers use consignment to minimise in-house stock (e.g. bulky products) most suppliers are reluctant to use consignment. The use of VMI and consigned inventory is often based on single supplier relations, and seen from a supplier's perspective the drawbacks of consignment, it is hoped, are counterbalanced by the benefits of VMI and improved information sharing. When considering consigned inventory special attention is needed regarding the functionality of the ERP system, as many mid-range ERP systems do not support consigned inventory.

## 3   VENDOR MANAGED INVENTORY SYSTEMS

Several standard VMI solutions exist from ERP-vendors such as Oracle and SAP and specific VMI-vendors such as Videlity (www.videlity.com) and Pipechain (www.pipechain.com). Both VMI-vendors use web-based clients to communicate with suppliers and customers. Videlity has focused on simplicity regarding use and implementation of their VMI solution, and their software has been well-accepted in Danish industry. Videlity's exchange module is able to interface a variety of sources such as Edifact, XML and printer-agents able to grab any print and extract the needed information. This enables integration regardless of the supplier's technical readiness (Videlity 2003) and makes integration with any ERP or order management system possible. Videlity also offers a large variety of notifications to the manufacturer and the supplier. This allows each partner to monitor the co-operation closely in the start of the process until the new procurement procedures have proven their value. An example of a notification is that the supplier has not confirmed production start even though the latest confirmation date is passed.

The large vendors of ERP and warehouse management systems have moved rapidly towards supply chain execution facilities, partly by acquisition of supply chain execution vendors (Trebilcock 2004). This development seems only natural considering the need for close integration with the ERP systems, and follows the previous steps of the ERP vendors incorporating finite capacity and product configuration functionality in their solutions.

VMI-ERP interface requirements are:

- Manage workflows and VMI buyer-supplier agreements based on events and KPI's;
- Monitor stock levels and supplies (needed versus actual, planned and initiated);
- Initiate actions and notifications in case of exceptions (e.g. a potential material shortage);
- Allow easy customisation of required notifications.

Regarding consigned inventory it is important that the IT-system is able to handle inventory holdings without a financial registration in the accounting system (as the inventory belongs to the supplier until the buyer pulls the item from the inventory). Otherwise more sub-systems (e.g. spreadsheets) are introduced in the organisation. Only a few systems are able to handle this at present (e.g. Oracle applications and SAP).

## 4   CASE STUDIES

Several examples of VMI initiatives can be found in Danish Industry. One is a medium-sized manufacturer of advanced products operating in a highly dynamic high-technology market with high demands for agility and flexibility (Hvolby et al. 2005). The starting point of their VMI considerations was primarily that the inventory levels of raw materials were far too high, the collaboration and integration to vendors were poor, and the average vendor service level and flexibility were also too low.

The objectives of the VMI-project (November 2003) was to:

- Introduce VMI agreements including consignment with top 25 vendors (more than 50% of total buy) within one year;
- Reduce total administrative acquisition costs by more than 50%;
- Reduce stock cover from over 8 weeks to 2 weeks;
- Reduce inventory of obsolete materials;
- Reduce number of slow moving items;
- Improve supplier delivery performance, agility and logistic attention.

The manufacturers' expected benefits for their suppliers by entering VMI collaboration were:

- Extended responsibility and long term relationship (preferred supplier);
- Improved long term planning and capacity utilisation:
  - Better possibilities of adjusting to changed demand without *rush-orders*
  - Savings on production change-over costs;
- Savings on own warehouse and insurance costs (consignment only);

The implementation was based on a Danish VMI-vendor, Videlity. The VMI initiative was reasonably well accepted by most vendors, and the results of the VMI-project (January 2005) have been:

- VMI agreements with 13 vendors, but this will be reduced to 11, as two of the vendors are not fulfilling their obligations of the VMI contracts. This is however expected to later increase to 15–20;
- Administrative acquisition workload are reduced, thought this has not led to reductions in the staff;
- Stock has been reduced with more than 50% for VMI items. Stock cover has only been slightly reduced as the production volume has decreased;
- The VMI project has to some degree reduced the number of slow moving items;
- Supplier delivery performance and agility have also been improved.

There were two main reasons for not obtaining the original goals so far. As for many other western companies, the manufacturer considered changing supply of some components to the Far East, and the current suppliers of these components obviously not have been involved in the VMI project. Further the manufacturer realised that their ERP system was not able to handle consigned inventory and found that the benefits of changed ownership would not match the drawback of manual stock handling. Unlike other companies' experiences on suppliers' reluctance towards implementing consigned inventory, the manufacturer has had several enquiries from suppliers specifically asking to enter consigned inventory agreements.

Gustafsson and Norrman (2001) also observed significant improvements at Ericsson Radio Systems by implementing VMI procedures and the VMI-system Pipechain. Over a period of four years Ericsson reduced their customer order lead-time from 15 days to 1, improved their delivery accuracy from 20% to 99.8% and improved their inventory turnover rate from 5 to 80. Alongside the lead-time and delivery accuracy improvements the costs were reduced by 20%. The pros and cons of the VMI implementation are listed in the following:

- Main benefits visible shortly after an implementation (months);
- The investment pays off shortly (months);
- The software tool is fast to implement (weeks-month);
- The customers and suppliers have gained a greater knowledge and understanding of one another's working processes and businesses;
- The users of the software tool rely on the system and find it logical and process oriented;
- The workload for the people working with operative logistics has been less fluctuating;
- The change of working procedures and shift of responsibility takes time;
- Time and resources is needed to adapt the ERP and VMI systems to the new processes.

Another Pipechain implementation at Volvo Powertrain reports a 50% reduction of average stock in 10 months without including consignment. Further, the service level with the suppliers also improved. The VMI implementation included 50 suppliers and during the project a missing comfort between Volvo managers and their suppliers was regained. (MSI/KeepMedia 2004).

Finally, pilot projects in the food sector in the Netherlands shows a decrease of stock level at food processing industries and also product quality improvements. Higher stock turnover and lower stock level lead to fresher food products – in a pilot project, up to three days were cut off the lead-time (Vorst et al. 2000).

## 5 CONCLUSION

VMI initiatives are well accepted in industry and the literature. Competitive advantages can be gained if the necessary trust and knowledge is present. As trust often is gained throughout a long relationship it might not always be present at project start. Therefore facilities to monitor the supply could ease a VMI project, especially in the starting phase.

Many collaborative initiatives by large manufacturers have not been received with great enthusiasm by their suppliers since the set-up has been solely on the premises of the manufacturer. The suppliers are thereby being squeezed by a number of manufacturers with each of their solutions. The chances of successful implementation of collaborative systems are fairly high, when the circumstances of both parties are included in the development process.

## REFERENCES

Alter, S. (2001). "Information Systems; a management perspective", Addison-Wesley Educational Publishers.
Bensaou M. (1999). "Portfolios of Buyer-Supplier Relationships, Sloan Management Review, Summer.
Blatherwick, A (1998). "Vendor Managed Inventory – fashion fad or important strategy?", Supply Chain Management, Volume 3, Number 1.
Cooper, M.C.; Lambert, D.M. & Pagh, J.D. (1997). "Supply Chain Management: more than a new name for logistics". International Journal of Logistics Management, Vol. 8, No. 1.
Disney, S.M. & Towill, D.R. (2003). "Vendor-managed inventory and bullwhip reduction in a two-level supply chain", International Journal of Operations & Production Management, Vol. 23 No. 6.
Erenguc, S.; Simpson, N.C. & Vakharia, A.J. (1999). "Integrated production/distribution planning in supply chains: an invited review", European Journal of Operational Research, 115, 219–236.
Gustafsson, J. & Norrman, A. (2001). "Network managed supply – execution of real-time replenishment in supply networks", Proceedings of International Symposium of Logistics, Austria.
Hvolby, H.-H.; Steger-Jensen, K.; Thorstenson, A. & Baastrup, J.-J. (2005). "Internet-based information exchange in the supply chain" (in Danish), Aalborg University Press, ISBN: 87-91200-46-6.
Hvolby, H.-H.& Trienekens, J.H. (2002). "Supply chain planning opportunities for small and medium sized companies", Computers in Industry, Vol. 49, No.1
Kaplan, S. & Sawhney, M. (2000). "E-Hubs: the new B2B market places", Harvard Business Review, May–June
Kim, J. & Shunk, D. (2004). "Matching indirect procurement process with different B2B e-procurement systems", Computers in Industry, Vol. 54 No. 2.
Lambert, M.D. & Cooper, M.C. (2000). "Issues in supply chain management", Industrial Marketing Management, 29, 65–83.
Lim, T.; Kim, H.; Kim, M. & Kang, S. (2003). "Object oriented XML document meta-model for B2B collaboration", Production Planning & Control, Vol. 14, No. 8, 810–826.
Lee, H.L. & Whang, S. (2000). "Information sharing in a supply chain", International Journal of Manufacturing Technology and Management, Vol. 1 No. 1.
MSI/KeepMedia (2004). "VMI gains elusive for some, but not for Volvo Powertrain", Reed Elsevier Business Information, July (www.keepmedia.com/pubs/msi/2004/393375 last accessed 9 August 2004)
O'Brian, K. (2003). "Value-Chain Report – Vendor Managed Inventory in Low-Volume Environments", Industry Week Publications (www.industryweek.com accessed 22 July 2004).
Steger-Jensen, K. & Hvolby, H.H. (2003). "Portfolios of Buyer-Supplier Relationships based on Supplier Order De-couple Point and Investment". In proceedings of the 6th SMESME International Conference, Greece, ISBN: 960-87716-0-9.

Thaler, K (1999). "Examination and measurement of delivery precision and its impact on the supply chain", 15th International Conference on Production Research, Limerick, Ireland.

Thoben, K.D.; Jagdev, H.S. (2001). Anatomy of enterprise collaborations, Production planning and control, Vol. 12, No. 5, 437–451.

Trebilcock, B (2004). "The world's top supply chain execution suppliers", Modern Materials Handling, Reed Elsevier Business Information, July.

Videlity (2003). "Videlity Supply Chain Connector – Technical concept and framework", whitepaper available at www.videlity.com.

Womack, J.P. & Jones, D.T. (1994). "From lean production to the lean enterprise". Harvard Business Review, Vol. 73, March-April.

Vorst, van der J.G.A.J., A.J.M. Beulens, P. van Beek (2000). Modelling and simulating multi-echelon food systems, European Journal of Operational Research, Vol. 122 No. 2, 354–366.

*Advanced Manufacturing – An ICT and Systems Perspective – Taisch,*
*Thoben & Montorio (eds)*
© *2007 Taylor & Francis Group, London, ISBN 978-0-415-42912-2*

# E-business solutions as supply chain enablers

Kenn Steger-Jensen & Hans-Henrik Hvolby
*Manufacturing Information Systems Group, Department of Production,*
*Aalborg University, Denmark*

ABSTRACT: This paper presents an overview of different web-based integration technologies for e-business solutions. Focus is on supply chain integration and related areas such as integrated application and integration of different enterprise information systems. The web-based integration technologies discussed are Business-to-Business (B2B) and Enterprise Application Integration (EAI). Advantages and disadvantages of these web-based integration technologies are discussed and reviewed from a business perspective. The conclusion is that all contributions are important to make an effective enterprise solution. Unfortunately, today, none of these two web-based technologies fully supports the enterprise integration needs, but further research is needed on EAI interfaces, which contains adapters for automatic mapping. Furthermore, these adapters need to be able to handle the different session's state across different types of asynchronous or synchronous systems running batch and online processing environments, based on industry standards.

*Keywords*: Enterprise application integration (EAI), Business-to-Business (B2B), web-based supply chain integration, integrated applications.

## 1 NEW BUSINESS MODELS

The global reach and interconnectivity of the Internet have spawned new business models and radically transformed the existing ones. In a recent survey, by Baer (1998), a third of the 100 respondents said that their organisations use the Internet in 10–25% of their business critical applications. It is expected that within the next three years the Internet will be the dominant platform for business applications in almost all organisations. The attractiveness of Internet-centric e-business models resides in their efficiency in:

- reducing search costs by facilitating comparison of price, products, and services;
- reducing lead times;
- improving production and supply capability;
- managing demand;
- improving personalisation and customisation of product offerings (Bakos 1998).

The Internet and the web technologies allow not only automation of inter-organisational processes but also individual users to interact with organisational information systems in novel ways and at very low cost. The advances in IT that have made this possible have been described as the second economic revolution (Essig and Arnold 2001).

E-business information systems (e-bis) are computer applications that use the Internet technology and the universal connectivity and the capabilities of the web browser to integrate business processes within and beyond an enterprise. The use of Internet technologies to manage information is a substantial improvement of traditional information systems and conventional uses of the web (Applegate 1995, Hsu and Pant 2000, Venkatraman 1994). E-business information systems allow transactions to be conducted in an integrated and enlarged information space by removing constraints imposed by diverse computing platforms, networks, and applications (Isakowitz and Fabio 1998, Lederer et al. 1998).

E-business technology facilitates companies to perform inter and intra-organisational business processes across the Internet. It has the power to transform the business process because it pervades all the business process steps. Simple information exchange, such as customer and purchase orders (traditional e-commerce), and in-dept business logic interaction and execution of business logic within customers and suppliers information systems (traditional e-collaboration) can be undertaken.

Here e-business is broadly defined to include also the design of the e-commerce and the e-collaboration across heterogeneous and homogeneous instances of the information system. This includes the development stage, the daily operation stage, and the daily tactical stage of adjusting the e-business model (e.g. switch e-commerce partners).

An indication of the benefits of e-business is found in projections for the growth of expected corporate buying on the Internet. For example, the Boston Consulting Group estimated that business-to-business Internet purchases would reach $2 trillion by 2003, up from $92 billion in 1998 (Whyte 2000). Although projections vary, they are generally in this range, demonstrating the clear move toward technology-facilitated purchasing in the new millennium.

## 2 WEB-BASED INTEGRATION TECHNOLOGIES

Companies which leverage Internet business practices throughout their business processes prerequisite, that the user applications support the business needs such as a) work together in the same transaction, b) commit to security boundaries, c) share the same session state, d) be remotely accessible and e) not suffer from code overload. This requires an e-business that contains:

- an infrastructure that hosts such components for secure remote and local access;
- an infrastructure that hides infrastructure code;
- an implementation in an enterprise solution framework;

These prerequisites and requirements are still a challenge for most businesses' IS/IT-systems. Furthermore, to fully participate in this new highly connected world and capitalise on the enormous potential benefits, it is becoming increasingly critical that organisations implement comprehensive solutions addressing the two following key classes of integration, which supports the prerequisites mentioned above and which drives further discussion:

- Enterprise Application Integration (EAI): is the sharing of data and business process logic across heterogeneous and homogeneous instances through message-oriented-middleware (MOM). EAI may be managed by SAP (NetWeaver), Oracle (Oracle Application Sever which contains Inter-Connect, PartnerConnect and ProcessConnect), or through solutions provided by private vendors (e.g., IBM, webMethods, etc.). EAI is sometimes called Application-Centric Interfacing or interoperability.

First, internal applications and business processes must be integrated and automated, before sharing business logic with external applications to the collaboration partners. As customers and partners grow to expect increasing information and at near real-time speeds, the organisation must prepare its systems and business processes to meet these needs. Secondly, true e-collaboration partners needs access to internal applications.

- B2B Integration: is the passing of data (not business process logic) through agreed-upon implementation conventions e.g., Extensible Mark-up Language (XML), and Electronic Data Interchange (EDI), etc. This may be done directly with trading partners, which use standards as e.g. RosettaNet (2004) and ebXML (2004). B2B integration is sometimes called Data-Centric Interfacing or Connectivity.

Once the internal applications and processes are integrated and automated to support the business objectives, the organisation is ready to implement systematic processes to link with trading partners to support the e-commerce processes. In the following sections these two mentioned web-based integration technologies are discussed.

## 3 BUSINESS-TO-BUSINESS (B2B) INTEGRATION

Business-to-Business (B2B) integration is the primary reason why e-commerce has reached high attention during the last decade. Although there is confusion between the B2B and B2C integration paradigms, there are differences that go much deeper than the differences between retail and wholesale purchasing. B2B integration is fundamentally about co-ordinating information among businesses and their information systems.

In today's world with companies operating in a global business environment, B2B integration is a prerequisite for them to remain competitive. They need to interact with their (worldwide) suppliers, partners and customers. B2B integration enables a company to focus on its core competencies and offload other services to partners to gain efficiency and reduce cost.

After a decade of implementing expensive ERP, CRM, and e-commerce applications in a departmental manner, companies are turning their attention to integrating these *information silos*. However, if this integration is done on a dedicated point-to-point basis, these companies end up using a large portion of their software budget on simply maintaining these connections. Connecting applications on a point-to-point basis is not enough. Without a thoroughly integrated internal infrastructure, B2B initiatives are sure to provide little value in the best-case scenario, or no value in the worst.

In connection with the Pathfinder project, Gulledge and Sommer (2001) found that baseline solutions for e-business and B2B where not as easy to establish as the different vendors' claim. One of the reasons was that many baseline solutions where not supported by other vendors' technologies and standards well enough. To obtain a satisfactory set-up, several baseline solutions had to be combined with customer-specific B2B programming.

Many B2B standards and flavours have arisen during the last decade, which have made the interoperability among these standards a mess. Furthermore, the authors believe that many companies are not aware of the differences in these standards, which also leads to expensive implementations for the companies. Normalisation among B2B-solutions has today become the essential element in implementing cost-effective B2B solutions.

Standardisation of B2B integration can be organised in four tiers (Steger-Jensen and Hvolby 2002): Meta-data tier, Content tier, Transport and Process tier, which all must be covered for successful B2B integration. Some elements of these tiers are shared with EAI. There are some pitfalls related to B2B integration that companies have to be aware of, as not all providers and standard B2B integration components cover all four tiers, which is necessary. Many different standards exist for each component of B2B integration. This complexity in B2B standards makes it difficult for companies to select the right standards for their purpose. In Figure 1, the different tiers and the main contributors are listed, and following on from this there is a brief description of each layer.

### 3.1 *Meta-data and content tiers*

The Meta-data tier describes data and how data are defined within the syntax of the XML document. Focus is on representation of data in terms of document type definitions (DTDs) and XML-schemas.

| Framework for B2B-integration based on the four main tiers | | | |
|---|---|---|---|
| Meta-data Tier | XML with Schemas or DTD, W3C | | |
| Content Tier | XML Messaging Model, OAG and CPFR | ebXML | RosettaNet |
| Process Tier | Workflow and business processes, WfMC, BPEL | | |
| Transport Tier | SOAP an XML-based protocol, W3C | | |

Figure 1. Framework for B2B-integration and main standard contributors to the tiers (Steger-Jensen and Hvolby 2002).

The Content tier focuses on the XML-message and the way XML-messages are described. Often the description of, for example, a purchase order differs from company to company, which is why standards for XML-messages are necessary.

The primary candidates for XML-message standards are Open Applications Group and Collaborative Planning, Forecasting and Replenishment which are both supported by the main vendors of B2B and ERP systems ( OAG 2004, CPFR 2004).

## 3.2   Process tier

The Process tier focuses on information flows in companies and among companies. Workflows and workflow systems, in general, control information flow. A workflow system is a tier, within an ERP system, placed between the logic data model and the end-user. Workflows representing the business processes are often different from vendor to vendor. For example, a workflow (e.g. customer order) used in SAP R/3 is not the same as in ebXML or Navision.

The primary candidate for process tier is the Business Process Execution Language (BPEL) based on OASIS (2004), which is the most mature process/workflow language standard to date and draws upon the rich history of its predecessor languages (specifically XLANG from Microsoft and WSFL from IBM) to provide rich process flow capabilities.

## 3.3   Transport tier

The transport tier is necessary to exchange XML-messages from one ERP system to another, which often is based on SOAP. The transport mechanism is Internet based such as HTTP, SMTP, TCP/IP, etc., and supports synchronous and asynchronous exchange. Most standardisation initiatives, which cover several tiers, have explicitly defined a mechanism for secure exchange of XML messages via a HTTP based infrastructure using the Internet standard MIME-based encoding and related technologies. In these standards (ebXML, RosettaNet, Open Application Group) this component is often referred to as implementation framework. These standardisation initiatives, which cover several tiers such as RosettaNet and ebXML, are discussed in the next section.

## 3.4   Multi-tier solutions

ebXML (electronic business with XML, 2004) is a complete B2B framework which enables business collaboration through the sharing of web-based business services. It is an effort of UN/CEFACT and OASIS.[1] The framework supports the definition and execution of B2B business processes expressed as choreographed sequences of business service exchanges. The framework includes specification for a Message Service, Collaborative Partner Agreements (CPA), Core Components, Business Process Methodology, and Registry and Repository.

Furthermore, ebXML and OAG, the contents tier within the B2B-framework collaborates as well. The OAG plans to use the ebXML Transport work, the ebXML Business Process Collaboration Schema, and the ebXML Collaboration Partner Protocol (CPP) as the Collaboration Partner Agreement (CPA) specifications. The OAG is also working and watching the Registry and Repository work as well and watching and contributing to the Core Components work and will work for convergence with these bodies of work as it makes sense to their constituency, (OAG 2001).

Rosettanet (2004) is dedicated to the development and deployment of open electronic commerce standards that align the business processes among partners in high-technology supply chains. RosettaNet standardisation efforts can be divided into three areas. 1) A dictionary, defining a common set of properties for use by 2) the business process (PIP) specifications and associated business documents and guidelines. The Partner Interface Processes (PIPs) (2004) define the specific sequence of steps required to execute business processes among supply chain partners, also known as the

---

[1] UN/CEFACT is the United Nations Centre for Trade Facilitation and Electronic Business, and OASIS is an Organisation for the Advancement of Structured Information Standards, a non-profit, global consortium that drives the development, convergence and adoption of e-business standards.

business process choreography. 3) The RosettaNet Implementation Framework, RNIF (2004), provides the fundamental plumbing required to execute business processes among trading partners in an open, interoperable, secure, platform and implementation independent way.

B2B integration is heavily dependent and based on XML and primarily with XML schemas from the Meta-data tier, which gives the capability to exchange XML messages, the core of the content tier. XML defines values for the information (focus on content, the data). XML also lets users create their own tags. XML is a way to enable data to have structure. The structure is not contained in the XML file itself but in a related text file (called a schema). Companies that adopt the same structure for, say, purchase orders can easily exchange orders via XML. Unfortunately, multiple versions of descriptions for such things as purchase orders have been created. Until schemas have been unified it is necessary to map the various schemas to one another to allow free and unconstrained communication. The process tier is a core element according to business processes and interaction based on workflow and the BPEL. A typical B2B integration solution is the discrete system model, in this example between OAGI and RosettaNet, which is as follows:

$$ERP(OAGI) - B2B(RNIF) - - - -B2B(RNIF) - ERP(OAGI)$$

Unfortunately, a one-to-one relationship does not exist between OAGI and RosettaNet messages, or for that matter among other content tier vendors and vendors who cover several tiers.

Most ERP vendors are members of OAGI and therefore use OAGI messages. At the same time, all XML-message vendors and the multi-tier standardisation initiatives such as RosettaNet and ebXML have their own workflow and exception handling method related to the XML-message. It is not only the workflow and XML-message that have to be mapped, but also the exception handling method. If two companies have implemented different B2B solutions, mapping between these two solutions could be necessary. Therefore, if the contribution from BPEL becomes a success, mapping among different workflow systems and different B2B solutions will be reduced drastically.

On the transport tier, SOAP and HTTP are the backbone. SOAP will presumably in future be the primary XML-message format, because it supports the connection among businesses in general and allows sending and receiving documents, making remote procedure calls and furthermore, it is firewall friendly.

## 4 ENTERPRISE APPLICATION INTEGRATION

Enterprise application integration (EAI) is the total integration of applications within an enterprise. EAI is a business computing term for the plans, methods, and tools aimed at modernising, consolidating, and co-ordinating the computer applications in an enterprise. EAI solutions ensure compliance and interoperability through open messaging, open queuing, open development tools, adapters (applications, web, e-commerce, communications, legacy, generic) and data standards (XML, EDI, HL7, Swift) across all leading platforms (Andrianopoulos 2002). Typically, an enterprise has existing legacy applications and databases and wants to continue to use them while adding or migrating to a new set of applications that exploit the Internet, e-commerce, Extranet, and other new technologies. EAI may involve developing a new total view of an enterprise's business and its applications, seeing how existing applications fit into the new view, and then devising ways to efficiently reuse what already exists while adding new applications and data. Unfortunately EAI considers the proprietary business logic and requires a detailed knowledge of the interfacing systems. As most information systems miss encapsulation, integration is often very difficult. The logical schema may comprise thousands of tables, which the information systems are based on. This means in practice that developers of outside components have to know a lot of details on which so-called Dynamic Link Libraries (DLL)[2] to choose, and what functionality this DLL provides. A better solution would be to publish standard methods that can be called on a limited number of business

---

[2] Dynamic Link Libraries (DLLs) is used to make the business logic code re-usable.

objects. For different application domains worldwide standards exist which define these standard messages (a/o ebXML and OAG), which are the core contribution of B2B-integration standards.

Today, most applications are implemented using a wide variety of programming languages, on an equally varied number of platforms, using many different (often non-standard) interfaces – APIs, messaging systems, batch file interfaces. Integration solutions must provide the ability to communicate with virtually any application or technology while providing a common framework for defining and managing the business process. Today, however, there are two fundamental limitations in the integration approaches being considered:

- **No Systematic Integration Architecture** – Many organisations have not approached integration from a systematic perspective within an architecture. Few have chosen an integration framework for the organisation into which all appropriate information systems, both new and legacy systems, are designed to fit. Instead, integration has been obtained by separate point-to-point solutions.
- **Components based on Middleware Solutions** – The lack of a systematic approach has also lead to the connection of many separate, piece part proprietary middleware solutions for integration purposes. For example, it is common to find solutions consisting of an enterprise application integration tool developed by one company coupled with a B2B communication product either licensed or acquired from another company. Further, multiple tools are used to address business process management – one tool for handling automated event processing and another for handling human interactions. Finally, when various middleware systems and security technologies are added, set-up and maintenance becomes very expensive.

Within state-of-art EAI-tools the underlying architecture is generally a *hub and spoke* model or *integration bus architecture*. These tools work on the control broker-adapter model wherein an application-specific adapter at the client interacts with the control broker to pass the messages/events to the target applications. There are two important aspects to understand about hub and spoke architectures:

- Hub-based deployment – The concept of a hub-and-spoke architecture is to eliminate point-to-point integration, which are less maintainable and have multiple points of failure. Instead, a central integration hub is used to which all the different systems are connected.
- Spokes – The spokes are typically the connectors that interface with various systems being integrated. A single hub may support many spokes – applications, B2B protocols, and web services.

Hub-and-spoke is a logical and a not a physical concept. For instance, the hub need not be deployed on a separate physical machine and can be co-located with one of the spokes. Further, hubs can be replicated and are often designed to be scalable and since the integration manager often is *stateless*, one or more hubs can be established to co-ordinate integration. Spokes may optionally run on remote hardware and may be deployed over the Internet, across firewalls. The *hub and spoke model* is best explained through an example where the application views are at the spokes and the common view is the hub within this model, hence the name. In general Oracle (2003) recommends the use of a hub-and-spoke topology because it is easy to maintain and manage and is suitable for high availability purposes.

An integration point is defined as an *event* that triggers communication between two or more participating applications in the integration scenario. Examples of integration points are, for example, *create customer*, *cancel purchase order*, or *get item info*. The common view consists of a list of such integration points, each with its own associated data. Applications participate in the integration by binding to one or more of these common view integration points. In the context of each binding, applications have their own application view of data that needs to be exchanged. Each binding involves mapping (also known as transformations) between the application view and the common view in the context of the integration point.

Most EAI-tools lack the features for B2B-integration, such as community management, trading partner profile management, sophisticated security mechanisms, and support for industry standards, such as open buying over the Internet (OBI), electronic business XML (ebXML) and

RosettaNet. On the other side, if the EAI-tool is from a leading vendor within IS/IT-systems there is a good chance that these standards are supported.

The difference between EAI and B2B are significant, even though they both may employ middleware, such as message brokers, to exchange information among various systems. Linthicum (2001) provides a good discussion of these differences:

- B2B typically resides outside of the integration domain, but functions in near real-time and with limited end-user influence.
- B2B typically passes information using accepted industry standards, such as XML or EDI, whereas EAI considers the proprietary business process configurations within enterprise software products.
- B2B allows users who understand little about internal business process logic to pass information across organisations, whereas EAI requires a detailed knowledge of the business processes as they are configured in the interfacing systems.
- B2B requires trading partners to agree on implementation conventions of industry standards. If agreement is reached, information can be easily passed.
- B2B assumes that the source and target enterprise system cannot be altered; hence, the passing of information is *non-intrusive* in the sense that the business process logic of the interfaced systems is not affected.
- B2B requires advanced security requirements, because the organisation is sharing information with external constituents.

## 5 CONCLUSION

E-business poses new challenges for an information system and ERP vendors must start to change their integration approaches. ERP systems are completely based on representation of data in tables (relational databases), but there are a growing demand to handle more complex documents, information objects and business objects. The notion of encapsulation is missing and application programmers are exposed nowadays to the precise way in which a complex object (such as a customer order) is mapped on the tables in logical schemas. DLLs are not specified as interfaces, but as navigation tools. Various ways of dealing with changeability exist. A common way is to provide *Application Programmer Interfaces* (APIs) to third-party programmers allowing them to insert additional logic.

Unfortunately this approach does not rectify the fact that ERP systems are interfaced by allowing DLL-calls to be made against the logical schemas and the APIs are seldom based on standards. The logical schema may comprise thousands of tables. This means in practice that developers of outside components have to know a lot of details on which DLLs to choose and what functionality the DLL provides. A better solution would be to publish standard methods that can be called on a limited number of business objects, as for example OAG does within the B2B-interation area.

B2B integration is used to pass information to external constituents, such as suppliers and customers. B2B integration could support any number of business requirements, such as sharing information with trading partners to support a supply chain or collaborating on product design. B2B-integration includes many features that are important requirements for interaction with external claimants, but typically does not include the deep business processes integration that is required when interfacing enterprise systems. This is where EAI comes in, as it supports some of the issues within B2B.

Unfortunately, lack of a systematic approach to integration makes connection of separate proprietary middleware solutions for integration purposes complex and expensive. This is probably one of the reasons why most SMEs have not entered the supply chain integration area yet, besides the fact that many SMEs do not have sufficient knowledge within the area. Supply chain integration has so far only been for the high-end integration market.

Further research is needed in development of standard EAI interfaces, which contain adapters for automatic mapping along with switching between different standards of B2B solutions as well as within the B2B tiers. Furthermore, these adapters need to be able to handle the different session states across different types of asynchronous or synchronous systems running batch and online processing environments. This new EAI-tool probably has to be based on a *hub and spoke* model.

## REFERENCES

Andrianopoulos, A (2002): 4 EAI -The framework behind WS Integration, WS Journal, Vol. 2, Issue 5, May, pp. 28–31.

Applegate, L.M. (1995), Electronics Commerce: Trends and Opportunities, Harvard Business School Case Series, Case 9-196-004.

Baer, T. (1998), "Premier 100 -making dollars and sense out of the Internet", Computer World, November 16, pp. 40–42.

Bakos, Y.J. (1998), "Towards friction-free markets: the emerging role of electronic marketplaces on the Internet", Communications of the ACM, Vol. 41 No. 8, pp. 35–42.

CPFR (2001), XML Messaging Model, June 25, 2001 Voluntary Interindustry Commerce Standards Association Copyright © 2001. http://www.cpfr.org/

CPFR (2004) - Collaborative Planning Forecasting and Replenishment Voluntary Guidelines. http://www.cpfr.org/

ebXML (2004), http://ebxml.org

Essig, M. and Arnold, U. (2001), "Electronic procurement in supply chain management: an information economics-based analysis of electronic markets", The Journal of Supply Chain Management, Fall, pp. 43–9.

Gullege, R. Thomas and Sommer, A. Ray (2001), Integration ebusiness Transaction Across Extended Enterprises, Proceedings of the IFIP 5.7 International Working Conference on Strategic Manufacturing, 26–29, Aalborg, Denmark, pp. 177–189.

Hsu, C. and Pant, S. (2000), Innovative Planning for Electronic Commerce and Enterprises: A Reference Model, Kluwer Academic Publishers, Norwall, MA.

Isakowitz, T., Michael B. and Fabio, V. (1998), "Web information systems introduction", Communications of the ACM, Vol. 41, pp. 78–80.

Lederer, A.L., Mirchandani, D.A. and Sims, K. (1998), "Using WISs to enhance competitiveness", Communications of the ACM, Vol. 41, pp. 94–5.

Linthicum, David, S. (2001), "Where EAI Meets B2B", Software Magazine, Vol.21 # 2 22–25.

OAG (2004), Open Applications Group, Plug and Play Business Software Integration, The Compelling Value of the Open Applications Group, http://www.openapplications.org/

OASIS (2004), http://www.oasis-open.org/committees/BPEL

Oracle Application Server -Integration Product Overview, (2003), An Oracle Technical White Paper, August 2003, p. 33.

RNIF (2004), RosettaNet Implementation Framework: Core Specification, Release 02.00.01, 6 March 2002, Version Identifier: V02.00, http://www.rosettanet.org/RNIF

RosettaNet (2004), http://www.rosettanet.org/HOME

RosettaNet PIP® Specification (2004), http://www.rosettanet.org/PIP

Steger-Jensen, K and Hvolby, H.-H. (2002), "Review of B2B integration", Proceedings of the APMS Conference on "Collaborative Systems for Production Management", Eindhoven, September 9–13, ISBN: 1-4020-7542-1.

WfMC (2004), Wf-XML 2.0,- XML Based Protocol for Run-Time Integration of Process Engines, Swenson K. D., et al., Draft October 3, 2003, http://www.wfmc.org/WorkflowStandard

Venkatraman, N. (1994), "IT-enabled business transformation: from automation to business scope redefinition", Sloan Management Review, Winter, pp. 73–87.

Whyte CK. (2000), E-procurement: the new competitive weapon. Purch Today April; 25.

*Advanced Manufacturing – An ICT and Systems Perspective – Taisch,*
*Thoben & Montorio (eds)*
*© 2007 Taylor & Francis Group, London, ISBN 978-0-415-42912-2*

# Web services as supply chain enablers

Kenn Steger-Jensen & Hans-Henrik Hvolby
*Manufacturing Information Systems Group, Department of Production, Aalborg University, Denmark*

ABSTRACT: The focus of this paper is on web services and integration of different enterprise information systems. During the last decade, B2B integration has obtained an extraordinary focus. Most companies have already implemented ERP systems for internal use, and are now ready to implement e-business and supply chain management solutions. B2B integration evolution today has reached a level that enables implementation of B2B solutions based on the Internet, standards, and technologies that are supported nearly by all leading hardware and software vendors.

B2B integration/connectivity only includes features that are required to interact with external claimants, and will typically not include the deep business process integration that is required when interfacing enterprise systems. This is where the use of web services becomes relevant.

The three important and emerging standards for web services are presented and discussed: Web Services Definition Language, Simple Object Access Protocol, and Universal Description, Discovery and Interoperability. Finally, business management through web services, and some pitfalls are discussed.

*Keywords*: Web Service, Enterprise application integration, Application integration technology.

## 1 INTRODUCTION

The Internet enables companies to extend their market reach through new customer channels. Internet technologies provide an efficient, low cost medium for business communication and exchange of data and, thus, enable multiple trading partners to collaborate on demand forecasts, production schedules, and inventory requirements. Companies that leverage Internet business practices throughout their business processes, such as collaboration with customers and suppliers, to obtain effective supply chain management, need to integrate their information systems. This collaboration and integration of information systems supports improvements such as increased visibility across the extended supply chain, reduced time to market, and reduced costs.

Web services is a term used for description of components and services that are addressable and available when using web technology. Web services are typically user-oriented and browser-based, application program interface (API) accessible, or system service functional. A web service could for example be a browser-based email program, a Simple Object Access Protocol (SOAP) monitoring service, or an Extensible Mark-up Language (XML) based integration with an enterprise application interface or legacy system.

There are at least three common types of web service deployment:

- One type enables adding of a web service interface on to an existing product. Examples of this approach include application servers, databases, messaging systems and enterprise resource planning tools. Generally, software vendors choose this approach.
- Another type is where customers can deploy vendor products to solve current integration needs. The customer then uses the new web service functionality to more easily integrate internal systems, or externally with partners.

- A third type is where an application service provider offers a web service interface, and customers access and use the service using web service standards.

An interesting difference between these deployment types is the different financial models they imply. Vendors make money from product sales, customers gain a return on their investment from increased efficiency and expanded customer revenue, and application service providers make money from recurring or rental revenue of the web service itself.

Web services are not especially new. Application Service Providers (ASPs) have provided end-user based web services for many years. While some have struggled, most ASPs – particularly the focused ones – have continued to grow in terms of customers and revenues over the past few years (Ekanayaka et al. 2002). Many ASPs have provided APIs for parts of their service for a long time. Many provide APIs for provisioning, security, billing, and aspects of the business process. Thus anyone who considers the web services provider model reliable, by implication should also consider ASPs as reliable since there is little difference between them. An examination of the issues and challenges faced by application service providers to secure their future success is discussed by (Sushil et al. 2002).

Web service standards are welcome enablers for efficient system-to-system connections. However, the current standards leave many questions unanswered when it comes to meeting enterprise integration needs, such as data transformation for interaction with complex back-end systems, guaranteed messaging, business process management, trading partner and protocol management, transactional integrity, and security.

Enterprise Application Integration (EAI) provides an open extensible framework for connecting applications within or among enterprises. EAI solutions ensure compliance and interoperability through open messaging, open queuing, open development tools, adapters (applications, web, e-commerce, communications, legacy, generic) and data standards (XML, EDI, HL7, Swift) across all leading platforms (Andrianopoulos 2002).

From a business perspective, a company gains extensive competitive advantages when all applications are integrated into a unified information system capable of sharing information and supporting business workflows. Only if the web services are satisfactorily connected with back-end information systems they can provide immediate and accurate information to business partners. The fact is that most web services will have to use existing and legacy applications as back-end solutions. Making the integration among such systems efficient is the key success factor. In particular, immediate response and immediate propagation of data to all related applications will be the important issue.

Information must often be gathered from several domains and integrated into a business process. Although the required information may well be available and exist in some form somewhere in an application, it is practically impossible for typical users to access it online without EAI. From the technical perspective, EAI refers to the process of integrating different applications and data to enable sharing of data and integration of business processes among applications with limited modification of the existing applications. Therefore, normalisation of the interfaces is the key to scaling into larger systems without creating expanding complexity and EAI must be performed using methods and activities which enables effectiveness in terms of costs and time.


## 2 CONCEPT OF WEB SERVICES

In this section, the concept of web services is presented. A web service represents a unit of business, application, or system functionality, which can be accessed via the web. Web services are applicable to any type of web environment, as for instance Internet, Intranet, or Extranet, with a focus on business-to-consumer, business-to-business, department-to-department, or peer-to-peer communication. A web service consumer could be a person who accesses the service through a desktop or wireless browser; it could be an application program; or it could be another web service (Sun 2001, Microsoft 2001).

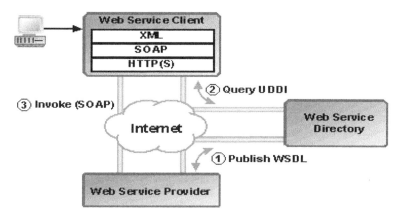

Figure 1.   Conceptual example of web services in action (Oracle web site).

A web service exhibits the following basic characteristics:

- is accessible via the web;
- exposes an XML interface;
- is registered and can be located through a web service directory;
- communicates, using XML messages based on standard web protocols;
- supports loosely coupled connections among systems.

The web services programming model is as follows: UDDI is used for publishing and querying web services, and SOAP is used for invoking the web service and, finally, WSDL is used for description of the web service.

An example of the high level protocol of discovery and access to a web service is shown in Figure 1. In this example operations proceed as follows:

1. First, the web service provider creates a service, defines the service in WSDL and then publishes the WSDL in the web service directory.
2. Next, a query from a web service client is made to the web service directory, perhaps as part of a request received from a browser. The reply will contain a WSDL descriptor of the requested service.
3. The web service client then invokes the service using the information in WSDL descriptors to structure the SOAP message, to determine the Internet address of the web service provider, and to understand the response to the request.

The above example illustrates a conceptual way in which web services might be discovered and accessed. These standards can be used together with the Internet and enable easy distribution of programming calls across the Internet, contrary to more internally focused protocols like J2EE RMI (Remote Method Invocation), Net9i (Oracle9i Database network protocol) and DCOM (Microsoft's distributed component model protocol).

Business logic performed by web service applications can be written in any language including Java and PL/SQL. Much of the development cycle including modelling, programming, security and the underlying component model does not change with web services. However, the additional steps made to describe it (WSDL), to access it (SOAP) and to publish it (UDDI) do change. These changes have some pitfalls, which are discussed later in the section on business management through web services.

Web services are often considered a radically new concept. The concept is very similar to other distributed programming models such as J2EE, CORBA and DCOM. One important difference is, however, that web service standards are just an XML meta-data layer on top of an application implementation, which describes the underlying application. Web services are not a standard for development of business logic or processes; they merely describe these.

The revolutionary aspect of web services is how easily they enable distributed component models to interact programmatically, particularly across the Internet using protocols like HTTP. By focusing initially on simplicity and interoperability, web services have gathered significant support, adoption and innovation across industry.

It is interesting to see that while simple web services now are well underway to become a popular development approach, much interest and impetus have been started in the area of complex web services. Complex web services take the foundation standards of SOAP, WSDL and UDDI and move them to higher level business processes, which have requirements for long running transactions, asynchronous interactions, authentication, encryption and non-repudiation. In this area, web service vendors and others leave the knowledge and standards to ebXML (Electronic Business XML) and to RosettaNet as well as to real life business process knowledge from the e-business suite line, where much work previously has been made to implement complex business processes (see e.g. ebXML (1) 2001, ebXML (2) 2001, RosettaNet web site).

## 3  WEB SERVICE COMPONENTS

Three components will be discussed in the following:

- Simple Object Access Protocol (SOAP)
- Universal Description, Discovery and Interoperability (UDDI)
- Web Services' Description Language (WSDL)

### 3.1  *Simple Object Access Protocol*

SOAP (WWWC (2) 2001) is an XML structure, which defines the XML formats of web service requests and responses. The benefits of SOAP as part of the definition of web services are that SOAP provides a way to leverage the huge industry-investment in XML. Furthermore, since SOAP is typically defined over *firewall friendly* protocols such as HTTP and SMTP, investment in firewall technology is leveraged as well. The result is the elimination of significant barriers to the production deployment of web services. Thus, by defining SOAP as an essential part of web services the industry is likely to enjoy volume production use of web services much sooner than if other strategies had been employed.

### 3.2  *Universal Description, Discovery and Interoperability*

UDDI creates a global, platform-independent, open framework to enable businesses to 1) discover one another, 2) define how they interact via the Internet, and 3) share information in a universal, web-based business directory called the UDDI Business Registry, which will more rapidly accelerate the global adoption of B2B e-commerce. The UDDI initiative is an industry consortium lead by Accenture, Ariba, Commerce One, Compaq, Edifices, Fujitsu, HP, I2, IBM, Intel, Microsoft, Oracle, SAP, Sun, Microsystems, and Verisign (ebXML (1) 2001). More than 130 companies have joined the UDDI initiative.

At first glance it seems simple to manage the process of web service *discovery*. After all, if a known business partner has a known electronic commerce gateway, what is left to discover? The tacit assumption, however, is that all the information is already known. When you want to find out which business partners have which services, the ability to discover the answers can quickly become difficult. One option is to call each partner on the phone, and then try to find the right person to talk with. It is difficult for a business, which espouses web services, to justify having highly technical staff to satisfy random discovery demand.

The core component of the UDDI project is the UDDI business registration in an XML file, which is used to describe a business entity and its web services. XML was chosen as it offers a platform-neutral view of data and allows hierarchical relationships to be described in a natural way. The emerging XML schema standard was chosen, instead of DTD, as it supports rich data types and easily describes and validates information based on information models represented in schemas.

Businesses can also locate potential partners through UDDI directly or, perhaps more likely, from online marketplaces and search engines, which use UDDI as a data source for their own value-added services. UDDI is designed to complement existing online marketplaces and search engines by providing them with standardised formats of programmatic business and service discovery. The ability to locate parties, which can provide a specific product or service at a given price or within a specific geographic boundary in a given timeframe, is not directly covered by UDDI specifications. These kinds of advanced discovery features require further collaboration and design work between buyer and seller. Instead, UDDI forms the basis for defining these services in a higher layer.

### 3.3   Web Services' Description Language

As communication protocols and message formats are standardised in the web community, it becomes possible and important to be able to describe the communication in a structured way. WSDL addresses this need and defines an XML grammar for description of network services as collections of communication endpoints capable of exchanging messages (WWWC (2) 2001). WSDL service definitions provide documentation for distributed systems and serves as a recipe for automating the details involved in applications communication. WDSL is a new specification, which describes networked XML-based services. It provides a simple language in which service providers can describe the basic format of requests to their systems regardless of the underlying protocol (such as SOAP or XML) or of encoding such as Multipurpose Internet Messaging Extensions (MIME). WSDL is a key part of the effort of the UDDI initiative to provide directories and descriptions of such online services for electronic business.

A WSDL document defines services as collections of network endpoints, or ports. In WSDL, the abstract definition of endpoints and messages is separated from their concrete network deployment or data format bindings. This allows the re-use of abstract definitions: Messages, which are abstract descriptions of the data being exchanged; and port types, which are abstract collections of operations. The concrete protocol and data format specifications for a particular port type constitute a reusable binding. A port is defined by associating a network address with a reusable binding, and a collection of ports defines a service.

It is important to observe that WSDL does not introduce a new type of definition language. WSDL recognises the need for rich type systems for description of message formats, and supports the XML schema specification (XSD) as its canonical type system. However, since it is unreasonable to expect a single type system grammar to be used for description of all message formats' at present and in future, WSDL allows use of other types of definition languages via extensibility.

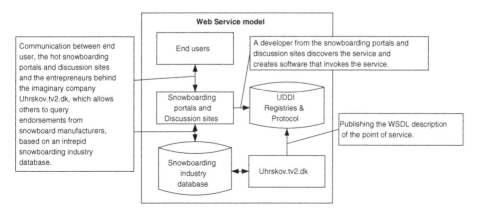

Figure 2.   Illustration of a web service model of an imaginary company (uhrskov.tv2.dk).

307

WSDL documents are divided into two types: *service interfaces* and *service implementations*. The service interface document is developed and published by the *service interface provider*. The service implementation document is created and published by the *service provider*. The roles of the service interface provider and service provider are logically separated, but they can be the same business entity, depending on the deployment type as mentioned in the introduction. If this is the case, they can use one document.

## 4   BUSINESS MANAGEMENT THROUGH WEB SERVICES

Web services pose new challenges for an information system. Today's requirements are very high – companies need to provide instant, online access to up-to-date information delivered with efficiency, reliability, and quality. Customers expect immediate response, and are not satisfied with days (or hours) of delay in confirmation of orders. Immediate response is only possible if the e-business application is backed by an efficiently integrated enterprise information system (EIS).

However, existing EIS applications have usually been developed with different, even legacy technologies, and are very heterogeneous. Accessing these systems directly from web services can be difficult because each existing system is unique and potentially requires a unique way to access it. This approach requires far too large an effort for development, and particularly for maintenance, as each change in the existing system will require an update in all related web services. Therefore, enterprise application integration (EAI) is the most important prerequisite for efficient web services. Only an integrated information system inside the company allows on-demand processing of e-business requests. The responsiveness of web services and the adequate quality of information provided through them can only be achieved by the tight integration of an enterprise information system in the back-end. EAI is therefore, the key success factor for the successful introduction of web services.

Research from leading consulting companies such as the Gartner Group confirms this thesis. It also shows that today there are very few web services (or other means of front-end systems to directly communicate with business partners) which are efficiently integrated with the back-end applications inside the company. Most non-integrated solutions fail to meet business expectations. As mentioned earlier, web services are not standards for developing business logic or processes, they merely describe these.

Many of the business and technical issues for service providers and consumers are incomplete. As very few service providers will offer a free service, the business conditions need to be laid down including payment of service. This typically involves a negotiated contract between the consumer and provider. Every time the service provider has a new consumer, and every time the consumer uses a different service provider, a new contract is required. If a service provider wants to reach fifty different consumers, fifty contracts are potentially needed. The problem is more difficult if the consumer and producer want different pricing models based upon different characteristics. Here, the issue of how the consumer and producer exchange usage and billing information skips over the fact that there are as yet no widely deployed standards for usage and billing (Steger-Jensen and Hvolby 2002). The service provider and consumer need a contract (in place) just as with any other non-technical customer-provider arrangement. As contract negotiation has many facets, this can be a difficult task to scale for either side.

Last, the issue of service version control emerges. Service providers change their interfaces and workflow. Usually the ASP will provide a new interface for the parameter, and the old interface will probably not be available. It is very difficult for ASPs to maintain backward compatibility, and they often do not. Their databases and business logic have all changed, so they do not want to keep the old stuff running. A typical example of this is the way that vendors seldom offer two different versions of an ERP system at the same time, using the same underlying data. What happens to the client when the interface changes? Generally, the client-code breaks. The ASP's interface is expecting a parameter, which the client software does not provide. In the worst scenario, the ASP simply changed the interface without warning the client in advance. Depending upon the complexity, the consumer's service will be down for an unknown period of time.

## 5 CONCLUSION

This paper has discussed that Enterprise Application Integration (EAI) is crucial for developing efficient and well-connected web services. Only if web services are satisfactorily connected with back-end information systems can they provide immediate and accurate information to business partners – a prerequisite for successful e-business collaboration. Making the integration between web services and legacy applications efficient will be the key success factor. In particular, immediate response and immediate propagation of data to all related applications will be the significant issue. Web services, which are not efficiently supported by back-end applications, will most likely fail to meet the requirements.

It is found that web services are a mature technology, supported by a large number of companies and organisations. Therefore web services are a recommendable technology for B2B integration taking the pitfalls of availability (payment) and maintenance (service version control) into consideration.

## REFERENCES

Andrianopoulos, A: 4 EAI – The framework behind Web Services Integration, Web Services Journal, Vol. 2, Issue 5, May 2002, pp.28–31.
ebXML (1): http://ebxml.org, June 2001.
ebXML (2): Technical Architecture Team – "ebxml Technical Architecture Specification v1.0.4", http://ebxml.org/specs/ebTA.pdf, June2001.
Ekanayaka Y; Currie, W & Seltsikas, P: "Delivering enterprise resource planning systems through service providers" Logistics Information Management, vol. 15, number 3, 2002.
Microsoft: Microsoft.Net, http://microsoft.com/net, June 2001.
Oracle, http://www.oracle.com/
RosettaNet, http://www.rosettanet.org/
Steger-Jensen, K. & Hvolby, H-H: "Review of B2B integration". The APMS conference on "Collaborative Systems for Production Management" Eindhoven, September 2002.
Sun Microsystems Open Net Environment (SunONE), http://www.sun.sonone, June 2001.
Sushil, K; Sharma; Jatinder & Gupta: Application service providers: issues and challenges, Logistics Information Management, vol. 15, number 3, 2002, pp.160–169.
UDDI: "Universal Description, Discovery, and Integration of business for the Web", http://www.uddi.org/, June 2001.
WWWC (1): World Wide Web Consortium – "simple Object Access Protocol (SOAP) v1.1, W3C Note", http://www.w3.org/TR/soap, May 2001.
WWWC (2): World Wide Web Consortium – "Web Services Description Language (WSDL) v1.1, W3C Note", http://www.w3.org/TR/wsdl, March 2001.

*Advanced Manufacturing – An ICT and Systems Perspective – Taisch,*
*Thoben & Montorio (eds)*
*© 2007 Taylor & Francis Group, London, ISBN 978-0-415-42912-2*

# From Enterprise Resource Planning (ERP) to Open Resource Planning (ORP): The openfactory project

Martin Meyer

*Forschungsinstitut für Rationalisierung (FIR) an der RWTH-Aachen, Pontdriesch, Aachen, Germany*

ABSTRACT:   A dynamic business network is the common type of organisation within the mechanical engineering industry. Owing to the complex structure and temporary business relationships of these networks today's ERP solutions do not sufficiently support efficient inter-organisational order processing. This paper describes the conceptual basis of Open Resource Planning (ORP), an innovative approach to realise seamless order management within dynamic business networks. Three important components are outlined: a data and process standard for inter-organisational order management within the mechanical engineering industry; a web-based information system; and an innovative Internet business model.

*Keywords*:   Enterprise Application Integration (EAI), Enterprise Resource Planning (ERP), Supply Chain Integration, Supply Chain Management

## 1   INTRODUCTION

For many decades the design, control and improvement of production facilities was focused on individual locations (Figure 1). Over the last years, however, this perspective has been extended to supply chains that represent all organisations involved in producing and distributing a product from raw material supply up to the ultimate consumer (Stadler 2000). Supply chain management (SCM) aims at the co-ordination and optimisation of all business processes within a supply chain (Christopher 1998, Chopra 2001) The implementation of a successful SCM requires the design of effective organisational structures as well as the development of appropriate software tools.

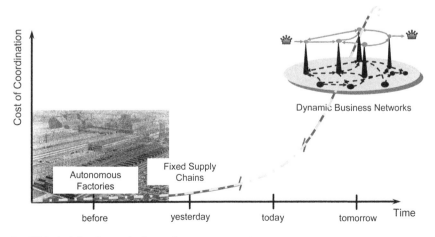

Figure 1.   Historical development of operations management.

As a recent survey of Fraunhofer ISI, Karlsruhe, Germany demonstrates, enterprises involved in inter-organisational corporations are significantly more successful than others. According to this survey co-operation correlates with striking advantages regarding sales growth, capacity utilisation and productivity (Eggers 2002).

However, most success stories in the field of supply chain management are reported from automotive and consumer goods industry. Fixed and linear supply chains designed for a product lifecycle of several years and dominated by a leading OEM (Original Equipment Manufacturer) seem to provide excellent conditions for today's concepts and tools of supply chain integration.

In the future, however, the scope of supply chain management must be extended to dynamic business networks designed for collaboration in development, manufacturing and distribution of customer specific products or small batches. The project network structure is the dominant type of organisation within the mechanical and plant engineering industry. While ERP systems such as mySAP primarily support order management within a single enterprise and SCM systems such as i2 particularly support the design and control of fixed supply chains, there is a lack of organisational concepts as well as information and communication technology (ICT) adequate to dynamic business networks.

## 2 CHALLENGES OF ORDER MANAGEMENT WITHIN DYNAMIC BUSINESS NETWORKS

In the first instance, challenges of order management within dynamic business networks result from their complex structure and short lifecycle. Up to several-hundred business partners all over the world can be involved in one single project (Figure 2). The prime contractor setting up this business network might not previously have closed a deal with many of them and maybe will never do ever after. Above all, most organisations within dynamic business networks – often including the prime contractor – are small or medium enterprises possessing limited resources in respect of staff and ICT infrastructure. Therefore, they will usually not intend to spend too much time and money on the set-up of temporary point-to-point connections (Cox 2001). Nonetheless, thousands of interactions among all business partners must be processed to jointly develop a product or procure material from suppliers and sub-contractors.

An efficient inter-organisational process management within dynamic business networks is at first constrained by the current disregard of order processing compared to manufacturing

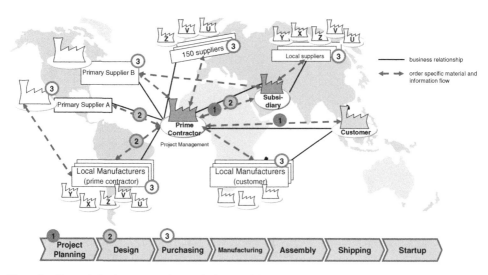

Figure 2.  Dynamic business network (practical example).

processes. While most companies have reduced stocks, thoroughly defined process standards and co-ordinated various departments within their factories, intra- as well as inter-organisational order management is still characterised by idle times, unstructured processes and improvisation rather than standardisation.

Above all, today's information and communication technology do not support integrated order management within dynamic business networks (Akkermans 2003). In this scope, the heterogeneity of data models must be understood to be a core problem. For example, it is hardly possible to achieve seamless data exchange among any of more than one hundred ERP systems available on the German market, without setting up an individual project. Current standards of Electronic Data Interchange (EDI) are not appropriate to the requirements of the mechanical engineering industry and require the customisation of each point-to-point connection (Meyer, Lücke, and Schmidt 2004).

This problem is reinforced by the fact that most companies apply various types of information systems beside a backbone system represented by ERP, e.g. PDM (Product Data Management), CRM (Customer Relationship Management) or SCM (Supply Chain Management). Thus, even small enterprises are required to spend a lot of money just to achieve internal enterprise application integration. As a result, between 80 and 95% of inter-organisational transactions within dynamic business networks are processed via telephone, fax and email (Meyer and Schweicher 2005). Even several production sites of one single enterprise are rarely able to achieve seamless data exchange among their ERP systems.

## 3   OPEN RESOURCE PLANNING (ORP)

The circumstance described above has led to an initiative of several machine tool manufacturers, some of their suppliers, and leading software vendors, funded by Germany's Ministry of Research and Education (BMBF) and co-ordinated by the Research Institute for Operations Management (FIR) at Aachen University of Technology. The project aims at the establishment of a *de facto* standard for data structures and processes within mechanical engineering industry (Figure 3). This *de facto* standard will enable software vendors to connect to the standardised data model by an individual interface. Via a web-based information system the partners of a dynamic business network will then be enabled to seamlessly exchange all necessary data such as enquiries, orders or capacity information (Open Resource Planning). The web-based information system, named *myOpenFactory.com*, will also allow for small enterprises without any advanced ICT infrastructure to participate in this scenario of integrated order management via a simple web cockpit providing

Figure 3.   Open resource planning (ORP).

the basic functionality of an ERP system. The ultimate aim is to enable small and medium-size enterprises to *plug-in and do business.*

The OpenFactory standard is based on existing approaches such as EDIFACT, ODETTE and Rosetta.Net. With regard to semantics the standard is geared to EDIFACT, which can be seen as the most common standard in various industrial sectors. However, in contrast to EDIFACT, the standard is based on XML (Extended Mark-up Language) and limited to 20 messages and 185 elements compared to EDIFACTS's more than 550 elements within 100 segments. The aim of the OpenFactory standard is to provide users with only 20% of EDIFACT's breadth that will, nevertheless, enable them to perform 80% of their transactions. A feature based structure of the OpenFactory standard will allow for its simple and individual extension according to a user's specific requirements.

Beside the data standard, a process standard for inter-organisational order management was designed within the project. The standard is differentiated according to the level of co-operation that is intended in a particular business relationship. For example, a customer and supplier that has not previously closed a deal might select a very formal process and only exchange basic messages such as enquiry, order, and order confirmation. In contrast to this scenario, two production sites of one enterprise might be allowed to view each other's capacity circumstances or even to reserve capacity of the other production site for a specific project. An inter-organisational workflow management will support the selection and realisation of an appropriate standard process.

The information system is based on the Service-Oriented Architecture (SOA) and applies web services as the main technological component. This is to ensure a high level of integration with regard to ERP systems and to take the constrictive ICT infrastructure of small enterprises into account. The technological basis of the information system is built by a web sever (Apache), an application server (Tomcat) and a database (MySQL).

Beside the conceptual and technological basis described above, Open Resource Planning (ORP) requires an innovative business model. Therefore, the participating software vendors will establish a legal entity to collaboratively manage marketing, sales, and further development of the information system (*OpenFactory Community*). To promote ORP as a *de facto* standard within the mechanical engineering industry, the standard as well as the community will be open to all software vendors worldwide. Users of the web-based information system will have to submit a basic charge and additional transaction fees to the community. According to current cost estimations and a thorough analysis of measurable benefits, *myOpenFactory.com* will generate a positive return on investment for users as well as software vendors within only a few months. Moreover, according to pilot users of *myOpenFactory.com*, significant competitive advantages will result from hard to measure benefits such as a gain of transparency and flexibility, rather than from cost reduction in itself.

## 4 CONCLUSIONS

In the future, efficient co-operation within dynamic business networks will be a precondition to gain competitive advantages within mechanical and plant engineering industry. Dynamic business networks require innovative organisational concepts and technological solutions to improve efficiency and flexibility. Therefore, various research institutes as well as software vendors and industrial enterprises combined to set up the OpenFactory initiative. The ultimate ambition of OpenFactory is to establish a *de facto* data and process standard for mechanical and plant engineering industry. Companies in this industrial sector, as well as software vendors, are invited to join the initiative. More information is to be found on the project homepage www.myopenfactory.com. The project is funded by PFT (Produktion und Fertigungstechnologien, Karlsruhe, Germany), a project execution organisation of BMBF (Germany's Ministry of Research and Education).

## REFERENCES

Stadtler, H., Kilger, Ch.: Supply Chain Management and Advanced Planning – Concepts, Models, Software and Case Studies (Springer Verlag, Berlin, Heidelberg, New York, 2000).

Christopher, M.: Logistics and Supply Chain Management – Strategies for Reducing Costs and Improving Service (Financial Time, Prentice Hall, New Jersey, 1998).

Chopra, Meindl, P.: Supply Chain Management – Strategy, Planning and Operation (Prentice Hall, New Jersey, 2001).

Eggers, Kinkel, S.: Die "virtuelle Fabrik" in weiter Ferne – Verbreitung und Nutzen von Produktionskooperation und Produktnetzwerken im Verarbeitenden Gewerbe (Fraunhofer ISI, Karlsruhe, 2002).

Cox, Chicksand, L., Ireland, P.: E-Supply Applications: The Inappropriateness of Certain Internet Solutions for SME's (10th International Annual IPSERA Conference, Jönköpping/ Sweden, 2001), pp. 189–200.

Akkermanns, H.A., Bogerd, P., Yücesan, E., van Wassenhove, L.: The impact of ERP on Supply Chain Management: Exploratory Findings from a European Delphi Study, European Journal of Operational Research 146 (1), 284–301 (2003).

Meyer, M., Lücke, T., Schmidt, C.: Plug and Do Business – ERP of the Next Generation for Efficient Order Processing in Dynamic Business Networks, International Journal of Internet and Enterprise Management 2(2), 152–162 (2004).

Meyer, M., Schweicher, B.: Innovative Koordination durch OpenFactory. Quasi-Standard für die überbetriebliche Auftragsabwicklung, isreport 9 (4), 42–44 (2005).

# Subject index

*Advanced Manufacturing – An ICT and Systems Perspective – Taisch,*
*Thoben & Montorio (eds)*
*© 2007 Taylor & Francis Group, London, ISBN 978-0-415-42912-2*

# Author index

Printed and bound by CPI Group (UK) Ltd, Croydon, CR0 4YY

01/11/2024

01782636-0010